Stanley Dalton Dodge

INTRODUCTION

TO THE

NATIONAL ARITHMETIC,

ON THE

INDUCTIVE SYSTEM,

COMBINING THE

ANALYTIC AND SYNTHETIC METHODS;

IN WHICH THE PRINCIPLES OF THE SCIENCE ARE FULLY EXPLAINED AND ILLUSTRATED.

DESIGNED FOR COMMON SCHOOLS AND ACADEMIES.

By BENJAMIN GREENLEAF, A.M.

AUTHOR OF THE "NATIONAL ARITHMETIC," "ALGEBRA," ETC.

NEW STEREOTYPE EDITION,
WITH ADDITIONS AND IMPROVEMENTS.

BOSTON:
PUBLISHED BY ROBERT S. DAVIS & CO.
NEW YORK: D. APPLETON & CO., AND MASON BROTHERS.
PHILADELPHIA: J. B. LIPPINCOTT AND COMPANY.
CHICAGO: WILLIAM B. KEEN.
1860.

OFFICE OF THE CONTROLLERS OF PUBLIC SCHOOLS,
FIRST SCHOOL DISTRICT OF PENNSYLVANIA.
PHILADELPHIA, December 14, 1859.

At a Meeting of the Controllers of Public Schools, First District of Pennsylvania, held at the CONTROLLERS' CHAMBER, on Tuesday, December 13th, 1859, the following Resolution was adopted:—

Resolved: That GREENLEAF'S COMMON SCHOOL AND NATIONAL ARITHMETICS be introduced to be used in the Public Schools of this District.

ROBERT J. HEMPHILL, *Secretary.*

GREENLEAF'S SERIES OF MATHEMATICS.

1. NEW PRIMARY ARITHMETIC; OR, MENTAL ARITHMETIC, upon the Inductive Plan; with Easy Exercises for the Slate. Designed for Primary Schools. 72 pp.

2. INTELLECTUAL ARITHMETIC, upon the Inductive Plan; being an advanced Intellectual Course, for Common Schools and Academies. Improved edition. 154 pp.

3. COMMON SCHOOL ARITHMETIC; OR, INTRODUCTION TO THE NATIONAL ARITHMETIC. Improved stereotype edition. 324 pp.

4. THE NATIONAL ARITHMETIC, being a complete course of Higher Arithmetic, for advanced scholars in Common Schools, High Schools, and Academies. New electrotype edition, with additions and improvements. 444 pp.

5. PRACTICAL TREATISE ON ALGEBRA, for Academies and High Schools, and for advanced Students in Common Schools. Improved stereotype edition. 360 pp.

6. ELEMENTS OF GEOMETRY; with Practical Applications to Mensuration. Designed for Academies and High Schools. Electrotype edition. 320 pp.

COMPLETE KEYS TO THE INTELLECTUAL, COMMON SCHOOL, AND NATIONAL ARITHMETICS, THE PRACTICAL TREATISE ON ALGEBRA, AND GEOMETRY, containing Solutions and Explanations, *for Teachers only.* In 5 volumes.

☞ Two editions of the NATIONAL ARITHMETIC, and also of the COMMON SCHOOL ARITHMETIC, one containing the ANSWERS to the examples, and the other without them, are published. Teachers are requested to state in their orders *which edition* they prefer.

PREFACE.

The present edition of this work has been thoroughly revised and re-written, and also improved by the addition of much valuable new material, rendering it a sufficiently complete practical treatise for the majority of learners.

The arrangement is strictly progressive; the aim having been to introduce subjects in an order most in accordance with the laws governing the proper development of mind. The rules have generally been deduced from the analysis of one or more questions, so that the reasons for the methods of solution adopted are rendered intelligible to the pupil; no knowledge of a principle being required, that has not been previously illustrated and explained. In this respect, it is believed the work will be found to differ from most other arithmetics.

In preparation of the rules, definitions, and illustrations, the utmost care has been taken to express them in language simple, precise, and accurate.

The examples are of a practical character, and adapted not only to fix in the mind the principles, which they involve, but also to interest the pupil, exercise his ingenuity, and inspire a love for mathematical science.

The reasons for the operations are explained, and an attempt is made to secure to the learner a knowledge of the philosophy of the subject, and prevent the too prevalent practice of merely performing, mechanically, operations, which he does not understand.

Analysis has been made a prominent subject, and employed in the solution of questions under most of the rules, in which it could be used with any practical advantage; and it cannot be too strongly recommended to the pupil to make use of this mode of operation, where it is recommended by the author.

All the most important methods of abridging operations, applicable to business transactions, have been given a place in the work, and, so introduced, as not to be regarded as mere blind mechanical expedients, but as rational labor-saving processes.

Old rules and distinctions, which modern improvements have rendered unnecessary, and which, deservedly, are becoming obsolete, have been avoided.

Rules for finding the greatest common divisor of fractions, and for finding the least common multiple of fractions; methods of equating accounts; division of duodecimals; exchange, foreign and inland; and several important tables, are among the new features of this edition, which will be found, it is believed, very valuable.

The articles on money, weights, measures, interest, and duties are the results of extensive correspondence and much laborious research, and are strictly conformable to present usage, and recent legislation, state and national.

Questions have been inserted at the bottom of each page, designed to direct the attention of teachers and pupils to the most important principles of the science, and fix them in the mind. It is not intended, however, nor is it desirable, that the teacher should servilely confine himself to these questions; but vary their form, and extend them at pleasure, and invariably require the pupil thoroughly to understand the subject, and give the reasons for the various steps in the operation, by which he arrives at any result in the solution of a question.

The object of studying mathematics is not only to acquire a knowledge of the subject, but also to secure mental discipline, to induce a habit of close and patient thought, and of persevering and thorough investigation. For the attainment of this object, the examples for the exercise of the pupil are numerous, and variously diversified, and so constructed as necessarily to require careful thought and reflection for the right application of principles.

The author would respectfully suggest to teachers, who may use this book, to require their pupils to become familiar with each rule before they proceed to a new one; and, for this purpose, a frequent review of rules and principles will be of service, and will greatly facilitate their progress. If the pupil has not a clear idea of the principles involved in the solution of questions, he will find but little pleasure in the study of the science; for no scholar can be pleased with what he does not understand.

BENJAMIN GREENLEAF.

BRADFORD, MASS., *August 1st*, 1856.

NOTICE.

☞ Two editions of this work, and also of the NATIONAL ARITHMETIC, one containing the ANSWERS to the examples, and the other without them, are now published.

CONTENTS.

ARITHMETIC.

ARTICLE 1. QUANTITY is anything that can be measured.

A *unit* is a single thing, or one.

A *number* is either a unit or a collection of units.

An *abstract number* is a number, whose units have no reference to any particular thing or quantity; as two, five, seven.

A *concrete number* is a number, whose units have reference to some particular thing or quantity; as two books, five feet, seven gallons.

ARITHMETIC is the science of numbers, and the art of computing by them.

A *rule* of arithmetic is a direction for performing an operation with numbers.

The *introductory* and *principal rules* of arithmetic are Notation and Numeration, Addition, Subtraction, Multiplication, and Division.

The last four are called the *fundamental rules*, because upon them depend all other arithmetical processes.

§ I. NOTATION AND NUMERATION.

NOTATION.

ART. 2. NOTATION is the art of expressing numbers by figures or other symbols.

There are two methods of notation in common use; the *Roman* and the *Arabic*.

QUESTIONS.—Art. 1. What is quantity? What is a unit? What is a number? What is an abstract number? What is a concrete number? What is arithmetic? What is a rule? Which are the introductory rules? What are the last four called?—Art. 2. What is notation? How many kinds of notation in common use? What are they?

ART. 3. The Roman notation, so called from its originating with the ancient Romans, employs in expressing numbers *seven* capital letters, viz.: I for *one;* V for *five;* X for *ten;* L for *fifty;* C for *one hundred;* D for *five hundred;* M for *one thousand.*

All the other numbers are expressed by the use of these letters, either in repetitions or combinations; as, II expresses *two;* IV, *four;* VI, *six,* &c.

By a *repetition* of a letter, the value denoted by the letter is represented as repeated; as, XX represents *twenty;* CCC, *three hundred.*

By writing a letter denoting a less value *before* a letter denoting a greater, their *difference* of value is represented; as, IV represents *four;* XL, *forty.* By writing a letter denoting a less value *after* a letter denoting a greater, their *sum* is represented; as, VI represents *six;* XV, *fifteen.*

A dash (—) placed over a letter increases the value denoted by the letter a *thousand* times; as, $\overline{\text{V}}$ represents *five thousand;* $\overline{\text{IV}}$, *four thousand.*

TABLE OF ROMAN LETTERS.

I	one.	LXXX	eighty.
II	two.	XC	ninety.
III	three.	C	one hundred.
IV	four.	CC	two hundred.
V	five.	CCC	three hundred.
VI	six.	CCCC	four hundred.
VII	seven.	D	five hundred.
VIII	eight.	DC	six hundred.
IX	nine.	DCC	seven hundred.
X	ten.	DCCC	eight hundred.
XX	twenty.	DCCCC	nine hundred.
XXX	thirty.	M	one thousand.
XL	forty.	MD	fifteen hundred.
L	fifty.	MM	two thousand.
LX	sixty.	$\overline{\text{X}}$	ten thousand.
LXX	seventy.	$\overline{\text{M}}$	one million.

QUESTIONS. — Art. 3. Why is the Roman notation so called? By what are numbers expressed in the Roman notation? What effect has the repetition of a letter? What is the effect of writing a letter expressing a less value before a letter denoting a greater? What of writing the letter after another denoting a greater value? How many times is the value denoted by a letter increased by placing a dash over it? Repeat the table.

The Roman notation is now but little used, except in numbering sections, chapters, and other divisions of books.

EXERCISES IN ROMAN NOTATION.

The learner may write the following numbers in letters:

1. Ninety-six. Ans. XCVI.
2. Eighty-seven.
3. One hundred and ten.
4. One hundred and sixty-nine.
5. Two hundred and seventy-five.
6. Five hundred and forty-two.
7. One thousand three hundred and nineteen.
8. One thousand eight hundred and fifty-eight.

ART. 4. The Arabic notation, so called from its having been made known through the Arabs, employs in expressing numbers *ten* characters or figures, viz.:

1,	2,	3,	4,	5,	6,	7,	8,	9,	0.
one,	two,	three,	four,	five,	six,	seven,	eight,	nine,	cipher.

The first *nine* are sometimes called *digits*, from *digitus*, the Latin signifying a finger, because of the use formerly made of the fingers in reckoning. The *cipher*, also, has sometimes been called *naught*, or *zero*, from its expressing the *absence* of a number, or *nothing*, when standing alone.

ART. 5. The particular position a figure occupies with regard to other figures is called its PLACE; as in 32, counting from the right, the 2 occupies the first place, and the 3 the second place, and so on for any other like arrangement of figures.

The digits have been denominated *significant figures*, because each expresses of itself a positive value, always representing so many *units*, or *ones*, as its name indicates. But the *size* or *value* of the units represented by a figure differs with the place occupied by the figure.

For example, there are written together to represent a number three figures, thus, 366 (three hundred and sixty-six). Each of the figures, without regard to its place, expresses units, or ones; but these units, or ones, differ in value. The 6 occupying the first place represents 6 single units; the 6 occupying the second place repre-

QUESTIONS. — What use is now made of Roman notation? — Art. 4. How many characters are employed in the Arabic notation? What are the first nine called, and why? What is the cipher sometimes called? What does it represent when standing alone? — Art. 5. What is meant by the place of a figure? What have the digits been denominated? Why? How does the size or value of units represented by figures differ?

sents 6 tens, or 6 units each *ten times* the size or value of a unit of the first place; and the 3 occupying the third place represents 3 hundreds, or 3 units each *one hundred times* the size or value of a unit of the first place.

ART. 6. The *cipher* becomes significant when connected with other figures, by filling a place that otherwise would be vacant; as in 10 (ten), where it occupies the vacant place of units; in 120 (one hundred and twenty), where it also occupies the vacant place of units; and in 304 (three hundred and four), where it fills the vacant place of tens.

ART. 7. The *simple* value of a unit is the value expressed by a figure standing alone; or, in a collection, when standing in the right-hand place. Thus 6 alone, or in 26, expresses a simple value of six single units, or ones.

The *local* value of a unit is the value expressed by a figure when it is used in combination with another figure or figures, and depends upon the place the figure occupies.

The local values expressed by figures will be made plain by the following table and its explanation.

Millions.	Hundreds of Thousands.	Tens of Thousands.	Thousands.	Hundreds.	Tens.	Units.	The figures in this table are read thus:
						9	Nine.
					9	8	Ninety-eight.
				9	8	7	Nine hundred eighty-seven.
			9	8	7	6	Nine thousand eight hundred seventy-six.
		9	8	7	6	5	Ninety-eight thousand seven hundred sixty-five.
	9	8	7	6	5	4	Nine hundred eighty-seven thousand six hundred fifty-four.
9	8	7	6	5	4	3	Nine millions eight hundred seventy-six thousand five hundred forty-three.

QUESTIONS. — Art. 6. When does a cipher become significant? — Art. 7. What is the simple value of a unit? What is the local value of a unit? What is the design of this table?

It will be noticed in the preceding table, that each figure in the right-hand or units' place expresses the local value of so many *units;* but when standing in the second place, it expresses the local value of so many *tens,* each of the value of ten ones; when in the third place, the local value of so many *hundreds,* each of the value of ten tens; when in the fourth place, the local value of so many *thousands,* each of the value of ten hundreds; *the value expressed by any figure being always made tenfold by each removal of it one place to the left hand.*

NUMERATION.

ART. **8.** NUMERATION is the art of reading numbers when expressed by figures.

ART. **9.** There are two methods of numeration in common use: the *French* and the *English.*

ART. **10.** The French method is that in general use on the continent of Europe and in the United States. It separates figures into groups, called *periods,* of three places each, and gives a distinct name to each period.

FRENCH NUMERATION TABLE.

Hundreds of Sextillions.	Tens of Sextillions.	Sextillions.	Hundreds of Quintillions.	Tens of Quintillions.	Quintillions.	Hundreds of Quadrillions.	Tens of Quadrillions.	Quadrillions.	Hundreds of Trillions	Tens of Trillions.	Trillions.	Hundreds of Billions.	Tens of Billions.	Billions.	Hundreds of Millions.	Tens of Millions.	Millions.	Hundreds of Thousands.	Tens of Thousands.	Thousands.	Hundreds.	Tens.	Units.
1	2	7,	8	9	4,	2	3	7,	8	6	7,	1	2	3,	6	7	8,	4	7	8,	6	3	8.
Period of Sextillions.			Period of Quintillions.			Period of Quadrillions.			Period of Trillions.			Period of Billions.			Period of Millions.			Period of Thousands.			Period of Units.		

QUESTIONS. — Art. 7. What value is expressed by a figure standing in the right-hand or units' place? What in the second place? What in the third? How do figures increase from the right towards the left? — Art. 8. What is numeration? What are the two methods of numeration in common use? Where is the French method more generally used? Repeat the French Numeration Table. What are the names of the different periods in the table? What is the value of the numbers represented in the table expressed in words?

The value of the numbers represented in this table, expressed in words, is, One hundred twenty-seven sextillions, eight hundred ninety-four quintillions, two hundred thirty-seven quadrillions, eight hundred sixty-seven trillions, one hundred twenty-three billions, six hundred seventy-eight millions, four hundred seventy-eight thousand, six hundred thirty-eight.

The names of the periods above Sextillions, in their order, are, Septillions, Octillions, Nonillions, Decillions, Undecillions, Duodecillions, Tredecillions, Quatuordecillions, Quindecillions, Sexdecillions, Septendecillions, Octodecillions, Novemdecillions, Vigintillions, &c.

ART. 11. The successive places occupied by figures are often called *orders*. Hence, a figure in the right-hand or units' place is called a figure of the *first* order, or of the order of *units*; a figure in the second place is a figure of the *second* order, or of the order of *tens*; in the third place, of the order of *hundreds*, and so on. Thus, in the number 1847, the 7 is of the order of *units*, 4 of the order of *tens*, 8 of the order of *hundreds*, and 1 of the order of *thousands*, each figure expressing as many units as its name indicates of that order to which it belongs; so that we read the whole number, *one thousand eight hundred and forty-seven.*

ART. 12. From the preceding table and explanation, we deduce the following rule for numerating and reading numbers expressed by figures according to the French method.

RULE. — *Begin at the right hand, and point off the figures into periods of* THREE *places each.*

Then, commencing at the left hand, read the figures of each period, adding the name of each period excepting that of units.

EXERCISES IN FRENCH NUMERATION.

The learner may read orally, or write in words, the numbers represented by the following figures:

1.	152	5.	2254	9.	84093	13.	610711
2.	276	6.	4384	10.	98612	14.	3031671
3.	998	7.	7932	11.	592614	15.	4869021
4.	1057	8.	42198	12.	400619	16.	637313789

QUESTIONS. — Art. 10. What are the names of the periods above sextillions? — Art. 11. What are the successive places of the figures in the table called? Of what order is the first or right-hand figure? The second? The third? &c. — Art. 12. What is the rule for *numerating* and *reading* numbers according to the French method?

17.	39461928	24.	37617001377706717
18.	427143271	25.	242173562357421
19.	6301706716	26.	870037637471078635
20.	14377670033	27.	8216243812706381
21.	2046316248613	28.	2403172914376931
22.	6382102471180	29.	376170613770616713
23.	4477063014767	30.	610167637896430607761607

ART. **13.** To write numbers by figures according to the French method, we have the following

RULE. — *Begin at the left hand, and write in each successive order the figure belonging to it.*

If any intervening order would otherwise be vacant, fill the place by a cipher.

EXERCISES IN FRENCH NOTATION AND NUMERATION.

The learner may represent by figures, and read, the following numbers:

1. Forty-seven.
2. Three hundred fifty-nine.
3. Six thousand five hundred seventy-five.
4. Nine hundred and eight.
5. Nineteen thousand.
6. Fifteen hundred and four.
7. Twenty-seven millions five hundred.
8. Ninety-nine thousand ninety-nine.
9. Forty-two millions two thousand and five.
10. Four hundred eight thousand ninety-six.
11. Five thousand four hundred and two.
12. Nine hundred seven millions eight hundred five thousand and seventy-four.
13. Three hundred forty-seven thousand nine hundred and fifteen.
14. Eighty-nine thousand forty-seven.
15. Fifty-one thousand eighty-one.
16. Seven thousand three hundred ninety-five.
17. Fifty-seven billions fifty-nine millions ninety-nine thousand and forty-seven.

QUESTIONS. — Art. 13. What is the rule for *writing* numbers according to the French method? At which hand do you begin to numerate figures? Where do you begin to read them? At which hand do you begin to write numbers? Why?

ART. 14. The following table exhibits the English method of numeration, in which it will be observed that the figures are separated by commas into periods of six figures each. The first or right-hand period is regarded as units and thousands of units, the second, as millions and thousands of millions; and so on, six places being assigned to each period designated by a distinct name.

ENGLISH NUMERATION TABLE.

Hund. of Thousands of Trillions.	Tens of Thousands of Trillions.	Thousands of Trillions.	Hundreds of Trillions.	Tens of Trillions.	Trillions.	Hund. of Thousands of Billions.	Tens of Thousands of Billions.	Thousands of Billions.	Hundreds of Billions.	Tens of Billions.	Billions.	Hund. of Thousands of Millions.	Tens of Thousands of Millions.	Thousands of Millions.	Hundreds of Millions.	Tens of Millions.	Millions.	Hundreds of Thousands.	Tens of Thousands.	Thousands.	Hundreds.	Tens.	Units
1	3	7	8	9	0,	7	1	1	7	1	6,	3	7	1	7	1	2,	4	5	6	7	1	1.
Period of Trillions.						Period of Billions.						Period of Millions.						Period of Units.					

The value of the figures in the above table, expressed in words according to the English method, is, One hundred thirty-seven thousand eight hundred ninety trillions; seven hundred eleven thousand seven hundred sixteen billions; three hundred seventy-one thousand seven hundred twelve millions; four hundred fifty-six thousand seven hundred eleven.

Although there is the same number of figures in the English and in the French table, yet it will be observed that in the French table we have the names of three periods other than in the English. It will also be observed that the variation commences after the ninth place, or the place of hundreds of millions. If, therefore, we would know the value of numbers

---

QUESTIONS. — Art. 14. How many figures in each period in the English method of numeration? What orders are found in the English method that are not in the French? Give the names of the periods in the English Numeration Table, beginning with the period of units. Repeat the table, giving the names of all the orders or places. What is the value of the numbers in the table expressed in words? How do the figures in the English and French table compare as to numbers? How as to periods? Why is this difference? Has a million the same value reckoned by the French table as when reckoned by the English?

higher than hundreds of millions, when we see them written in words, or hear them read, we need to know whether they are expressed according to the French or the English method of numeration.

The English method of numeration is that generally used in Great Britain, and in the British Provinces.

ART. **15.** To numerate and read numbers expressed by figures according to the English method, we have the following

RULE. — *Begin at the right hand, and point off the figures into periods of* SIX *places each. Then, commencing at the left hand, read the figures of each period, adding the name of each period, excepting that of units.*

EXERCISES IN ENGLISH NUMERATION.

The learner may read orally, or write in words, the following numbers:

1.	125	5.	27306387903
2.	1063	6.	531470983712
3.	25842	7.	4230578032765038
4.	904357	8.	716756378807370767086389706473

ART. **16.** To write numbers in figures, according to the English method, we have the following

RULE. — *Begin at the left hand, and write in each successive order the figure belonging to it.*

If any intervening order would otherwise be vacant, fill the place by a cipher.

EXERCISES IN ENGLISH NOTATION AND NUMERATION.

The learner may write in figures, and read, the following numbers:

1. Three hundred twenty-five thousand four hundred and twelve.

2. Two hundred fourteen thousand, one hundred sixty-five millions, seventy-eight thousand and fifty-six.

3. Forty-two billions, six hundred seventeen thousand thirty-one millions, forty-one thousand three hundred forty-two.

4. Two thousand eight billions, nine thousand eighty-two millions, seven hundred one thousand, nine hundred and eight.

QUESTIONS. — Has the billion the same value as that by the French table? Why not? By which table has it the greater value? Where is the English method of numerating in use? — Art. 15. What is the rule for numerating and reading numbers by the English method? — Art. 16. What is the rule for *writing* numbers according to the English method?

§ II. ADDITION.

Mental Exercises.

Art. 17. When it is required to find a single number to express the sum of the units contained in several smaller numbers, the process is called *Addition*.

Ex. 1. James has 3 pears, and his younger brother has 4, how many have both?

Illustration. — To solve this question, the 3 pears and 4 pears must be added together; thus, 3 added to 4 makes 7. Therefore James and his brother have 7 pears.

ADDITION TABLE.

2 and 0 are 2	3 and 0 are 3	4 and 0 are 4	5 and 0 are 5
2 and 1 are 3	3 and 1 are 4	4 and 1 are 5	5 and 1 are 6
2 and 2 are 4	3 and 2 are 5	4 and 2 are 6	5 and 2 are 7
2 and 3 are 5	3 and 3 are 6	4 and 3 are 7	5 and 3 are 8
2 and 4 are 6	3 and 4 are 7	4 and 4 are 8	5 and 4 are 9
2 and 5 are 7	3 and 5 are 8	4 and 5 are 9	5 and 5 are 10
2 and 6 are 8	3 and 6 are 9	4 and 6 are 10	5 and 6 are 11
2 and 7 are 9	3 and 7 are 10	4 and 7 are 11	5 and 7 are 12
2 and 8 are 10	3 and 8 are 11	4 and 8 are 12	5 and 8 are 13
2 and 9 are 11	3 and 9 are 12	4 and 9 are 13	5 and 9 are 14
2 and 10 are 12	3 and 10 are 13	4 and 10 are 14	5 and 10 are 15
2 and 11 are 13	3 and 11 are 14	4 and 11 are 15	5 and 11 are 16
2 and 12 are 14	3 and 12 are 15	4 and 12 are 16	5 and 12 are 17
6 and 0 are 6	7 and 0 are 7	8 and 0 are 8	9 and 0 are 9
6 and 1 are 7	7 and 1 are 8	8 and 1 are 9	9 and 1 are 10
6 and 2 are 8	7 and 2 are 9	8 and 2 are 10	9 and 2 are 11
6 and 3 are 9	7 and 3 are 10	8 and 3 are 11	9 and 3 are 12
6 and 4 are 10	7 and 4 are 11	8 and 4 are 12	9 and 4 are 13
6 and 5 are 11	7 and 5 are 12	8 and 5 are 13	9 and 5 are 14
6 and 6 are 12	7 and 6 are 13	8 and 6 are 14	9 and 6 are 15
6 and 7 are 13	7 and 7 are 14	8 and 7 are 15	9 and 7 are 16
6 and 8 are 14	7 and 8 are 15	8 and 8 are 16	9 and 8 are 17
6 and 9 are 15	7 and 9 are 16	8 and 9 are 17	9 and 9 are 18
6 and 10 are 16	7 and 10 are 17	8 and 10 are 18	9 and 10 are 19
6 and 11 are 17	7 and 11 are 13	8 and 11 are 19	9 and 11 are 20
6 and 12 are 18	7 and 12 are 19	8 and 12 are 20	9 and 12 are 21
10 and 0 are 10	11 and 0 are 11	12 and 0 are 12	13 and 0 are 13
10 and 1 are 11	11 and 1 are 12	12 and 1 are 13	13 and 1 are 14
10 and 2 are 12	11 and 2 are 13	12 and 2 are 14	13 and 2 are 15
10 and 3 are 13	11 and 3 are 14	12 and 3 are 15	13 and 3 are 16
10 and 4 are 14	11 and 4 are 15	12 and 4 are 16	13 and 4 are 17
10 and 5 are 15	11 and 5 are 16	12 and 5 are 17	13 and 5 are 18
10 and 6 are 16	11 and 6 are 17	12 and 6 are 18	13 and 6 are 19
10 and 7 are 17	11 and 7 are 18	12 and 7 are 19	13 and 7 are 20
10 and 8 are 18	11 and 8 are 19	12 and 8 are 20	13 and 8 are 21
10 and 9 are 19	11 and 9 are 20	12 and 9 are 21	13 and 9 are 22
10 and 10 are 20	11 and 10 are 21	12 and 10 are 22	13 and 10 are 23
10 and 11 are 21	11 and 11 are 22	12 and 11 are 23	13 and 11 are 24
10 and 12 are 22	11 and 12 are 23	12 and 12 are 24	13 and 12 are 25

Question. — Art. 17. What is the process called by which we find the sum of several numbers?

2. How many are 2 and 3? 2 and 5? 2 and 7? 2 and 9? 2 and 4? 2 and 2? 2 and 8? 2 and 6?

3. How many are 3 and 3? 3 and 5? 3 and 7? 3 and 9? 3 and 4? 3 and 6? 3 and 8? 3 and 3?

4. How many are 4 and 3? 4 and 5? 4 and 8? 4 and 9? 4 and 1? 4 and 2? 4 and 4? 4 and 7?

5. How many are 5 and 3? 5 and 4? 5 and 7? 5 and 8? 5 and 9? 5 and 2? 5 and 5? 5 and 6? 5 and 1?

6. How many are 6 and 2? 6 and 4? 6 and 3? 6 and 5? 6 and 7? 6 and 0? 6 and 1? 6 and 6? 6 and 8?

7. How many are 7 and 3? 7 and 5? 7 and 7? 7 and 6? 7 and 8? 7 and 9? 7 and 2? 7 and 4? 7 and 10?

8. How many are 8 and 2? 8 and 4? 8 and 5? 8 and 7? 8 and 9? 8 and 8? 8 and 1? 8 and 3? 8 and 6?

9. How many are 9 and 1? 9 and 3? 9 and 5? 9 and 4? 9 and 6? 9 and 8? 9 and 9? 9 and 2?

10. James had 4 apples, Samuel gave him 5 more, and John gave him 6; how many had he in all?

11. Gave 7 dollars for a barrel of flour, 5 dollars for a hundred weight of sugar, and 8 dollars for a tub of butter; what did I give for the whole?

12. Paid 5 dollars for a pair of boots, 12 dollars for a coat, and 6 dollars for a vest; what was the whole cost?

13. Gave 25 cents for an arithmetic, and 67 for a geography; what was the cost of both?

ILLUSTRATION. — We may divide the cents into tens and units. Thus, 25 equals 2 tens and 5 units; 67 equals 6 tens and 7 units; 2 tens and 6 tens are 8 tens; and 5 units and 7 units are 12 units, or 1 ten and 2 units; 1 ten and 2 units added to 8 tens make 9 tens and 2 units, or 92. Therefore the arithmetic and geography cost 92 cents.

14. On the fourth of July 20 cents were given to Emily, 15 cents to Betsey, 10 cents to Benjamin, and none to Lydia; what did they all receive?

15. Bought four loads of hay; the first cost 15 dollars, the second 12 dollars, the third 20 dollars, and the fourth 17 dollars; what was the price of the whole?

16. Gave 55 dollars for a horse, 40 dollars for a wagon, and 17 dollars for a harness; what did they all cost?

17. Sold 3 loads of wood for 17 dollars, 6 tons of timber for 19 dollars, and a pair of oxen for 60 dollars; what sum did I receive?

ART. 18. From the solution of the preceding questions, the learner will perceive, that

ADDITION is the process of finding the sum of two or more numbers. The result obtained, is called their *amount*.

Addition is commonly represented by this character, +, which signifies *plus*, or added to. The expression 7+5 is read, 7 plus 5, or 7 added to 5.

This character, =, is called the sign of equality, and signifies *equal to*. The expression 7+5=12 is read, 7 plus 5, or 7 added to 5, is equal to 12.

EXERCISES FOR THE SLATE.

ART. 19. The method of operation when the numbers are large, and the sum of each column is less than 10.

Ex. 1. A man bought a watch for 42 dollars, a coat for 16 dollars, and a set of maps for 21 dollars; what did he pay for the whole? Ans. 79 dollars.

OPERATION.

	Dollars.
	42
	16
	21
Amount	79

Having arranged the numbers so that all the units of the same order shall stand in the same column, we first add the column of *units;* thus, 1 and 6 are 7, and 2 are 9 (units), and write down the amount under the column of units. We then add the column of *tens;* thus, 2 and 1 are 3, and 4 are 7 (tens), which we write under the column of tens, and thus find the amount of the whole to be 79 dollars.

ART. 20. *First Method of Proof.* — Begin at the top and add the columns downward in the same manner as they were before added upward, and if the two sums agree the work is presumed to be right.

The reason of this proof is, that, by adding downward, the order of the figures is inverted; and, therefore, any error made in the first addition would probably be detected in the second.

NOTE. — This method of proof is generally used in business.

QUESTIONS. — Art. 18. What is addition? What is the sign of addition, and what does it signify? What is the sign of equality, and what does it signify? — Art. 19. How are numbers arranged for addition? Which column must first be added? Why? Where do you place its sum? Where must the sum of each column be placed? What is the whole sum called? — Art. 20. How is addition proved? What is the reason for this method of proof? Is this method in common use?

EXAMPLES FOR PRACTICE.

	2.	3.	4.	5.
	Miles.	Furlongs.	Days.	Weeks.
	151	234	472	121
	212	423	315	516
	321	321	102	361
Ans.	684			

6. What is the sum of 231, 114, and 324? Ans. 669.

7. Required the sum of 235, 321, and 142. Ans. 698.

8. What is the sum of 11, 22, 505, and 461? Ans. 999.

9. Sold twelve ploughs for 104 dollars, two wagons for 214 dollars, and one chaise for 121 dollars; what was the amount of the whole? Ans. 439 dollars.

10. A drover bought 125 sheep of one man, 432 of another, and of a third 311; how many did he buy? Ans. 868 sheep.

ART. 21. Method of operation when the sum of any column is equal to or exceeds 10.

Ex. 1. I have three lots of wild land; the first contains 246 acres, the second 764 acres, and the third 918 acres. I wish to know how many acres are in the three lots. Ans. 1928 acres.

OPERATION.

	Acres.
	246
	764
	918
Amount	1928

Having arranged the numbers as in the preceding examples, we first add the units; thus, 8 and 4 are 12, and 6 are 18 units, equal 1 ten and 8 units. We write the 8 units under the column of units, we *carry* or add the 1 ten to the column of tens; thus, 1 added to 1 makes 2, and 6 are 8, and 4 are 12 (tens), equal to 1 hundred and 2 tens. We write the 2 tens under the column of tens, and add the 1 hundred to the column of hundreds; thus, 1 added to 9 makes 10, and 7 are 17, and 2 are 19 (hundreds), equal to 1 thousand and 9 hundreds. We write the 9 under the column of hundreds; and there being no other column to be added, we set down the 1 thousand in thousands' place, and find the amount of the several numbers to be 1928.

NOTE.—A more concise way, in practice, is to omit calling the name of each figure as added, and name only results.

QUESTIONS.—Art. 21. When the sum of any column exceeds ten, where are the units written? What is done with the tens? Why do you carry, or add, one for every 10? How is the sum of the last column written?

ART. 22. From the preceding examples and illustrations in addition, we deduce the following general

RULE. — *Write the numbers so that all the figures of the same order shall stand in the same column.*

Add, upward, all the figures in the column of units, and, if the amount be less than ten, write it underneath. But, if the amount be ten or more, write down the unit figure only, and add in the figure denoting the ten or tens with the next column.

Proceed in this way with each column, until all are added, observing to write down the whole amount of the last column.

ART. 23. *Second Method of Proof.* — Separate the numbers to be added into two parts, by drawing a horizontal line between them. Add the numbers below the line, and set down their sum. Then add this sum and the number, or numbers, above the line together; and, if their sum is equal to the first amount, the work is presumed to be right.

The reason of this proof depends on the principle, *That the sum of all the parts into which any number is divided is equal to the whole.*

EXAMPLES FOR PRACTICE.

2. OPERATION.	2. OPERATION AND PROOF.	3. OPERATION.	3. OPERATION AND PROOF.
526	526	241	241
317	———	532	———
529	317	207	532
132	529	913	207
	132		913
Ans. 1504	First am't 1504	Ans. 1893	First am't 1893
	978		1652
	Ans. 1504		Ans. 1893

4. Dollars.	5. Miles.	6. Pounds.	7. Rods.	8. Inches.	9. Feet.
11	47	127	678	789	1769
23	87	396	971	478	7895
97	58	787	147	719	7563
86	83	456	716	937	8765
217	275	1766	2512	2923	25992

QUESTIONS. — Art 22. What is the general rule for addition? — Art. 23. What is the second method of proving addition? What is the reason of this method of proof?

10.	11.	12.	13.	14.
Ounces.	Drams.	Cents.	Eagles.	Degrees.
876	789	123	471	1234
376	567	478	617	3456
715	743	716	871	6544
678	435	478	317	7891
910	678	127	899	8766

15.	16.	17.	18.
Feet.	Inches.	Hours.	Minutes.
78956	71678	71123	98765
37667	12345	45678	12345
12345	67890	34680	67111
67890	34567	56777	33333
78999	89012	67812	71345
13579	78917	71444	99999

19.	20.	21.	22.
Days.	Years.	Months.	Hogsheads
17875897	789567	37	30176
7167512	7613	1378956	31
876567	123123	700714	8601
98765	70071	367	11
7896	475	76117	9911
789	1069	4611779	89120
78	374176	9171	710
7	761	131765	4325

23. Add 1001, 76, 10078, 15, 8761, 7, and 1678.
Ans. 21616.

24. Add 49, 761, 3756, 8, 150, 761761, and 18.
Ans. 766503.

25. Required the sum of 3717, 8, 7, 10001, 58, 18, and 5.
Ans. 13814.

26. Add 19, 181, 5, 897156, 81, 800, and 71512.
Ans. 969754.

27. What is the sum of 999, 8081, 9, 1567, 88, 91, 7, and 878?
Ans. 11720.

28. Add 71, 18765, 9111, 1471, 678, 9, 1446, and 71.
Ans. 31622.

29. Add 51, 1, 7671, 89, 871787, 61, and 70001.
Ans. 949661.

30. What is the sum of 71, 8956, 1, 785, 587, and 76178? Ans. 86578.

31. Add 9999, 8008, 8, 81, 4777, and 516785. Ans. 539658.

32. Add 5, 7, 8911, 467, 47895, and 87. Ans. 57372.

33. Add 123456, 71, 8005, 21, and 716787. Ans. 848340.

34. Add 47, 911111, 717, 81, 88767, and 56. Ans. 1000779.

35. What is the sum of 71, 8899, 4, 7111, and 678679? Ans. 694764.

36. Add 81, 879, 41, 76789, 42, 1, and 78967. Ans. 156800.

37. Add 917658, 75, 876789, 46, and 8222. Ans. 1802790.

38. Add 91, 76756895, 76, 14, 3, and 76378. Ans. 76833457.

39. Add 10, 100, 1000, 10000, 100000, and 1000000. Ans. 1111110.

40. What is the sum of 9, 99, 99, 1111, 8000, and 5? Ans. 9323.

41. Add 41, 7651, 7678956, 43, 15, and 6780. Ans. 7693486.

42. Add 1234, 7891, 3146751, 27, 9, and 5. Ans. 3155917.

43. What is the sum of 19, 91, 1, 1, 1478, 1007, and 46? Ans. 2643.

44. Add four hundred seventy-six, seventy-one, one hundred five, and three hundred eighty-seven. Ans. 1039.

45. Add fifty-six thousand seven hundred eighty-five, seven hundred five, thirty-six, one hundred seventy thousand and one, and four hundred seven. Ans. 227934.

46. Add fifty-six thousand seven hundred eleven, three thousand seventy-one, four hundred seventy-one, sixty-one, and three thousand and one. Ans. 63315.

47. What is the sum of the following numbers: seven hundred thousand seven hundred one, seventeen thousand nine, one million six hundred thousand seven hundred six, forty-seven thousand six hundred seventy-one, seven thousand forty-seven, four hundred one, and nine? Ans. 2373544.

48. Gave 73 dollars for a watch, 15 dollars for a cane, 119 dollars for a horse, 376 dollars for a carriage, and 7689 dollars for a house; how much did they all cost? Ans. 8272 dollars.

49. In an orchard, 15 trees bear plums, 73 trees bear apples, 29 trees bear pears, and 14 trees bear cherries; how many trees are there in the orchard? Ans. 131 trees.

50. The hind quarters of an ox weighed 375 pounds each, the fore quarters 315 pounds each; the hide weighed 96 pounds, and the tallow 87 pounds. What was the whole weight of the ox? Ans. 1563 pounds.

51. A man bought a farm for 1728 dollars, and sold it so as to gain 375 dollars; how much did he sell it for?
Ans. 2103 dollars.

52. A merchant bought five pieces of cloth. For the first he gave 376 dollars, for the second 198 dollars, for the third 896 dollars, for the fourth 691 dollars, and for the fifth 96 dollars. How much did he give for the whole? Ans. 2257 dollars.

53. A merchant bought five hogsheads of molasses for 375 dollars, and sold it so as to gain 25 dollars on each hogshead; for how much did he sell it? Ans. 500 dollars.

54. John Smith's farm is worth 7896 dollars; he has bank stock valued at 369 dollars, and he has in cash 850 dollars. How much is he worth? Ans. 9115 dollars.

55. Required the number of inhabitants in the New England States. By the census of 1850 there were in Maine 583,169, in New Hampshire 317,976, in Massachusetts 994,514, in Rhode Island 147,545, in Connecticut 370,792, and in Vermont 314,120. Ans. 2,728,116.

56. Required the number of inhabitants in the Middle States, including the District of Columbia. In 1850 there were in New York 3,097,394, in New Jersey 489,555, in Pennsylvania 2,311,786, in Delaware 91,532, in Maryland 583,034, and in the District of Columbia 51,687. Ans. 6,624,988.

57. Required the number of inhabitants in the Southern States. In 1850 there were in Virginia 1,421,661, in North Carolina 869,039, in South Carolina 668,507, in Georgia 906,185, and in Florida 87,445. Ans. 3,952,837.

58. Required the number of inhabitants in the South-Western States. In 1850 there were in Alabama 771,623, in Mississippi 606,526, in Louisiana 517,762, in Texas 212,592, in Arkansas 209,897, and in Tennessee 1,002,917. Ans. 3,321,317.

59. Required the number of inhabitants in the North-Western States and Territories. In 1850 there were in Missouri 682,044, in Kentucky 982,405, in Ohio 1,980,329, in Indiana 988,416, in Illinois 851,470, in Michigan 397,654, in Wisconsin 305,391,

in Iowa 192,214, in California 92,597, and in the Territories 92,298. Ans. 6,564,818.

ART. **24.** Method of adding two or more columns at a single operation.

Ex. 1. Washington lived 68 years; John Adams, 91 years Jefferson, 83 years; Madison, 85 years. What is the sum of the years they all lived? Ans. 327.

OPERATION.

	Years.
	6 8
	9 1
	8 3
	8 5
Amount	3 2 7

Beginning with the number last written down, we add the units and tens, thus: 85 and 3 equal 88, and 80 equal 168, and 1 equal 169, and 90 equal 259, and 8 equal 267, and 60 equal 327, the sum sought. In like manner may be added more than two columns at one operation.

NOTE. — The examples that follow can be performed as the above, or by the common method, or by both, as the teacher may advise.

2.	3.	4.	5.	6.
Ounces.	Yards.	Feet.	Inches.	Chaldrons.
1234	2345	3456	7891	5678
5678	6789	7891	1356	3215
9012	1023	3456	7891	6789
3456	4456	7891	2345	3214
7890	7890	3456	6789	1234
1345	1234	7890	1234	3789
6789	5678	1378	5678	1379
3216	9012	8123	9123	9008
7890	3456	4567	4567	1071
1030	7890	8912	8912	7163
7055	1345	3456	3456	6781
5678	6789	7891	7812	1780
1234	3456	3456	3456	3007
5678	7890	7891	7812	5617
9001	5678	3783	3713	4456
2345	9012	1237	7891	3456
6789	3456	7891	1357	7891
1030	7890	1007	9009	3070
7816	1234	5670	8765	4567
1781	5678	1234	4321	3456

§ III. SUBTRACTION.

Mental Exercises.

Art. 25. When it is required to find the difference between two numbers, the process is called *Subtraction*. The operation is the reverse of addition.

Ex. 1. John has 7 oranges, and his sister but 4; how many more has John than his sister?

Illustration. — We first inquire what number added to 4 will make 7. From addition we learn that 4 and 3 are 7; consequently, if 4 oranges be taken from 7 oranges, 3 will remain Hence John has 3 oranges more than his sister.

SUBTRACTION TABLE.

1 from	1 leaves	0	2 from	2 leaves	0	3 from	3 leaves	0	4 from	4 leaves	0
1 from	2 leaves	1	2 from	3 leaves	1	3 from	4 leaves	1	4 from	5 leaves	1
1 from	3 leaves	2	2 from	4 leaves	2	3 from	5 leaves	2	4 from	6 leaves	2
1 from	4 leaves	3	2 from	5 leaves	3	3 from	6 leaves	3	4 from	7 leaves	3
1 from	5 leaves	4	2 from	6 leaves	4	3 from	7 leaves	4	4 from	8 leaves	4
1 from	6 leaves	5	2 from	7 leaves	5	3 from	8 leaves	5	4 from	9 leaves	5
1 from	7 leaves	6	2 from	8 leaves	6	3 from	9 leaves	6	4 from	10 leaves	6
1 from	8 leaves	7	2 from	9 leaves	7	3 from	10 leaves	7	4 from	11 leaves	7
1 from	9 leaves	8	2 from	10 leaves	8	3 from	11 leaves	8	4 from	12 leaves	8
1 from	10 leaves	9	2 from	11 leaves	9	3 from	12 leaves	9	4 from	13 leaves	9
1 from	11 leaves	10	2 from	12 leaves	10	3 from	13 leaves	10	4 from	14 leaves	10
1 from	12 leaves	11	2 from	13 leaves	11	3 from	14 leaves	11	4 from	15 leaves	11
1 from	13 leaves	12	2 from	14 leaves	12	3 from	15 leaves	12	4 from	16 leaves	12
5 from	5 leaves	0	6 from	6 leaves	0	7 from	7 leaves	0	8 from	8 leaves	0
5 from	6 leaves	1	6 from	7 leaves	1	7 from	8 leaves	1	8 from	9 leaves	1
5 from	7 leaves	2	6 from	8 leaves	2	7 from	9 leaves	2	8 from	10 leaves	2
5 from	8 leaves	3	6 from	9 leaves	3	7 from	10 leaves	3	8 from	11 leaves	3
5 from	9 leaves	4	6 from	10 leaves	4	7 from	11 leaves	4	8 from	12 leaves	4
5 from	10 leaves	5	6 from	11 leaves	5	7 from	12 leaves	5	8 from	13 leaves	5
5 from	11 leaves	6	6 from	12 leaves	6	7 from	13 leaves	6	8 from	14 leaves	6
5 from	12 leaves	7	6 from	13 leaves	7	7 from	14 leaves	7	8 from	15 leaves	7
5 from	13 leaves	8	6 from	14 leaves	8	7 from	15 leaves	8	8 from	16 leaves	8
5 from	14 leaves	9	6 from	15 leaves	9	7 from	16 leaves	9	8 from	17 leaves	9
5 from	15 leaves	10	6 from	16 leaves	10	7 from	17 leaves	10	8 from	18 leaves	10
5 from	16 leaves	11	6 from	17 leaves	11	7 from	18 leaves	11	8 from	19 leaves	11
5 from	17 leaves	12	6 from	18 leaves	12	7 from	19 leaves	12	8 from	20 leaves	12
9 from	9 leaves	0	10 from	10 leaves	0	11 from	11 leaves	0	12 from	12 leaves	0
9 from	10 leaves	1	10 from	11 leaves	1	11 from	12 leaves	1	12 from	13 leaves	1
9 from	11 leaves	2	10 from	12 leaves	2	11 from	13 leaves	2	12 from	14 leaves	2
9 from	12 leaves	3	10 from	13 leaves	3	11 from	14 leaves	3	12 from	15 leaves	3
9 from	13 leaves	4	10 from	14 leaves	4	11 from	15 leaves	4	12 from	16 leaves	4
9 from	14 leaves	5	10 from	15 leaves	5	11 from	16 leaves	5	12 from	17 leaves	5
9 from	15 leaves	6	10 from	16 leaves	6	11 from	17 leaves	6	12 from	18 leaves	6
9 from	16 leaves	7	10 from	17 leaves	7	11 from	18 leaves	7	12 from	19 leaves	7
9 from	17 leaves	8	10 from	18 leaves	8	11 from	19 leaves	8	12 from	20 leaves	8
9 from	18 leaves	9	10 from	19 leaves	9	11 from	20 leaves	9	12 from	21 leaves	9
9 from	19 leaves	10	10 from	20 leaves	10	11 from	21 leaves	10	12 from	22 leaves	10
9 from	20 leaves	11	10 from	21 leaves	11	11 from	22 leaves	11	12 from	23 leaves	11
9 from	21 leaves	12	10 from	22 leaves	12	11 from	23 leaves	12	12 from	24 leaves	12

Questions. — Art. 25. What does subtraction teach? Of what is it the reverse?

2. Thomas had five oranges, and gave two of them to John how many had he left?

3. Peter had six marbles, and gave two of them to Samuel· how many had he left?

4. Lydia had four cakes; having lost one, how many had she left?

5. Daniel, having eight cents, gives three to Mary; how many has he left?

6. Benjamin had ten nuts; he gave four to Jane, and three to Emily; how many had he left?

7. Moses gives eleven oranges to John, and eight to Enoch; how many more has John than Enoch?

8. Paid seven dollars for a pair of boots, and two dollars for shoes; how much did the boots cost more than the shoes?

9. How many are 4 less 2? 4 less 1? 4 less 4?

10. How many are 4 less 3? 5 less 1? 5 less 5?

11. How many are 5 less 2? 5 less 3? 5 less 4?

12. How many are 6 less 1? 6 less 2? 6 less 4? 6 less 5?

13. How many are 7 less 2? 7 less 3? 7 less 4? 7 less 6?

14. How many are 8 less 6? 8 less 5? 8 less 2? 8 less 4? 8 less 1?

15. How many are 9 less 2? 9 less 4? 9 less 5? 9 less 7? 9 less 3?

16. How many are 10 less 8? 10 less 7? 10 less 5? 10 less 3? 10 less 1?

17. How many are 11 less 9? 11 less 7? 11 less 5? 11 less 3? 11 less 4?

18. How many are 12 less 10? 12 less 8? 12 less 6? 12 less 4? 12 less 7?

19. How many are 13 less 11? 13 less 10? 13 less 7? 13 less 9? 13 less 5?

20. How many are 14 less 11? 14 less 9? 14 less 8? 14 less 6? 14 less 7? 14 less 3?

21. How many are 15 less 2? 15 less 4? 15 less 5? 15 less 7? 15 less 9? 15 less 13?

22. How many are 16 less 3? 16 less 4? 16 less 7? 16 less 9? 16 less 11? 16 less 15?

23. How many are 17 less 1? 17 less 3? 17 less 5? 17 less 7? 17 less 8? 17 less 12?

24. How many are 18 less 2? 18 less 4? 18 less 7? 18 less 8? 18 less 10? 18 less 12?

25. How many are 19 less 1? 19 less 3? 19 less 5? 19 less 7? 19 less 9? 19 less 16?

26. How many are 20 less 5? 20 less 8? 20 less 9? 20 less 12? 20 less 15? 20 less 19?

27. Bought a horse for 60 dollars, and sold him for 90 dollars; how much did I gain?

ILLUSTRATION. — We may divide the two prices of the horse into tens, and subtract the greater from the less. Thus 60 equals 6 tens, and 90 equals 9 tens; 6 tens from 9 tens leave 3 tens, or 30. Therefore I gained 30 dollars.

28. Sold a wagon for 70 dollars, which cost me 100 dollars; how much did I lose?

29. John travels 30 miles a day, and Samuel 90 miles; what is the difference?

30. I have 100 dollars, and after I shall have given 20 to Benjamin, and paid a debt of 30 dollars to J. Smith, how many dollars have I left?

31. John Smith, Jr., had 170 dollars; he gave his oldest daughter, Angeline, 40 dollars, his youngest daughter, Mary, 50 dollars, his oldest son, James, 30, and his youngest son, William, 20 dollars; he also paid 20 dollars for his taxes; how many dollars had he remaining?

ART. **26.** The pupil, having solved the preceding questions, will perceive, that

SUBTRACTION is the taking of one number from another to find the difference.

When the two numbers are unequal, the larger is called the *Minuend*, and the less number the *Subtrahend*. The answer, or number found by the operation, is called the *Difference*, or *Remainder*.

NOTE. — The words *minuend* and *subtrahend* are derived from two Latin words; the former from *minuendum*, which signifies *to be diminished* or *made less*, and the latter from *subtrahendum*, which means *to be subtracted* or *taken away*.

ART. **27.** SIGNS. — Subtraction is denoted by a short horizontal line, thus —, signifying *minus*, or *less*. It indicates that the number *following* is to be taken from the one that *precedes* it. The expression $6 - 2 = 4$ is read, 6 minus, or less, 2 is equal to 4.

QUESTIONS. — Art. 26. What is subtraction? What is the greater number called? What is the less number called? What the answer? — Art. 27. What is the sign of subtraction? What does it signify and indicate?

EXERCISES FOR THE SLATE.

ART. 28. Method of operation, when the numbers are large and each figure in the subtrahend is less than the figure above it in the minuend.

Ex. 1. Let it be required to take 245 from 468, and to find their difference. Ans. 223.

	OPERATION.
Minuend	4 6 8
Subtrahend	2 4 5
Remainder	2 2 3

We place the less number under the greater, units under units, tens under tens, &c., and draw a line below them. We then begin at the right hand, and say, 5 units from 8 units leave 3 units, and write the 3 in units' place below. We then say, 4 tens from 6 tens leave 2 tens, and write the 2 in tens' place below; and proceed with the next figure, and say, 2 hundreds from 4 hundreds leave 2 hundreds, which we write in hundreds' place below. We thus find the difference to be 223.

ART. 29. *First Method of Proof.* — Add the remainder and the subtrahend together, and their sum will be equal to the minuend, if the work is right.

This method of proof depends on the principle, *That the greater of any two numbers is equal to the less added to the difference between them.*

EXAMPLES FOR PRACTICE.

	2. OPERATION.	2. OPERATION AND PROOF.	3. OPERATION.	3. OPERATION AND PROOF.
Minuend	5 4 7	5 4 7	9 8 6	9 8 6
Subtrahend	2 3 5	2 3 5	7 6 3	7 6 3
Remainder	3 1 2	3 1 2	2 2 3	2 2 3
		Min. 5 4 7		Min. 9 8 6

	4.	5.	6.	7.
From	6 8 4	7 3 5	8 6 4	9 4 8
Take	4 6 2	5 2 3	6 5 1	7 4 6

8. A farmer paid 539 dollars for a span of fine horses, and sold them for 425 dollars; how much did he lose?

Ans. 114 dollars.

9. A farmer raised 896 bushels of wheat, and sold 675 bushels of it; how much did he reserve for his own use?

Ans. 221 bushels.

QUESTIONS. — Art. 28. How are numbers arranged for subtraction? Where do you begin to subtract? Why? Where do you write the difference? — Art. 29. What is the first method of proving subtraction? What is the reason of this proof, or on what principle does it depend?

10. A gentleman gave his son 3692 dollars, and his daughter 1212 dollars less than his son; how much did his daughter receive? Ans. 2480 dollars.

ART. **30.** Method of operation when any figure in the subtrahend is greater than the figure above it in the minuend.

Ex. 1. If I have 624 dollars, and lose 342 of them, how many remain? Ans. 282.

	OPERATION.
Minuend	6 2 4
Subtrahend	3 4 2
Remainder	2 8 2

We first take the 2 units from the 4 units, and find the difference to be 2 units, which we write under the figure subtracted. We then proceed to take the 4 tens from the 2 tens above it; but we here find a difficulty, since the 4 is greater than 2, and cannot be subtracted from it. We therefore add 10 to the 2 tens, which makes 12 tens, and then subtract the 4 from 12, and 8 tens remain, which we write below. Then, to compensate for the 10 thus added to the 2 in the minuend, we add one to the 3 hundreds in the next higher place in the subtrahend, which makes 4 hundreds, and subtract the 4 from 6 hundreds, and 2 hundreds remain. The remainder, therefore, is 282.

The reason of this operation depends upon the self-evident truth, *That, if any two numbers are equally increased, their difference remains the same.* In this example 10 tens, equal to 1 hundred, were added to the 2 tens in the upper number, and 1 was added to the 2 hundreds in the lower number. Now, since the 3 stands in the hundreds' place, the 1 added was in fact 1 hundred. Hence, the two numbers being equally increased, the difference is the same.

NOTE. — This addition of 10 to the minuend is sometimes called *borrowing* 10, and the addition of 1 to the subtrahend is called *carrying* 1.

ART. **31.** From the preceding examples and illustrations in subtraction, we deduce the following general

RULE. — *Place the less number under the greater, so that units of the same order shall stand in the same column.*

Commencing at the right hand, subtract each figure of the subtrahend from the figure above it.

If any figure of the subtrahend is larger than the figure above it in the minuend, add 10 *to that figure of the minuend before subtracting, and then add* 1 *to the next figure of the subtrahend.*

QUESTIONS. — Art. 30. How do you proceed when a figure of the subtrahend is larger than the one above it in the minuend? How do you compensate for the 10 which is added to the minuend? What is the reason for this addition to the minuend and subtrahend? How does it appear that the 1 added to the subtrahend equals the 10 added to the minuend? What is the addition of 10 to the minuend sometimes called? The addition of 1 to the subtrahend? — Art. 31. What is the general rule for subtraction?

ART. 32. *Second Method of Proof.* — Subtract the remainder or difference from the minuend, and the result will be like the subtrahend if the work is right.

This method of proof depends on the principle, *That the smaller of any two numbers is equal to the remainder obtainea by subtracting their difference from the greater.*

EXAMPLES FOR PRACTICE.

	2. OPERATION.	2. OPERATION AND PROOF.	3. OPERATION.	3. OPERATION AND PROOF
Minuend	376	376	531	531
Subtrahend	167	167	389	389
Remainder	209	209	142	142
		Sub. 167		Sub. 389

	4. Tons.	5. Gallons.	6. Pecks.	7. Feet.
From	978	67158	14711	100000
Take	199	14339	9197	90909
Ans.	779	52819	5514	9091

	8. Miles.	9. Dollars.	10. Minutes.	11. Seconds.
From	67895	456798	765321	555555
Take	19999	190899	177777	177777

	12. Rods.	13. Acres.
From	1002003004005 00	1000000000000
Take	908070605040 30	999999999999

14. From 671111 take 199999. Ans. 471112.
15. From 1789100 take 808088. Ans. 981012.
16. From 1000000 take 999999. Ans. 1.
17. From 9999999 take 1607. Ans. 9998392.
18. From 6101507601061 take 3806790989. Ans. 6097700810072.

QUESTIONS. — Art. 32. What is the second method of proving subtraction? What is the reason for this method of proof, or on what principle does it depend?

19. From 8054010657811 take 76909748598.
Ans. 7977100909213.

20. From 7100071641115 take 10071178.
Ans. 7100061569937.

21. From 501505010678 take 794090589.
Ans. 500710920089.

22. Take 99999999 from 100000000. Ans. 1.

23. Take 44444444 from 500000000. Ans. 455555556.

24. Take 1234567890 from 9987654321.
Ans. 8753086431.

25. From 800700567 take 1010101. Ans. 799690466.

26. Take twenty-five thousand twenty-five from twenty-five millions. Ans. 24974975.

27. Take nine thousand ninety-nine from ninety-nine thousand. Ans. 89901.

28. From one hundred one millions ten thousand one hundred one take ten millions one hundred one thousand and ten.
Ans. 90909091.

29. From one million take nine. Ans. 99991.

30. From three thousand take thirty-three. Ans. 2967.

31. From one hundred millions take five thousand.
Ans. 99995000.

32. From 1,728 dollars, I paid 961 dollars; how many remain? Ans. 767 dollars.

33. Our national independence was declared in 1776; how many years from that period to the close of the last war with Great Britain, in 1815? Ans. 39 years.

34. The last transit of Venus was in 1769, and the next will be in 1874; how many years will intervene?
Ans. 105 years.

35. The State of New Jersey contains 6851 square miles, and Delaware 2120. How many more square miles has the former State than the latter? Ans. 4731.

36. In 1840 the number of inhabitants in the United States was 17,069,453, and in 1850 it was 23,191,876; what was the increase? Ans. 6,122,423.

37. In 1850 there were raised in the State of Ohio 56,619,608 bushels of corn, and in 1853, 73,436,690 bushels; what was the increase? Ans. 16,817,082 bushels.

38. By the census of 1850, 13,121,498 bushels of wheat were raised in New York, and 15,367,691 bushels in Pennsylvania; how many bushels in the latter State more than in the former?
Ans. 2,246,193 bushels.

39. The city of New York owes 13,960,856 dollars, and Boston owes 7,779,855 dollars; how much more does New York owe than Boston? Ans. 6,181,001 dollars.

40. From five hundred eighty-one thousand take three thousand and ninety-six. Ans. 577,904.

41. It was ascertained by a transit of Venus, June 3, 1769 that the mean distance of the earth from the sun was ninety-five millions one hundred seventy-three thousand one hundred twenty-seven miles, and that the mean distance of Mars from the sun was one hundred forty-five millions fourteen thousand one hundred forty-eight miles. Required the difference of their distances from the sun. Ans. 49,841,021 miles.

ART. 33. Method of subtracting when there are two or more subtrahends.

Ex. 1. A man owing 767 dollars, paid at one time 190 dollars, at another time 131 dollars, at another time 155 dollars; how much did he then owe? Ans 291 dollars.

FIRST OPERATION.

	Dollars.
Minuend	7 6 7
	1 3 1
	1 9 0
	1 5 5
Subtrahend	4 7 6
Remainder	2 9 1

SECOND OPERATION.

	Dollars.
Minuend	7 6 7
Subtrahends	1 3 1
	1 9 0
	1 5 5
Remainder	2 9 1

In the first operation, the *several subtrahends*, having been properly placed, *are added for a single subtrahend*, to be taken from the minuend. In the second, *the subtrahends are subtracted, as they are added, at one operation*, thus: beginning with units, 5 and 1 equal 6, which from 7 units leaves 1 unit; passing to tens, 5 and 9 and 3 equal 17 tens; reserving the left-hand figure to add in with the figures of the subtrahends in the next column, the right-hand figure, 7, being larger than 6 tens of the minuend, we add 10 to the 6, and, subtracting, have left 9 tens; and, passing to hundreds, we add in the left-hand figure 1 reserved from the 17 tens, and also add 1, equal 10 tens, to compensate for the 10 added to the minuend, and with the other figures, 1 and 1 and 1 equal 5 hundreds, which, taken from 7 hundreds, leave 2 hundreds; and 291 as the answer sought.

2. E. Webster owned 6,765 acres of land, and he gave to his oldest brother 2,196 acres, and his uncle Rollins 1,981 acres; how much has he left? Ans. 2,588 acres.

3. The real estate of James Dow is valued at 3,769 dollars, and his personal estate at 2,648 dollars; he owes John Smith 1,728 dollars, and Job Tyler 1,161 dollars; how much is Dow worth? Ans. 3,528 dollars.

§ IV. MULTIPLICATION.

MENTAL EXERCISES.

ART. 34. WHEN any number is to be added to itself several times, the operation may be shortened by a process called *Multiplication.*

Ex. 1. If a man can earn 8 dollars in 1 week, what will he earn in 4 weeks?

ILLUSTRATION. — It is evident, since a man can earn 8 dollars in 1 week, in 4 weeks he will earn 4 times as much, and the result may be obtained by addition; thus, $8 + 8 + 8 + 8 = 32$; or, by a more convenient process, by setting down the 8 but once, and multiplying it by 4, the number of times it is to be repeated; thus, 4 times 8 are 32. Hence in 4 weeks he will earn 32 dollars.

MULTIPLICATION TABLE.

2 times 1 are 2	3 times 1 are 3	4 times 1 are 4	5 times 1 are 5
2 times 2 are 4	3 times 2 are 6	4 times 2 are 8	5 times 2 are 10
2 times 3 are 6	3 times 3 are 9	4 times 3 are 12	5 times 3 are 15
2 times 4 are 8	3 times 4 are 12	4 times 4 are 16	5 times 4 are 20
2 times 5 are 10	3 times 5 are 15	4 times 5 are 20	5 times 5 are 25
2 times 6 are 12	3 times 6 are 18	4 times 6 are 24	5 times 6 are 30
2 times 7 are 14	3 times 7 are 21	4 times 7 are 28	5 times 7 are 35
2 times 8 are 16	3 times 8 are 24	4 times 8 are 32	5 times 8 are 40
2 times 9 are 18	3 times 9 are 27	4 times 9 are 36	5 times 9 are 45
2 times 10 are 20	3 times 10 are 30	4 times 10 are 40	5 times 10 are 50
2 times 11 are 22	3 times 11 are 33	4 times 11 are 44	5 times 11 are 55
2 times 12 are 24	3 times 12 are 36	4 times 12 are 48	5 times 12 are 60
6 times 1 are 6	7 times 1 are 7	8 times 1 are 8	9 times 1 are 9
6 times 2 are 1[illegible]	7 times 2 are 14	8 times 2 are 16	9 times 2 are 18
6 times 3 are 18	7 times 3 are 21	8 times 3 are 24	9 times 3 are 27
6 times 4 are 24	7 times 4 are 28	8 times 4 are 32	9 times 4 are 36
6 times 5 are 30	7 times 5 are 35	8 times 5 are 40	9 times 5 are 45
6 times 6 are 36	7 times 6 are 42	8 times 6 are 48	9 times 6 are 54
6 times 7 are 42	7 times 7 are 49	8 times 7 are 56	9 times 7 are 63
6 times 8 are 48	7 times 8 are 56	8 times 8 are 64	9 times 8 are 72
6 times 9 are 54	7 times 9 are 63	8 times 9 are 72	9 times 9 are 81
6 times 10 are 60	7 times 10 are 70	8 times 10 are 80	9 times 10 are 90
6 times 11 are 66	7 times 11 are 77	8 times 11 are 88	9 times 11 are 99
6 times 12 are 72	7 times 12 are 84	8 times 12 are 96	9 times 12 are 108
10 times 1 are 10	10 times 11 are 110	11 times 8 are 88	12 times 4 are 48
10 times 2 are 20	10 times 12 are 120	11 times 9 are 99	12 times 5 are 60
10 times 3 are 30		11 times 10 are 110	12 times 6 are 72
10 times 4 are 40	11 times 1 are 11	11 times 11 are 121	12 times 7 are 84
10 times 5 are 50	11 times 2 are 22	11 times 12 are 132	12 times 8 are 96
10 times 6 are 60	11 times 3 are 33		12 times 9 are 108
10 times 7 are 70	11 times 4 are 44	12 times 1 are 12	12 times 10 are 120
10 times 8 are 80	11 times 5 are 55	12 times 2 are 24	12 times 11 are 132
10 times 9 are 90	11 times 6 are 66	12 times 3 are 36	12 times 12 are 144
10 times 10 are 100	11 times 7 are 77		

QUESTION. — Art. 34. How may the process of adding a number to itself several times be shortened?

2. What cost 5 barrels of flour at 6 dollars per barrel?

ILLUSTRATION. — If a barrel of flour cost 6 dollars, 5 barrels will cost 5 times as much; 5 times 6 are 30. Hence 5 barrels of flour at 6 dollars per barrel will cost 30 dollars.

3. What cost 6 bushels of beans at 2 dollars per bushel?
4. What cost 5 quarts of cherries at 7 cents per quart?
5. What will 7 quarts of vinegar cost at 12 cents per quart?
6. What cost 9 acres of land at 10 dollars per acre?
7. If a pint of currants cost 4 cents, what cost 9 pints?
8. If in 1 penny there are 4 farthings, how many in 9 pence? In 7 pence? In 8 pence? In 4 pence? In 3 pence?
9. If 12 pence make a shilling, how many pence in 3 shillings? In 5 shillings? In 7 shillings? In 9 shillings?
10. If one pound of raisins cost 6 cents, what cost 4 pounds? 5 pounds? 6 pounds? 7 pounds? 8 pounds? 9 pounds? 10 pounds? 12 pounds?
11. In one acre there are four roods; how many roods in 2 acres? In 3 acres? In 4 acres? In 5 acres? In 6 acres? In 9 acres?
12. A good pair of boots is worth 5 dollars; what must I give for 5 pairs? For 6 pairs? For 7 pairs? For 8 pairs? For 9 pairs?
13. A cord of good walnut wood may be obtained for 8 dollars; what must I give for 4 cords? For 6 cords? For 9 cords?
14. What cost 4 quarts of milk at 5 cent a quart, and 8 gallons of vinegar at 10 cents a gallon?
15. If a man earn 7 dollars a week, how much will he earn in 3 weeks? In 4 weeks? In 5 weeks? In 6 weeks? In 7 weeks? In 9 weeks?
16. If 1 thousand feet of boards cost 12 dollars, what cost 4 thousand? 5 thousand? 6 thousand? 7 thousand? 9 thousand? 12 thousand?
17. If 3 pairs of shoes buy 1 pair of boots, how many pairs of shoes will it take to buy 7 pairs of boots?
18. If 5 bushels of apples buy 1 barrel of flour, how many bushels of apples are equal in value to 12 barrels of flour?
19. If 1 yard of canvas cost 25 cents, what will 12 yards cost?

ILLUSTRATION. — The number 25 is composed of 2 tens and 5 units; 12 times 2 tens are 24 tens; and 12 times 5 units are

60 units, or 6 tens, 24 tens added to 6 tens make 30 tens, or 300. Therefore, 12 yards will cost 300 cents, or 3 dollars.

20. In 1 pound there are 20 shillings; how many shillings in 3 pounds? In 4 pounds? In 6 pounds?

21. A gallon of molasses is worth 25 cents; what is the value of 2 gallons? Of 3 gallons? Of 4 gallons? Of 5 gallons? Of 6 gallons? Of 9 gallons?

22. If 1 man earn 12 dollars in 16 days, how much would 10 men earn in the same time?

23. If a steam-engine runs 28 miles in 1 hour, how far will it run in 4 hours? In 6 hours? In 9 hours?

24. If the earth turns on its axis 15 degrees in 1 hour, how far will it turn in 7 hours? In 11 hours? In 12 hours?

25. In a certain regiment there are 8 companies, in each company 6 platoons, and in each platoon 12 soldiers; how many soldiers are there in the regiment?

26. If 1 man walk 7 miles in 1 hour, how far will he walk in 8 hours? In 9 hours? In 11 hours? In 12 hours? In 20 hours? In 30 hours?

ART. **35.** The learner, having performed the foregoing questions, will perceive that

MULTIPLICATION is the process of taking a number as many times as there are units in another number.

In multiplication three terms are employed, called the *Multiplicand*, the *Multiplier*, and the *Product*.

The *multiplicand* is the number to be multiplied or taken.

The *multiplier* is the number by which we multiply, and denotes the number of times the multiplicand is to be taken.

The *product* is the result, or number produced by the multiplication.

The multiplicand and multiplier are often called FACTORS.

SIGNS. — The sign of multiplication is formed by two short lines crossing each other obliquely; thus, $\times$. It shows that the numbers between which it is placed are to be multiplied together; thus, the expression $7 \times 5 = 35$ is read, 7 multiplied by 5 is equal to 35.

QUESTIONS. — Art. 35. What is multiplication? What three terms are employed? What is the multiplicand? What is the multiplier? What is the product? What are the multiplicand and multiplier often called? What is the sign of multiplication? What does it show?

EXERCISES FOR THE SLATE.

ART. **36.** Method of operation when the multiplier does not exceed 12.

Ex. 1. Let it be required to multiply 175 by 7.

Ans. 1225.

	OPERATION.
Multiplicand	1 7 5
Multiplier	7
Product	1 2 2 5

Having written the multiplier under the unit figure of the multiplicand, we multiply the 5 units by 7, obtaining 35, and set down the 5 units directly under the 7, and reserve the 3 tens for the tens' column. We then multiply the 7 tens by 7, obtaining 49, and, adding the 3 tens which were reserved, we have 52 tens, or 5 hundreds and 2 tens. Writing down the 2 tens, and reserving the 5 hundreds, we multiply 1 by 7; and, adding the reserved 5 hundreds, we have 12 hundreds, which we write down in full, and the product is 1225.

EXAMPLES FOR PRACTICE.

	2.	3.	4.
Multiply	8 7 5 6	4 5 6 7	7 8 9 6
By	4	3	5
Ans.	3 5 0 2 4	1 3 7 0 1	3 9 4 8 0

5.	6.	7.	8.
5 6 8 0 7	4 7 8 9 3	6 1 6 5 7	8 9 7 6 5
5	6	7	9
2 8 4 0 3 5	2 8 7 3 5 8	4 3 1 5 9 9	8 0 7 8 8 5

9. Multiply 767853 by 9. Ans. 6910677.
10. Multiply 876538765 by 8. Ans. 7012310120.
11. Multiply 7654328 by 7. Ans. 53580296.
12. Multiply 4976387 by 5. Ans. 24881935.
13. Multiply 8765448 by 12. Ans. 105185376.
14. Multiply 4567839 by 11. Ans. 50246229.
15. What cost 8675 barrels of flour at 7 dollars per barrel? Ans. 60725 dollars.

QUESTIONS. — Art. 36. How must numbers be written for multiplication? At which hand do you begin to multiply? Why? Where do you write the first or right-hand figure of the product of each figure in the multiplicand? Why? What is done with the tens or left-hand figure of each product? How, then, do you proceed when the multiplier does not exceed 12?

16. What cost 25384 tons of hay at 9 dollars per ton?
Ans. 228456 dollars.

17. If, on 1 page in this book, there are 2538 letters, how many are there on 11 pages? Ans. 27918 letters.

ART. 37. Method of operation when the multiplier exceeds 12.

Ex. 1. Let it be required to multiply 763 by 24.
Ans. 18312.

OPERATION.

Multiplicand	7 6 3
Multiplier	2 4
	3 0 5 2
	1 5 2 6
Product	1 8 3 1 2

We write the multiplier under the multiplicand, and proceed to multiply the multiplicand by 4, the unit figure of the multiplier, precisely as in Art. 36. We then, in like manner, multiply the multiplicand by the 2 tens in the multiplier, taking care to write the first figure obtained by this multiplication in tens' column, directly under the 2 of the multiplier; and, adding together these *partial* products obtained by the two multiplications, and placed as in the operation, we have the full product of 763 multiplied by 24, which is 18312.

ART. 38. The preceding examples sufficiently illustrate the principle and method of multiplication; hence the following general

RULE. — *Write the multiplier under the multiplicand, arranging units under units, tens under tens, &c.*

Multiply each figure of the multiplicand by each figure of the multiplier, beginning with the right-hand figure, writing the right-hand figure of each product under the figure multiplied, and adding the left-hand figure or figures, if any, to the succeeding product.

If the multiplier consists of more than one figure, the right-hand figure of each partial product must be placed directly under the figure of the multiplier that produces it. The sum of the partial products will be the whole product required.

NOTE. — When there are ciphers between the significant figures of the multiplier, pass over them in the operation, and multiply by the significant figures only, remembering to set the first figure of the product directly under the figure of the multiplier that produces it.

QUESTIONS. — Art. 37. How do you proceed when the multiplier exceeds 12? Where do you set the first figure of each partial product? Why? How is the true product found? — Art. 38. What is the general rule for multiplication? When there are ciphers between the significant figures of the multiplier, how do you proceed?

ART. **39.** *First Method of Proof.* — Multiply the multiplier by the multiplicand, and if the result is like the first product the work is supposed to be right.

The reason of this proof depends on the principle, *That, when two or more numbers are multiplied together, the product is the same, whatever the order of multiplying them.*

Ex. 2. Multiply 7895 by 56. Ans. 442120.

OPERATION.

```
Multiplicand    7895
Multiplier        56
               -----
               47370
              39475
              ------
Product       442120
```

PROOF.

```
           56
         7895
         ----
          280
         504
        448
       392
       ------
Product 442120
```

NOTE. — The common mode of proof in business is to divide the product by the multiplier, and, if the work is right, the quotient will be like the multiplicand. This mode of proof anticipates the principles of division, and therefore cannot be employed without a previous knowledge of that rule.

ART. **40.** *Second Method of Proof.* — Begin at the left hand of the multiplicand, and add together its successive figures toward the right till the sum obtained equals or exceeds the number nine. If it equals it, drop the nine, and begin to add again at this point, and proceed till you obtain a sum equal to, or greater than, nine. If it exceeds nine, drop the nine as before, and carry the excess to the next figure, and then continue the addition as before. Proceed in this way till you have added all the figures in the multiplicand and rejected all the nines contained in it, and write the final excess at the right hand of the multiplicand.

Proceed in the same manner with the multiplier, and write the final excess under that of the multiplicand. Multiply these excesses together, and place the excess of nines in their product at the right.

Then proceed to find the excess of nines in the product obtained by the original operation; and, if the work is right,

---

QUESTIONS. — Art. 39. How is multiplication proved by the first method? What is the reason for this method of proof? What is the common mode of proof in business? — Art. 40. What is the second method of proving multiplication?

the excess thus found will be equal to the excess contained in the product of the above excesses of the multiplicand and multiplier.

Ex. 3.

OPERATION.

Multiplicand	12345 =	6 excess.	
Multiplier	2231 =	8 excess.	
	12345	48 = 3	Proof.
	37035		
	24690		
	24690		
Product	27541695 =	3	

Note. — This method of proof, though perhaps sufficiently sure for common purposes, is not always a test of the correctness of an operation. If two or more figures in the work should be transposed, or the value of one figure be just as much too great as another is too small, or if a nine be set down in the place of a cipher, or the contrary, the excess of nines will be the same, and still the work may not be correct. Such a balance of errors will not, however, be likely to occur.

Examples for Practice.

	4.	5.
Multiply	67895	78956
By	36	47
	407370	552692
	203685	315824
Ans.	2444220	3710932

	6.	7.
Multiply	89325	47896
By	901	2008
	89325	383168
	803925	95792
Ans.	80481825	96175168

8. What cost 47 hogsheads of molasses at 13 dollars per hogshead? Ans. 611 dollars.

9. What cost 97 oxen at 29 dollars each? Ans. 2813 dollars.

Questions. — Is this method of proof always a true test of the correctness of an operation? What is the reason for this method of proof?

10. Sold a farm containing 367 acres, at 97 dollars per acre what was the amount? Ans. 35599 dollars.

11. An army of 17006 men receive each 109 dollars as theii annual pay; what is the amount paid the whole army? Ans. 1853654 dollars.

12. If a mechanic deposit annually in the Savings Bank 407 dollars, what will be the sum deposited in 27 years? Ans. 10989 dollars.

13. If a man travel 37 miles in 1 day, how far will he travel in 365 days? Ans. 13505 miles.

14. If there be 24 hours in 1 day, how many hours in 365 days? Ans. 8760 hours.

15. How many gallons in 87 hogsheads, there being 63 gallons in each? Ans. 5481 gallons.

16. If the expenses of the Massachusetts Legislature be 1839 dollars per day, what will be the amount in a session of 109 days? Ans. 200451 dollars.

17. If a hogshead of sugar contains 368 pounds, how many pounds in 187 hogsheads? Ans. 68816 pounds.

18. Multiply 675 by 476. Ans. 321300.

19. Multiply 679 by 763. Ans. 518077.

20. Multiply 899 by 981. Ans. 881919.

21. Multiply 7854 by 1234. Ans. 9691836.

22. Multiply 3001 by 6071. Ans. 18219071.

23. Multiply 7117 by 9876. Ans. 70287492.

24. Multiply 376546 by 407091. Ans. 153288487686.

25. Multiply 7001009 by 7007867. Ans. 49062139937803.

26. Multiply five hundred and eighty-six by nine hundred and eight. Ans. 532088.

27. Multiply three thousand eight hundred and five by one thousand and seven. Ans. 3831635.

28. Multiply two thousand and seventy-one by seven hundred and six. Ans. 1462126.

29. Multiply eighty-eight thousand and eight by three thousand and seven. Ans. 264640056.

30. Multiply ninety thousand eight hundred and seven by one thousand and ninety-one. Ans. 99070437.

31. Multiply ninety thousand eight hundred and seven by nine thousand one hundred and six. Ans. 826888542.

32. Multiply fifty thousand and one by five thousand eight hundred and seven. Ans. 290355807.

33. Multiply eighty thousand and nine by nine thousand and sixteen. Ans. 721361144.

34. Multiply forty-seven thousand and thirteen by eighty thousand eight hundred and seven. Ans. 3798979491.

ART. 41. A COMPOSITE number is one produced by multiplying together two or more numbers greater than unity or one; thus, 12 is a composite number, since it is the product of 3×4; and also 24 is a composite number, since it is the product of $2 \times 3 \times 4$.

A FACTOR of any number is a name given to one of two or more numbers, which, being multiplied together, produce that number; thus, 3 and 4 are factors of 12, since $3 \times 4 = 12$.

ART. 42. To multiply by a composite number.

EX. 1. A merchant bought fifteen pieces of broadcloth, at 96 dollars per piece; how much did he pay for the whole?

Ans. 1440 dollars.

OPERATION.

```
  9 6 dolls., price of  1 piece.
    3
 ------
2 8 8 dolls., price of  3 pieces.
    5
 ------
1 4 4 0 dolls., price of 15 pieces.
```

The factors of 15 are 3 and 5. Now, if we multiply the price of one piece by the factor 3, we get the cost of 3 pieces; and then, by multiplying the cost of 3 pieces by the factor 5, it is evident we obtain the cost of 15, the number of pieces bought, since 15 is equal to 5 times 3. Hence we adopt the following

RULE. — *Multiply the multiplicand by one of the factors of the multiplier, and the product thus obtained by another, and so on until each of the factors has been used as a multiplier. The last product will be the answer.*

NOTE. — The product of any number of factors is the same in whatever order they are multiplied. Thus, $3 \times 4 = 12$, and $4 \times 3 = 12$.

EXAMPLES FOR PRACTICE.

2. Multiply 30613 by $25 = 5 \times 5$. Ans. 765325.
3. Multiply 1469 by $84 = 7 \times 12$. Ans. 123396.
4. Multiply 7546 by 81, using its factors. Ans. 611226.
5. Multiply 7901 by 125, using its factors. Ans. 987625.
6. In 1 mile there are 63360 inches; how many inches in 45 miles? Ans. 2851200 inches.
7. If in one year there are 8766 hours, how many hours in 72 years? Ans. 631152 hours.

QUESTIONS. — Art. 41. What is a composite number? What is a factor of any number? — Art. 42. What are the factors of 15? How do we multiply by a composite number? Repeat the rule. In what order must the factors of a composite number be multiplied?

8. If sound moves 1142 feet in a second, how far will it move in one minute? Ans. 68520 feet.

ART. 43. When the multiplier is 1 with one or more ciphers annexed to it, as 10, 100, &c.

Ex. 1. In 1 day there are 24 hours; how many hours in 10 days? In 100 days? Answers. 240, 2400.

OPERATION.

Multiplicand	2 4	2 4
Multiplier	1 0	1 0 0
Product	2 4 0	2 4 0 0

Or thus, 240, 2400.

The removal of a figure one place to the left increases its value *ten times*. (Art. 7.) If, then, we annex a cipher to the right of 24, the multiplicand, each figure is removed one place to the left, and its value is increased ten times, or multiplied by 10. If two ciphers are annexed, each figure is removed two places to the left, and its value is increased 100 times, or multiplied by 100; every additional cipher increasing the value *ten times*. Hence the following

RULE. — *Annex to the multiplicand as many ciphers as has the multiplier. The number thus formed will be the product required.*

EXAMPLES FOR PRACTICE.

2. Multiply 2356 by 10. Ans. 23560.
3. Multiply 5873 by 100. Ans. 587300.
4. Multiply 7964 by 1000. Ans. 7964000.
5. Multiply 98725 by 100000. Ans. 9872500000.

ART. 44. When there are ciphers on the right hand of the multiplier or multiplicand, or both.

Ex. 1. What will 600 acres of land cost at 20 dollars per acre? Ans 12,000 dollars.

OPERATION.

Multiplicand	6 0 0
Multiplier	2 0
Product	1 2 0 0 0

The multiplicand may be resolved into the factors 6 and 100, and the multiplier into the factors 2 and 10. Now, it is evident (Art. 42), if these several factors be multiplied together, they will produce the same product as the original factors 600 and 20. Thus $6 \times 2 = 12$, and $12 \times 100 = 1200$, and $1200 \times 10 = 12000$, the same result as in the operation. Hence the following

QUESTIONS. — Art. 43. What is the effect of removing a figure one place to the left? What is the effect of annexing a cipher to any figure or number? Two ciphers? &c. What is the rule when the multiplier is 1 with ciphers annexed? — Art 44. How do you arrange the figures for multiplication, when there are ciphers on the right hand of either the multiplier or multiplicand, or both? Why does multiplying the significant figures and annexing the ciphers produce the true product?

RULE. — *Write the significant figures of the multiplier under those of the multiplicand, and multiply them together. To their product annex as many ciphers as there are on the right of both multiplicand and multiplier.*

EXAMPLES FOR PRACTICE.

	2.	3.
Multiply	8785324	713378900
By	3200	70080
	17570648	57070312
	26355972	49936523
Ans.	28113036800	49993593312000

4. Multiply 8010700 by 9000909.
Ans. 72103581726300.
5. Multiply 700110000 by 700110000
Ans. 490154012100000000.
6. Multiply 4070607 by 7007000.
Ans. 28522743249000.
7. Multiply 4110000 by 1017010.
Ans. 4179911100000.

8. Multiply twenty-nine millions two thousand nine hundred and nine by four hundred and four thousand.
Ans. 11717175236000.

9. Multiply eighty-seven millions by eight hundred thousand seven hundred. Ans. 69660900000000.

10. Multiply one million one thousand one hundred by nine hundred nine thousand and ninety. Ans. 910089999000.

11. Multiply forty-nine millions and forty-nine by four hundred and ninety thousand. Ans 24010024010000.

12. Multiply two hundred millions two hundred by two millions two thousand and two. Ans. 400400800400400.

13. Multiply four millions forty thousand four hundred by three hundred three thousand. Ans. 1224241200000.

14. Multiply three hundred thousand thirty by forty-seven thousand seventy. Ans. 14122412100.

15. Multiply fifteen millions one hundred by two thousand two hundred. Ans. 33000220000.

16. Multiply one billion twenty thousand by one thousand one hundred. Ans. 1100022000000.

QUESTION. — What is the rule?

§ V. DIVISION.

MENTAL EXERCISES.

ART. 45. WHEN it is required to find how many times one number contains another, the process is called *Division.*

Ex. 1. A boy has 32 cents, which he wishes to give to 8 of his companions, to each an equal number; how many must each receive?

ILLUSTRATION. — It is evident that each boy must receive as many cents as the number 8 is contained times in 32. We therefore inquire what number 8 must be multiplied by to make 32. By trial, we find that 4 is the number; because 4 times 8 make 32. Hence 8 is contained in 32 4 times, and the boys receive 4 cents apiece.

The following table should be studied by the learner to aid him in solving questions in division:

DIVISION TABLE.

2 in 2 1 time	3 in 3 1 time	4 in 4 1 time	5 in 5 1 time
2 in 4 2 times	3 in 6 2 times	4 in 8 2 times	5 in 10 2 times
2 in 6 3 times	3 in 9 3 times	4 in 12 3 times	5 in 15 3 times
2 in 8 4 times	3 in 12 4 times	4 in 16 4 times	5 in 20 4 times
2 in 10 5 times	3 in 15 5 times	4 in 20 5 times	5 in 25 5 times
2 in 12 6 times	3 in 18 6 times	4 in 24 6 times	5 in 30 6 times
2 in 14 7 times	3 in 21 7 times	4 in 28 7 times	5 in 35 7 times
2 in 16 8 times	3 in 24 8 times	4 in 32 8 times	5 in 40 8 times
2 in 18 9 times	3 in 27 9 times	4 in 36 9 times	5 in 45 9 times
2 in 20 10 times	3 in 30 10 times	4 in 40 10 times	5 in 50 10 times
2 in 22 11 times	3 in 33 11 times	4 in 44 11 times	5 in 55 11 times
2 in 24 12 times	3 in 36 12 times	4 in 48 12 times	5 in 60 12 times
6 in 6 1 time	7 in 7 1 time	8 in 8 1 time	9 in 9 1 time
6 in 12 2 times	7 in 14 2 times	8 in 16 2 times	9 in 18 2 times
6 in 18 3 times	7 in 21 3 times	8 in 24 3 times	9 in 27 3 times
6 in 24 4 times	7 in 28 4 times	8 in 32 4 times	9 in 36 4 times
6 in 30 5 times	7 in 35 5 times	8 in 40 5 times	9 in 45 5 times
6 in 36 6 times	7 in 42 6 times	8 in 48 6 times	9 in 54 6 times
6 in 42 7 times	7 in 49 7 times	8 in 56 7 times	9 in 63 7 times
6 in 48 8 times	7 in 56 8 times	8 in 64 8 times	9 in 72 8 times
6 in 54 9 times	7 in 63 9 times	8 in 72 9 times	9 in 81 9 times
6 in 60 10 times	7 in 70 10 times	8 in 80 10 times	9 in 90 10 times
6 in 66 11 times	7 in 77 11 times	8 in 88 11 times	9 in 99 11 times
6 in 72 12 times	7 in 84 12 times	8 in 96 12 times	9 in 108 12 times
10 in 10 1 time	10 in 110 11 times	11 in 88 8 times	12 in 48 4 times
10 in 20 2 times	10 in 120 12 times	11 in 99 9 times	12 in 60 5 times
10 in 30 3 times		11 in 110 10 times	12 in 72 6 times
10 in 40 4 times	11 in 11 1 time	11 in 121 11 times	12 in 84 7 times
10 in 50 5 times	11 in 22 2 times	11 in 132 12 times	12 in 96 8 times
10 in 60 6 times	11 in 33 3 times		12 in 108 9 times
10 in 70 7 times	11 in 44 4 times	12 in 12 1 time	12 in 120 10 times
10 in 80 8 times	11 in 55 5 times	12 in 24 2 times	12 in 132 11 times
10 in 90 9 times	11 in 66 6 times	12 in 36 3 times	12 in 144 12 times
10 in 100 10 times	11 in 77 7 times		

2. A farmer received 8 dollars for 2 sheep; what was the price of each?

ILLUSTRATION. — It is evident, since he received 8 dollars for 2 sheep, for 1 sheep he must receive as many dollars as 2 is contained times in 8. 2 is contained in 8 4 times, because 4 times 2 are 8; hence 4 dollars was the price of each sheep.

3. A man gave 15 dollars for 3 barrels of flour; what was the cost of each barrel?

4. A lady divided 20 oranges among her 5 daughters; how many did each receive?

5. If 4 casks of lime cost 12 dollars, what costs 1 cask?

6. A laborer earned 48 shillings in 6 days; what did he receive per day?

7. A man can perform a certain piece of labor in 30 days; how long will it take five men to do the same?

8. When 72 dollars are paid for 8 acres of land, what costs 1 acre? What cost 3 acres?

9. If 21 pounds of flour can be obtained for 3 dollars, how much can be obtained for 1 dollar? How much for 8 dollars? How much for 9 dollars?

10. Gave 56 cents for 8 pounds of raisins; what costs 1 pound? What cost 7 pounds?

11. If a man walk 24 miles in 6 hours, how far will he walk in 1 hour? How far in 10 hours?

12. Paid 56 dollars for 7 hundred weight of sugar; what costs 1 hundred weight? What cost 10 hundred weight?

13. If 5 horses will eat a load of hay in 1 week, how long would it last 1 horse?

14. In 20, how many times 2? How many times 4? How many times 5? How many times 10?

15. In 24, how many times 3? How many times 4? How many times 6? How many times 8?

16. How many times 7 in 21? In 28? In 56? In 35? In 14? In 63? In 77? In 70? In 84?

17. How many times 6 in 12? In 36? In 18? In 54? In 60? In 42? In 48? In 72? In 66?

18. How many times 9 in 27? In 45? In 63? In 81? In 99? In 108?

19. How many times 11 in 22? In 55? In 77? In 88? In 110? In 132?

20. How many times 12 in 36? In 60? In 72? In 84? In 120? In 144?

ART. **46.** The pupil will now perceive that

DIVISION is the process of finding how many times one number is contained in another.

In division there are three principal terms: the *Dividend* the *Divisor*, and the *Quotient*, or *answer*.

The *dividend* is the number to be divided.

The *divisor* is the number by which we divide.

The *quotient* is the number of times the divisor is contained in the dividend.

NOTE. — *Quotient* is derived from the Latin word *quoties*, which signifies *how often*, or *how many times*.

When the dividend does not contain the divisor an exact number of times, the *excess* is called a *remainder*, and may be regarded as a *fourth* term in the *division*. The remainder, being part of the dividend, will always be of the same denomination or kind as the dividend, and must always be less than the divisor.

ART. **47.** SIGNS. — The sign of division is a short horizontal line, with a dot above it and another below; thus, $\div$. It shows that the number *before* it is to be divided by the number *after* it. The expression $6 \div 2 = 3$ is read, 6 divided by 2 is equal to 3.

Division is also indicated by writing the dividend above a short horizontal line and the divisor below; thus, $\frac{6}{2}$. The expression $\frac{6}{2} = 3$ is read, 6 divided by 2 is equal to 3.

There is a third method of indicating division, by a curved line placed between the divisor and dividend. Thus, the expression 6) 12 shows that 12 is to be divided by 6.

EXERCISES FOR THE SLATE.

ART. **48.** The method of operation by *Short Division*, or when the divisor does not exceed 12.

Ex. 1. Divide 8574 dollars equally among 6 men.

Ans. 1429 dollars.

QUESTIONS. — Art. 46. What is division? What are the three principal terms in division? What is the dividend? What is the divisor? What is the quotient? What the remainder? What will be the denomination of the remainder? How does it compare with the divisor? — Art. 47. What is the first sign of division, and what does it show? What is the second, and what does it show? What is the third, and what does it show? — Art. 48. What is short division?

OPERATION.

Divisor 6) 8574 Dividend.

1429 Quotient.

We first inquire how many times 6, the divisor, is contained in 8, the first figure of the dividend, which is thousands, and find it to be 1 time, and 2 thousands remaining. We write the 1 directly under the 8, its dividend, for the thousands' figure of the quotient. To 5, the next figure of the dividend, which is hundreds, we regard as prefixed the 2 thousands remaining, which equal 20 hundreds, and thus form the number 25 hundreds, in which we find the divisor 6 to be contained 4 times, and 1 hundred remaining. We write the 4 for the hundreds' figure in the quotient, and the 1 hundred remaining, equal 10 tens, we regard as prefixed to 7, the next figure of the dividend, which is tens, forming 17 tens, in which the divisor 6 is contained 2 times, and 5 tens remaining. We write the 2 for the tens' figure in the quotient, and the 5 tens remaining, equal 50 units, we regard as prefixed to 4, the last figure of the dividend, which is units, forming 54 units, in which the divisor 6 is contained 9 times. Writing the 9 for the units' figure of the quotient, we have 1429 as the entire quotient, or the number of times which the dividend contains the divisor 6.

ART. **49.** From the foregoing illustration we deduce the following

RULE. — *Write the divisor at the left hand of the dividend, with a curved line between them, and draw a horizontal line under the dividend.*

Then, beginning at the left, find how many times the divisor is contained in the fewest figures of the dividend that will contain it, and write the quotient under its dividend.

If there be a remainder, regard it as prefixed to the next figure of the dividend, and divide as before.

Should any dividend be less than the divisor, write a cipher in the quotient, and annex another figure, if any remains, for a new dividend.

NOTE 1. — When there is a remainder after dividing the last figure of the dividend, write it with the divisor underneath, with a line between them, at the right of the quotient.

NOTE 2. — *Prefix* means to place before; *annex*, to place after.

ART. **50.** *First Method of Proof.* — Multiply the divisor by the quotient, and to the product add the remainder, if any, and, if the work is right, the sum thus obtained will be equal to the dividend.

QUESTIONS. — How are the numbers arranged for short division? At which hand do you begin to divide? Why not begin at the right, where you begin to multiply? Where do you write the quotient? If there is a remainder after dividing a figure, what is done with it? — Art. 49. What is the rule for short division? Repeat the notes.

NOTE.—It will be seen, from this method of proof, that division is the reverse of multiplication. The *dividend* answers to the *product*, the *divisor* to one of the *factors*, and the *quotient* to the *other*.

EXAMPLES FOR PRACTICE

2. Divide 6375 by 5.

OPERATION.

Divisor 5) 6 3 7 5 Dividend.
1 2 7 5 Quotient.

PROOF.

1 2 7 5 Quotient.
5 Divisor.
6 3 7 5 Dividend.

3.	4.	5.
3) 7 8 9 3 7 6 2	4) 4 7 6 3 2 5 6	5) 3 7 8 9 5 6 5
2 6 3 1 2 5 4	1 1 9 0 8 1 4	

6.	7.	8.
6) 8 7 6 5 3 8 9	7) 9 8 7 6 3 5	8) 3 7 8 5 3 2

9.	10.	11.
9) 8 9 5 3 7 8 4	11) 7 6 7 8 9 0 3	12) 6 3 4 5 3 2 1

	Quotients.
12. Divide 479956 by 6.	$79992\frac{4}{6}$
13. Divide 385678 by 7.	$55096\frac{6}{7}$
14. Divide 438789 by 8.	$54848\frac{5}{8}$
15. Divide 1678767 by 9.	$186529\frac{6}{9}$
16. Divide 11497583 by 12.	$958131\frac{11}{12}$

	Quotients.	Rem
17. Divide 5678956 by 5.		1
18. Divide 1135791 by 7.		6
19. Divide 1622550 by 8.		6
20. Divide 2028180 by 9.		3
21. Divide 2253530 by 12.		2
22. Divide 1877940 by 11.		9
Sum of the quotients,	2084732	27

QUESTIONS.—Art. 50. How is short division proved? Of what is division the reverse? To what do the three terms in division answer in multiplication? What, then, is the reason for this proof of division?

23. Divide 944,580 dollars equally among 12 men, and what will be the share of each? Ans. 78,715 dollars.

24. Divide 154,503 acres of land equally among 9 persons. Ans. 17,167 acres.

25. A plantation in Cuba was sold for 7,011,608 dollars, and the amount was divided among 8 persons. What was paid to each person? Ans. 876,451 dollars.

26. A prize, valued at 178,656 dollars, is to be equally divided among 12 men; what is the share of each? Ans. 14,888 dollars.

27. Among 7 men, 67,123 bushels of wheat are to be distributed; how many bushels does each man receive? Ans. 9,589 bushels.

28. If 9 square feet make 1 square yard, how many yards in 895,347 square feet? Ans. 99,483 yards.

29. A township of 876,136 acres is to be divided among 8 persons; how many acres will be the portion of each? Ans. 109,517 acres.

30. Bought a farm for 5670 dollars, and sold it for 7896 dollars, and I divide the net gain among 6 persons; what does each receive? Ans. 371 dollars.

31. If 6 shillings make a dollar, how many dollars in 7890 shillings? Ans. 1315.

ART. **51.** The method of operation by *Long Division*, or, in general, when the divisor exceeds 12.

Ex. 1. A gentleman divided 896 dollars equally among his 7 children; how much did each receive? Ans. 128 dollars.

OPERATION.

```
           Dividend.
Divisor 7 ) 8 9 6 ( 1 2 8 Quotient.
            7
            ---
            1 9
            1 4
            -----
              5 6
              5 6
              ---
```

Having set down the divisor and dividend as in short division, we draw a curved line at the right of the dividend, to mark the place for the quotient. We then inquire how many times 7, the divisor, is contained in the 8 hundreds of the dividend; and, finding it to be 1 hundred times, we write the 1 in the quotient,

QUESTIONS. — Art. 51. What is long division? What is the difference between long division and short division? How do you arrange the numbers for long division? What do you first do after arranging the numbers for long division? Where do you place the figures of the quotient?

and multiply the divisor, 7, by it, writing the product, 7 hundreds under the 8 hundreds, from which we subtract it, and to the remainder, 1, annex the 9 tens of the dividend, making 19 tens. We now inquire how many times 7 is contained in 19 tens, and write the number, 2, at the right of the quotient figure before obtained. We then multiply the divisor by it, and place the product under the 19, and subtract as before; and to the remainder, 5, we annex 6 units, the next and last figure of the dividend, making 56 units. We proceed, as before, to find the next quotient figure, and, after subtracting the product of the divisor multiplied by it from 56, find there is no remainder left. Hence we learn that each one of the 7 children must receive 128 dollars.

Note. — The preceding example and the four that follow are usually performed by short division, but are here introduced to illustrate more clearly the method of operation by long division.

Examples for Practice.

2. Divide 1728 by 8. Ans. 216.
3. Divide 987656 by 11. Ans. $89786\frac{10}{11}$.
4. Divide 123456789 by 9. Ans. 13717421.
5. Divide 390413609 by 12. Ans. $32534467\frac{5}{12}$.

Ex. 6. A gentleman divided 4712 dollars equally among his 19 sons; what was the share of each? Ans. 248 dollars.

OPERATION.

```
            Dividend.
Divisor 19) 4712 (248 Quotient.
            38
            --
             91
             76
             --
             152
             152
             ---
```

We first inquire how many times 19, the divisor, is contained in 47, the two left-hand figures of the dividend; and, finding it to be 2 times, we write the 2 in the quotient, multiply the divisor by it, and subtract the product from the 47; and to the remainder, 9, annex 1, the next figure of the dividend, making 91. We next inquire how many times 19 is contained in 91, place the number, 4, in the quotient, then multiply and subtract as before, and to the remainder, 15, annex 2, the last figure of the dividend, and, proceeding as before, after finding the quotient figure, no remainder is left. Hence the share of each of the 19 sons is 248 dollars. This illustration, except in omissions, is essentially like the preceding one.

Questions. — After the quotient figure is found, what is the next thing you do? Where do you place the product? What do you next do? What is the next step? How do you then proceed? Is long division the same in principle as short division?

ART. **52.** From the preceding illustrations, the pupil will perceive the propriety of the following general

RULE. — *Write the divisor and dividend as in short division, and draw a curved line at the right hand of the dividend.*

Then inquire how many times the divisor is contained in the fewest figures on the left hand of the dividend that will contain it, and write the result at the right hand of the dividend for the first quotient figure.

Multiply the divisor by the quotient figure, and subtract the product from the figures of the dividend used, and to the remainder annex the next figure of the dividend.

Find how many times the divisor is contained in the number thus formed; write the figure denoting it at the right hand of the former quotient figure.

Thus proceed until all the figures of the dividend are divided.

NOTE 1. — The proper remainder is in all cases *less* than the divisor. If, in the course of the operation, it is at any time found to be as large as, or *larger* than, the divisor, it will show that there is an error in the work, and that the quotient figure should be increased.

NOTE 2. — If, at any time, the divisor, multiplied by the quotient figure, produces a product *larger* than the part of the dividend used, it shows that the quotient figure is too *large*, and must be diminished.

NOTE 3. — It will often happen that, when a figure is brought down, the number will not contain the divisor; and in that case a cipher must be placed in the quotient, and another figure of the dividend brought down, and so on until the number is large enough to contain the divisor.

NOTE 4. — If there is a remainder after dividing all the figures of the dividend, it must be written as directed in the preceding rule.

ART. **53.** *Second Method of Proof.* — Add together the remainder, if any, and all the products that have been produced by multiplying the divisor by the several quotient figures, and the result will be like the dividend, if the work is right.

ART. **54.** *Third Method.* — Subtract the remainder, if any, from the dividend, and divide the difference by the quotient. The result will be like the original divisor, if the work is right.

NOTE. — The first method of proof (Art. 50) is usually most convenient, and is most commonly employed.

QUESTIONS. — Art. 52. What is the general rule for long division? How may you know when the quotient figure is too small? How may you know when it is too large? What do you do when the part of the dividend used will not contain the divisor? — Art. 53. What is the second method of proof for division? — Art. 54. What is the third method? Can long division be proved by the first method of proof (Art. 50)?

EXAMPLES FOR PRACTICE.

Ex. 7. It is required to find how many times 48 is contained in 28618. Ans. 596.

OPERATION.

```
              Dividend.
Divisor 4 8 ) 2 8 6 1 8 ( 5 9 6 Quotient.
              2 4 0
              -----
              4 6 1
              4 3 2
              -------
                2 9 8
                2 8 8
                -----
                  1 0 Remainder.
```

PROOF BY MULTIPLICATION

```
    5 9 6 Quotient.
      4 8 Divisor.
  ---------
  4 7 6 8
2 3 8 4
---------
2 8 6 0 8
      1 0 Remainder
---------
2 8 6 1 8 Dividend.
```

8.

OPERATION.

```
             Dividend.
Divisor 2 6 ) 5 6 9 8 ( 2 1 9 Quotient.
            *+5 2
              ---
              4 9
             +2 6
              -----
              2 3 8
             +2 3 4
              -----
               +4 Remainder.
```

PROOF BY ADDITION.

```
5 2   )
  2 6 } Products.
2 3 4 )
    4   Remainder.
-------
5 6 9 8 Dividend.
```

9.

OPERATION.

```
               Dividend.
Divisor 1 4 4 ) 1 3 8 2 4 ( 9 6 Quotient.
                1 2 9 6
                -------
                  8 6 4
                  8 6 4
                  -----
```

PROOF BY DIVISION.

```
       Dividend.
9 6 ) 1 3 8 2 4 ( 1 4 4 Divisor.
      9 6
      -----
      4 2 2
      3 8 4
      -------
        3 8 4
        3 8 4
        -----
```

	Quotients.	Rem
10. Divide 3276 by 14.	234	
11. Divide 6205 by 17.	365	

* This sign of addition denotes the several products to be added.

	Quotients.	Rem.
12. Divide 3051 by 21.	145	6
13. Divide 190850 by 25.	7634	0
14. Divide 218579 by 42.	5204	11
15. Divide 9012345 by 31.	290720	25
16. Divide 6717890 by 98.	68549	88
17. Divide 4567890 by 19.	240415	5
18. Divide 1357901 by 87.	15608	5
19. Divide 9988891 by 77.	129725	66
20. Divide 9999999 by 69.	144927	36
21. Divide 867532 by 59.	14703	55
22. Divide 167008 by 87.	1919	55
23. Divide 345678 by 379.	912	30
24. Divide 3456789567 by 987.	3502319	714
25. Divide 8997744444 by 345.	26080418	234
26. Divide 4500700701 by 407.	11058232	277
27. Divide 6789563 by 1234.		95
28. Divide 78112345 by 8007.		4060
29. Divide 34533669 by 9999.		7122
30. Divide 99999999 by 3333.		0
31. Divide 47856712 by 1789.		962
32. Divide 345678901765 by 4007.	86268755	480
33. Divide 478656785178 by 56789.	8428688	22346
34. Divide 678957000107 by 10789561.	62927	2295060
35. Divide 990070171009 by 900700601.	1099	200210510

36. Divide three hundred twenty-one thousand three hundred dollars equally among six hundred seventy-five men.

Ans. 476 dollars.

37. Four hundred seventy-one men purchase a township containing one hundred eighty-six thousand forty-five acres; what is the share of each? Ans. 395 acres.

38. A railroad, which cost five hundred eighteen thousand seventy-seven dollars, is divided into six hundred seventy-nine shares; what is the value of each share? Ans. 763 dollars.

39. Divide forty-two thousand four hundred thirty-five bushels of wheat equally among one hundred twenty-three men.

Ans. 345 bushels each.

40. A prize, valued at one hundred eighty-four thousand seven hundred seventy-five dollars, is to be divided equally among four hundred seventy-five men; what is the share of each? Ans. 389 dollars.

41. A certain company purchased a valuable township for nine millions six hundred ninety-one thousand eight hundred

thirty-six dollars; each share was valued at seven thousand eight hundred fifty-four dollars; of how many men did the com pany consist? Ans. 1234 men.

42. A tax of thirty millions fifty-six thousand four hundred sixty-five dollars is assessed equally on four thousand five hundred ninety-seven towns; what sum must each town pay?

Ans. $6538\frac{1279}{4597}$ dollars.

ART. **55.** Method of operation when the divisor is a com posite number.

Ex. 1. A merchant bought 15 pieces of broadcloth for 1440 dollars; what was the value of each piece? Ans. 96 dollars.

OPERATION.

```
3 ) 1 4 4 0 dolls., cost of 15 pieces.
    -------
  5 ) 4 8 0 dolls., cost of 5 pieces.
      -----
        9 6 dolls., cost of 1 piece.
```

The factors of 15 are 3 and 5. Now, if we divide the 1440 dollars, the cost of 15 pieces, by 3, we obtain 480 dollars, which is evidently the cost of 5 pieces, because there are 5 times 3 in 15. Then, dividing 480 dollars, the cost of 5 pieces, by 5, we get the cost of 1 piece. Hence we deduce the following

RULE. — *Divide the dividend by one of the factors, and the quotient thus found by another, and thus proceed till every factor has been made a divisor. The last quotient will be the true quotient required.*

EXAMPLES FOR PRACTICE.

	Quotients.
2. Divide 765325 by 25 = 5 × 5.	30613
3. Divide 123396 by 84 = 7 × 12.	1469
4. Divide 611226 by 81, using its factors.	7546
5. Divide 987625 by 125, using its factors.	7901
6. Divide 17472 by 96, using its factors.	182
7. Divide 34848 by 132, using its factors.	264

ART. **56.** Method of finding the true remainder when there are several in the operation.

Ex. 1 How many months of 4 weeks each are there in 298 days, and how many days remaining?

Ans. 10 months and 18 days.

QUESTIONS. — Art. 55. What are the factors of 15? What do you get the cost of, in this example, when you divide by the factor 3? What, when you divide by 5? Why? What is the rule for dividing by a composite number?

OPERATION.

```
7 ) 2 9 8
  ————
 4 ) 4 2,   4 days  }
   ————             } 18 days.
   1 0,   2 weeks   }
```

Since there are 7 days in 1 week, we first divide the 298 days by 7, and have 42 weeks and a remainder of 4 days. Then, since 4 weeks make 1 month, we divide the 42 weeks by 4, and have 10 months and a remainder of 2 weeks. Now, to find the true remainder in days, it is evident that we must multiply the 2 weeks by 7, because 7 days make a week, and to the product add the 4 days; thus, $2 \times 7 = 14$, and $14 + 4 = 18$ days, for the remainder. Hence the following

RULE. — *Multiply each remainder, except the first, by all the divisors preceding the one which produced it; and the first remainder being added to the sum of the products, the amount will be the true remainder.*

NOTE. — There will be but one product to add to the first remainder when there are only two divisors and two remainders.

Ex. 2. Divide 789 by 36, using the factors 2, 3, and 6, and find the true remainder. Ans. 33.

OPERATION.

```
2 ) 7 8 9
3 ) 3 9 4,   1, 1st Rem.
6 ) 1 3 1,   1, 2d Rem.
      2 1,   5, 3d Rem.
```

FINDING THE TRUE REMAINDER.

```
5 × 3 × 2 = 30, 1st Product.
    1 × 2 =  2, 2d Product.
             1, 1st Remainder.
            ——
            33, true Rem.
```

EXAMPLES FOR PRACTICE.

3. Divide 934 by 55, using the factors 5 and 11, and find the true remainder. Ans. 54.

4. Divide 5348 by 48, using the factors 6 and 8, and find the true remainder. Ans. 20.

5. Divide 5873 by 84, using the factors 3, 4, and 7, and find the true remainder. Ans. 77.

6. Divide 249237 by 1728, using the factors 12, 6, 6, and 4, and find the true remainder. Ans. 405.

ART. **57.** When the divisor is 1, with one or more ciphers at the right; as 10, 100, &c.

Ex. 1. Divide 356 dollars equally among 10 men; what will each man have? Ans. $35\frac{6}{10}$ dollars.

QUESTIONS. — Art. 56. When there are several remainders, what is the rule for finding the true remainder? Will you give the reason for this rule?

OPERATION.

1 | 0) 3 5 | 6

Quotient 3 5, 6 Rem.

Or thus, 3 5 | 6.

It will be remembered, that to multiply by 10 we annex one cipher, which removes the figures one place to the left, and thus *increases* their value *ten times*. Now, it is obvious that if we *reverse* the process, and cut off the right-hand figure by a line, we remove the remaining figures *one* place to the *right*, and consequently *diminish* the value of each *ten times*, and thus divide the whole number by 10. The figures on the left of the line are the quotient, and the one on the right is the remainder, which may be written over the divisor, and annexed to the quotient. Hence the share of each man is $35\frac{6}{10}$ dollars.

EXAMPLES FOR PRACTICE.

	Quotient.	Rem.
2. Divide 6892 by 10.	689	2
3. Divide 4375 by 100.		75
4. Divide 24815 by 1000.		815
5. Divide 987654321123 by 100000000.		54321123

ART. **58.** When the divisor has ciphers on the right, and is not 10, 100, &c.

Ex. 1. If I divide 5832 pounds of bread equally among 600 soldiers, what is each one's share? Ans. $9\frac{432}{600}$ pounds.

OPERATION.

1 | 0 0) 5 8 | 3 2

6) 5 8, 3 2, 1st Rem.

9, 4, 2d Rem.

Or thus, 6 | 0 0) 5 8 | 3 2

9, 4 3 2

The divisor, 600, may be resolved into the factors 6 and 100. We first divide by the factor 100, by cutting off two figures at the right, and get 58 for the quotient and 32 for a remainder. We then divide the quotient, 58, by the other factor, 6, and obtain 9 for the quotient and 4 for a remainder. The last remainder, 4, being multiplied by the divisor, 100, and 32, the first remainder, added, we obtain 432 for the true remainder (Art. 56). Hence each soldier receives $9\frac{432}{600}$ pounds.

ART. **59.** From the preceding illustrations is deduced the general

RULE. — *Cut off the ciphers from the divisor, and the same number of figures from the right of the dividend.*

Then divide the remaining figures of the dividend by the remaining figures of the divisor.

QUESTIONS. — Art. 57. How do you divide by 10? How does it appear that this divides the number by 10? — Art. 58. How do you divide by 600 in the example? How does it appear that this divides the number? — Art. 59. What is the general rule?

NOTE. — When by the operation there is a last remainder, to it must be annexed the figures cut off from the dividend to form the true remainder Should there be no last remainder, then the significant figures, if any, cut off from the dividend, will form the true remainder.

EXAMPLES FOR PRACTICE.

	Quotients.	Rem.
2. Divide 3594 by 80.	44	74
3. Divide 79872 by 240.	332	192
4. Divide 467153 by 700.	667	253
5. Divide 13112297 by 8900.		2597
6. Divide 71897654325 by 700000000.		497654325
7. Divide 3456789123456787 by 990000.		306787
8. Divide 967231731328000 by 1020000000.		411328000
9. Divide 33166405115000 by 1600000000.		5115000
10. Divide 18191618562300 by 1000000000.		618562300
11. Divide 4766666000000 by 55550000000.		44916000000

§ VI. QUESTIONS INVOLVING FRACTIONS.

ART. 60. IF a unit or individual thing is divided into equal parts, each of the parts is called a *Fraction* of the number or thing divided. *Hence* a FRACTION *is one or more equal parts of a unit.*

ILLUSTRATIONS. — 1. When any number or thing is divided into *two* equal parts, *one* of the parts is called *one half*, and is written thus: $\frac{1}{2}$.

2. When any number or thing is divided into *three* equal parts, *one* of the parts is called *one third* ($\frac{1}{3}$); *two* of the parts are called *two thirds* ($\frac{2}{3}$).

3. When any number or thing is divided into *four* equal parts, *one* of the parts is called *one fourth* ($\frac{1}{4}$); *three* of the parts, *three fourths* ($\frac{3}{4}$).

4. When any number or thing is divided into *five* equal parts, *one* of the parts is called *one fifth* ($\frac{1}{5}$); *two* parts, *two fifths* ($\frac{2}{5}$); *three* parts, *three fifths* ($\frac{3}{5}$); and *four* parts, *four fifths* ($\frac{4}{5}$).

QUESTIONS. — Art. 60. What is a fraction? What is meant by one half of any number or thing? How is it written? What is meant by one third, and how is it written? What by one fourth, and how written? What by one fifth, and how written? What by four fifths, and how written? How do you find one half of any number? How one third? How one fourth? &c. How many halves make a whole one? How many thirds? How many fourths? How many fifths?

5. When any number or thing is divided into *six* equal parts, what is *one* of the parts called? Two parts? Five parts?

6. When a number or thing is divided into 7 equal parts, what is 1 part called? 2 parts? 3 parts? 4 parts? 5 parts? 6 parts?

7. When a number or thing is divided into 9 equal parts, what is 1 part called? 2 parts? 4 parts? 5 parts? 7 parts? 8 parts?

8. What is 1 *half* of 4? Of 8? Of 16? Of 20? Of 28? Of 32?

9. What is 1 *third* of 9? Of 12? Of 15? Of 27? Of 30? Of 36? Of 60?

10. What is 1 *fourth* of 8? Of 16? Of 20? Of 24? Of 40? Of 48? Of 100?

11. What is 1 *fifth* of 10? Of 25? Of 30? Of 35? Of 45? Of 50? Of 55? Of 65?

12. What is 1 *sixth* of 12? Of 18? Of 30? Of 42? Of 60? Of 72? Of 90?

13. How many fourths in 1 apple?

14. How many fourths in 2 apples? In 3 apples? In 8 apples? In 16 apples?

15. How many fifths in 1 barrel of flour? In 3 barrels? In 5 barrels? In 7 barrels? In 9 barrels?

16. How many sixths in 1 bushel of wheat? In 4 bushels? In 7 bushels? In 9 bushels? In 12 bushels?

17. James owns 3 fifths of a kite, and his brother Thomas the remainder. How many fifths does Thomas own?

ILLUSTRATION. — Since there are 5 fifths in the kite, if James owns 3 fifths, there will remain for Thomas 5 fifths ($\frac{5}{5}$) less 3 fifths ($\frac{3}{5}$) = 2 fifths. Ans. 2 fifths.

18. From a load of hay I sold 4 sevenths; how many sevenths remain?

19. John Jones found a large sum of money; he gave 5 eighths of it to the poor of the parish; how much did he reserve for himself?

20. John Smith gave 2 ninths of his farm to his son, 3 ninths to his daughter, and the remainder to his wife; how many ninths did his wife receive?

ILLUSTRATION. — Since he gave 2 ninths ($\frac{2}{9}$) to his son, and 3 ninths ($\frac{3}{9}$) to his daughter, he gave them both $\frac{2}{9} + \frac{3}{9} = \frac{5}{9}$; and since there are 9 ninths ($\frac{9}{9}$) in the farm, he must have given his wife $\frac{9}{9} - \frac{5}{9} = \frac{4}{9}$. Ans. $\frac{4}{9}$.

21. In a certain school $\frac{2}{11}$ of the pupils study grammar, $\frac{3}{11}$ study arithmetic, $\frac{4}{11}$ geography, and the remainder philosophy. What part of the school study philosophy?

22. J. Dow spends $\frac{1}{7}$ of his time in reading, $\frac{3}{7}$ in labor, and $\frac{2}{7}$ in visiting. How large a portion of his time remains for eating and sleeping?

23. If a yard of cloth cost $8, what cost $\frac{1}{4}$ of a yard? What cost $\frac{3}{4}$ of a yard?

ILLUSTRATION. — If 1 yard cost $8, $\frac{1}{4}$ of a yard will cost $\frac{1}{4}$ of $8 = $2; and if $\frac{1}{4}$ of a yard cost $2, $\frac{3}{4}$ will cost three times as much; 3 times $2 = $6. Ans. $6.

24. If an acre of land cost $24, what will $\frac{1}{8}$ of an acre cost? What will $\frac{5}{8}$ cost?

25. When 96 cents are paid for a bushel of rye, what cost $\frac{11}{12}$ of a bushel?

26. If $\frac{1}{5}$ of a barrel of flour cost 2 dollars, what cost $\frac{4}{5}$ of barrel?

ILLUSTRATION. — If $\frac{1}{5}$ cost 2 dollars, $\frac{4}{5}$ will cost 4 times 2 dollars = $8. Ans. $8.

27. If $\frac{1}{8}$ of an acre of land cost $24 dollars, what will $\frac{7}{8}$ of an acre cost?

28. If $\frac{1}{3}$ of a hogshead of molasses cost $11, what will a hogshead cost?

29. If $\frac{7}{8}$ of an acre of land cost $21, what cost $\frac{1}{8}$ of an acre? What cost an acre? What cost 10 acres?

ILLUSTRATION. — If $\frac{7}{8}$ cost $21, $\frac{1}{8}$ will cost $\frac{1}{7}$ of $21, and $\frac{1}{7}$ of $21 is $3; and $\frac{8}{8}$ will cost 8 times $3 = $24, and 10 acres will cost 10 times $24 = $240. Ans. $240.

30. If $\frac{9}{11}$ of a hogshead of sugar cost $18, what costs 1 hogshead? What cost 4 hogsheads?

31. If $\frac{5}{8}$ of a barrel of apples cost $1.50, what costs a barrel? What cost 10 barrels?

32. When $49 are paid for $\frac{7}{11}$ of a ton of potash, what must be paid for 2 tons?

33. How many half-barrels of flour are there in 2 and a half ($2\frac{1}{2}$) barrels?

ILLUSTRATION. — Since 1 barrel contains 2 halves, 2 barrels will contain 2 times 2 = 4 halves, and the 1 half added makes 5 halves. Ans. $\frac{5}{2}$.

34. How many half-bushels in $4\frac{1}{2}$ bushels of oats? In $5\frac{1}{2}$ bushels? In $7\frac{1}{2}$ bushels? In $9\frac{1}{2}$ bushels?

35. How many eighths of a dollar in $2\frac{1}{8}$ dollars? In $4\frac{3}{8}$ dollars? In $7\frac{5}{8}$ dollars? In $9\frac{7}{8}$ dollars? In $12\frac{1}{8}$ dollars?

36. How many tenths of an ounce in $4\frac{3}{10}$ ounces? In $5\frac{7}{10}$ ounces? In $8\frac{1}{10}$ ounces? In $10\frac{9}{10}$ ounces?

37. How many barrels of wine in 6 half ($\frac{6}{2}$) barrels?

Illustration. — Since it takes 2 halves to make one whole one, there will be as many whole barrels in 6 halves ($\frac{6}{2}$) as 2 is contained times in 6. 2 is contained in 6, 3 times.

Ans. 3 barrels.

38. How many firkins of butter in $\frac{8}{2}$ firkins? In $\frac{20}{2}$ firkins?

39. How many whole numbers in $\frac{10}{5}$? In $\frac{15}{5}$? In $\frac{25}{5}$?

40. How many whole numbers in $\frac{16}{8}$? In $\frac{8}{8}$? In $\frac{32}{8}$? In $\frac{48}{8}$?

41. If a skein of silk is worth $3\frac{1}{2}$ cents, what are 6 skeins worth?

Illustration. — If 1 skein is worth $3\frac{1}{2}$ cents, 6 skeins are worth 6 times as much; 6 times $3\frac{1}{2}$ are equal to 6 times 3 and 6 times $\frac{1}{2}$; 6 times $3 = 18$; 6 times $\frac{1}{2} = \frac{6}{2} = 3$; $18 + 3 = 21$.

Ans. 21 cents.

42. Bought one pair of boots for $\$6\frac{1}{2}$; what must I pay for 4 pairs? For 8 pairs? For 10 pairs? For 12 pairs?

43. Paid $12\frac{1}{2}$ cents for one pound of cloves; what will 6 pounds cost? 10 pounds? 12 pounds?

44. If one pound of butter is worth 12 cents, what are $4\frac{1}{2}$ pounds worth?

Illustration. — If 1 pound is worth 12 cents, $4\frac{1}{2}$ pounds are worth $4\frac{1}{2}$ times as much; $4\frac{1}{2}$ times 12 cents are equal to 4 times 12 and $\frac{1}{2}$ of 12; 4 times 12 are 48, and $\frac{1}{2}$ of 12 is 6; 48 cents and 6 cents are 54 cents. Ans. 54 cents.

45. When lard is sold for 9 cents per pound, what must be paid for $7\frac{1}{3}$ pounds? For $8\frac{1}{3}$ pounds? For $9\frac{1}{3}$ pounds?

46. Bought 1 pound of coffee at 16 cents; what will $5\frac{1}{2}$ pounds cost? $3\frac{1}{4}$ pounds? $5\frac{1}{4}$ pounds? $6\frac{1}{8}$ pounds?

47. If 1 yard of cloth is worth 20 cents, what is the value of $16\frac{1}{2}$ yards? $12\frac{1}{4}$ yards? $8\frac{1}{2}$ yards? $11\frac{1}{4}$ yards?

48. If $1\frac{1}{2}$ bushels of corn cost \$1.20, what will 1 bushel cost?

Illustration. — $1\frac{1}{2}$ bushels $= \frac{3}{2}$ bushels. Now, if $\frac{3}{2}$ cost \$1.20, $\frac{1}{2}$ will cost $\frac{1}{3}$ of $\$1.20 = \$0.40$; and $\frac{2}{2}$ or a whole bushel will cost 2 times $\$0.40 = \$0.80$. Ans. \$0.80.

49. If $2\frac{2}{5}$ pounds of coffee cost 60 cents, what will 1 pound cost?

ILLUSTRATION. — $2\frac{2}{5}$ pounds $= \frac{12}{5}$ pounds. If $\frac{12}{5}$ cost 60 cents, $\frac{1}{5}$ will cost $\frac{1}{12}$ of 60 cents $=$ 5 cents; and $\frac{5}{5}$, or a pound, will cost 5 times 5 cents $=$ 25 cents. Ans. $0.25.

50. How many times will 60 contain $2\frac{2}{5}$?

51. Paid $54 for $7\frac{5}{7}$ barrels of oil; what cost 1 barrel? Ans. $7.

52. How many times is $7\frac{5}{7}$ contained in 54?

53. How many cords of wood, at $$5\frac{1}{2}$ per cord, can be bought for $66?

54. How many times will 66 contain $5\frac{1}{2}$?

55. Gave $40 for $6\frac{2}{3}$ yards of broadcloth; what cost 1 yard?

56. How many times is $6\frac{2}{3}$ contained in 40?

57. The distance between two places is 110 rods. I wish to divide this distance into spaces of $5\frac{1}{2}$ rods each. Required the number of spaces.

---

§ VII. CONTRACTIONS IN MULTIPLICATION AND DIVISION.*

CONTRACTIONS IN MULTIPLICATION.

ART. **61.** To multiply by 25.

Ex. 1. Multiply 876581 by 25. Ans. 21914525.

OPERATION.

```
4 ) 8 7 6 5 8 1 0 0
    ---------------
    2 1 9 1 4 5 2 5  Product.
```

We multiply by 100, by annexing two ciphers to the multiplicand; and since 25, the multiplier, is only *one fourth* of 100, we divide by 4 to obtain the true product.

RULE. — *Annex two ciphers to the multiplicand, and divide it by* 4.

* If the principles on which these contractions depend are considered too difficult for the young pupil to understand at this stage of his progress, they may be omitted for the present, and attended to when he is further advanced.

QUESTIONS. — Art. 61. What is the rule for multiplying by 25? What is the reason for the rule?

EXAMPLES FOR PRACTICE.

2. Multiply 76589658 by 25. Ans. 1914741450.
3. Multiply 567898717 by 25. Ans. 14197467925.
4. Multiply 123456789 by 25. Ans. 3086419725.

ART. **62.** To multiply by $33\frac{1}{3}$.

Ex. 1. Multiply 87678963 by $33\frac{1}{3}$ Ans. 2922632100.

OPERATION.

3) 8 7 6 7 8 9 6 3 0 0

2 9 2 2 6 3 2 1 0 0 Product.

We multiply by 100, as before; and since $33\frac{1}{3}$, the multiplier, is only *one third* of 100, we divide by 3 to obtain the true product.

RULE. — *Annex two ciphers to the multiplicand, and divide it by* 3.

EXAMPLES FOR PRACTICE.

2. Multiply 356789541 by $33\frac{1}{3}$. Ans. 11892984700.
3. Multiply 871132182 by $33\frac{1}{3}$. Ans. 29037739400.
4. Multiply 583647912 by $33\frac{1}{3}$. Ans. 19454930400.

ART. **63.** To multiply by 125.

Ex. 1. Multiply 7896538 by 125. Ans. 987067250.

OPERATION.

8) 7 8 9 6 5 3 8 0 0 0

9 8 7 0 6 7 2 5 0 Product.

We multiply by 1000, by annexing three ciphers to the multiplicand; and since 125, the multiplier, is only *one eighth* of 1000, we divide by 8 to obtain the true product.

RULE. — *Annex three ciphers to the multiplicand, and divide it by* 8

EXAMPLES FOR PRACTICE.

2. Multiply 7965325 by 125. Ans. 995665625.
3. Multiply 1234567 by 125. Ans. 154320875.
4. Multiply 3049862 by 125. Ans. 381232750.

QUESTIONS. — Art. 62. What is the rule for multiplying by $33\frac{1}{3}$? What is the reason for this rule? — Art. 63. What is the rule for multiplying by 125? Give the reason for the rule.

ART. **64.** To multiply by any number of 9's.

Ex. 1. Multiply 4789653 by 99999. Ans. 478960510347.

OPERATION.

```
478965300000
       4789653
478960510347 Product.
```

By adding 1 to any number composed of nines, we obtain a number expressed by 1 with as many ciphers annexed as there are nines in the number to which 1 is added. Thus, $999 + 1 = 1000$. Therefore, annexing to the multiplicand as many ciphers as there are nines in the multiplier is the same thing as multiplying the number by a multiplier too large by 1, and subtracting the number to be multiplied from this enlarged product will give the true product.

RULE. — *Annex as many ciphers to the multiplicand as there are* 9's *in the multiplier, and from this number subtract the number to be multiplied.*

EAXMPLES FOR PRACTICE.

2. Multiply 1234567 by 999. Ans. 1233332433.
3. Multiply 876543 by 999999. Ans. 876542123457.
4. Multiply 999999 by 999999. Ans. 999998000001.

CONTRACTIONS IN DIVISION.

ART. **65.** To divide by 25.

Ex. 1. Divide 1234567 by 25. Ans. $49382\frac{68}{100}$.

OPERATION.

```
1234567
      4
49382|68 Quotient.
```

Multiplying the dividend by 4 makes it four times too great; therefore, to obtain the true quotient, we must divide by 100, a divisor four times greater than the true one. This we do by cutting off two figures on the right.

RULE. — *Multiply the dividend by* 4, *and divide the product by* 100.

EXAMPLES FOR PRACTICE.

2. Divide 9876525 by 25. Ans. 395061.
3. Divide 1378925 by 25. Ans. 55157.
4. Divide 899999 by 25. Ans. $35999\frac{96}{100}$.

QUESTIONS. — Art. 64. What is the rule for multiplying by any number of 9's? What is the reason for the rule? — Art. 65. What is the rule for dividing by 25? Give the reason for the rule.

ART. 66. To divide by $33\frac{1}{3}$.

Ex. 1. Divide 6789543 by $33\frac{1}{3}$. Ans. $203686\frac{29}{100}$.

OPERATION.

```
6 7 8 9 5 4 3
            3
─────────────
2 0 3 6 8 6|2 9 Quotient.
```

Multiplying the dividend by 3 makes it three times too great; therefore, to obtain the true quotient, we must divide by 100, a divisor three times greater than the true one. This is done by cutting off two figures on the right.

RULE. — *Multiply the dividend by* 3, *and divide the product by* 100

EXAMPLES FOR PRACTICE.

2. Divide 987654321 by $33\frac{1}{3}$. Ans. $29629629\frac{63}{100}$.
3. Divide 8712378 by $33\frac{1}{3}$. Ans. $261371\frac{34}{100}$.
4. Divide 4789536 by $33\frac{1}{3}$. Ans. $143686\frac{8}{100}$.
5. Divide 89676 by $33\frac{1}{3}$. Ans. $2690\frac{28}{100}$.
6. Divide 17854 by $33\frac{1}{3}$. Ans. $535\frac{62}{100}$.

ART. 67. To divide by 125.

Ex. 1. Divide 9874725 by 125. Ans. $78997\frac{8}{10}$.

OPERATION.

```
9 8 7 4 7 2 5
            8
─────────────
7 8 9 9 7|8 0 0 Quotient.
```

Multiplying the dividend by 8 makes it eight times too great; therefore, to obtain the true quotient, we must divide by 1000, a divisor eight times greater than the true one. We do this by cutting off three figures on the right.

RULE. — *Multiply the dividend by* 8, *and divide the product by* 1000.

EXAMPLES FOR PRACTICE.

2. Divide 1728125 by 125. Ans. 13825.
3. Divide 478763250 by 125. Ans. 3830106.
4. Divide 591234875 by 125. Ans. 4729879.
5. Divide 489648 by 125. Ans. $3917\frac{184}{1000}$.
6. Divide 836184 by 125. Ans. $6689\frac{472}{1000}$.

QUESTIONS. — Art. 66. What is the rule for dividing by $33\frac{1}{3}$? Give the reason for the rule. — Art. 67. What is the rule for dividing by 125? What is the reason for the rule?

§ VIII. MISCELLANEOUS EXAMPLES,

INVOLVING THE FOREGOING RULES.

1. A BOUGHT 73 hogsheads of molasses at 29 dollars per hogshead, and sold it at 37 dollars per hogshead; what did he gain? Ans. 584 dollars.

2. B bought 896 acres of wild land at 15 dollars per acre, and sold it at 43 dollars per acre; what did he gain? Ans. 25088 dollars.

3. N. Gage sold 47 bushels of corn at 57 cents per bushel, which cost him only 37 cents per bushel; how many cents did he gain? Ans. 940 cents.

4. A butcher bought a lot of beef weighing 765 pounds at 11 cents per pound, and sold it at 9 cents per pound; how many cents did he lose? Ans. 1530 cents.

5. A taverner bought 29 loads of hay at 17 dollars per load and 76 cords of wood at 5 dollars a cord; what was the amount of the hay and the wood? Ans. 873 dollars.

6. Bought 17 yards of cotton at 15 cents per yard, 46 gallons of molasses at 28 cents per gallon, 16 pounds of tea at 76 cents a pound, and 107 pounds of coffee at 14 cents a pound; what was the amount of my bill? Ans. 4257 cents.

7. A man travelled 78 days, and each day he walked 27 miles; what was the length of his journey? Ans. 2106 miles.

8. A man sets out from Boston to travel to New York, the distance being 223 miles, and walks 27 miles a day for 6 days in succession; what distance remains to be travelled? Ans. 61 miles.

9. What cost a farm of 365 acres at 97 dollars per acre? Ans. 35405 dollars.

10. Bought 376 oxen at 36 dollars per ox, 169 cows at 27 dollars each, 765 sheep at 4 dollars per head, and 79 elegant horses at 275 dollars each; what was paid for all? Ans. 42884 dollars.

11. J. Barker has a fine orchard, consisting of 365 trees, and each tree produces 7 barrels of apples, and these apples will bring him in market 3 dollars per barrel; what is the income of the orchard? Ans. 7665 dollars.

12. J. Peabody bought of E. Ames 7 yards of his best broadcloth at 9 dollars per yard, and in payment he gave Ames a

one hundred-dollar bill; how many dollars must Ames return to Peabody? Ans. 37 dollars.

13. Bought of P. Parker a cooking-stove for 31 dollars, 7 quintals of his best fish at 6 dollars per quintal, 14 bushels of rye at 1 dollar per bushel, and 5 mill-saws at 16 dollars each, in part payment for the above articles, I sold him eight thousand feet of boards at 15 dollars per thousand; how much must I pay him to balance the account? Ans. 47 dollars.

14. In 1 day there are 24 hours; how many in 57 days? Ans. 1368 hours.

15. In one pound avoirdupois weight there are 16 ounces; how many ounces are there in 369 pounds? Ans. 5904 ounces.

16. In a square mile there are 640 acres; how many acres are there in a town, which contains 89 square miles? Ans. 56960 acres.

17. What cost 78 barrels of apples at 3 dollars per barrel? Ans. 234 dollars.

18. Bought 500 barrels of flour at 5 dollars per barrel, 47 hundred weight of cheese at 9 dollars per hundred weight, and 15 barrels of salmon at 17 dollars per barrel; what was the amount of my purchase? Ans. 3178 dollars.

19. Bought 760 acres of land at 47 dollars per acre, and sold J. Emery 171 acres at 56 dollars per acre, J. Smith 275 acres at 37 dollars per acre, and the remainder I sold to J. Kimball at 75 dollars per acre; how much did I gain by my sales? Ans. 7581 dollars.

20. Bought a hogshead of oil containing 184 gallons, at 75 cents per gallon; but 28 gallons having leaked out, I sold the remainder at 98 cents per gallon; did I gain or lose by my bargain? Ans. 1488 cents, gain.

21. Bought a quantity of flour, for which I gave 1728 dollars, there being 288 barrels; I sold the same at 8 dollars per barrel; how much did I gain? Ans. 576 dollars.

22. Purchased a cargo of molasses for 9212 dollars, there being 196 hogsheads; I sold the same at 67 dollars per hogshead; how much did I gain on each hogshead? Ans. 20 dollars.

23. A farmer bought 5 yoke of oxen at 87 dollars a yoke; 37 cows at 37 dollars each; 89 sheep at 3 dollars apiece. He sold the oxen at 98 dollars a yoke; for the cows he received 40 dollars each; and for the sheep he had 4 dollars apiece. How much did he gain by his trade? Ans. 255 dollars.

24. The sum of two numbers is 5482, and the smaller number is 1962; what is the difference? Ans. 3520

25. The difference between two numbers is 125, and the smaller number is 1482; what is the greater? Ans. 1607.

26. The difference between two numbers is 1282, and the greater number is 6958; what is the smaller? Ans. 5676.

27. If the dividend is 21775, and the divisor 871, what is the quotient? Ans. 25.

28. If the quotient is 482, and the divisor 281, what is the dividend? Ans. 135442.

29. If 144 inches make 1 square foot, how many square feet in 20736 inches? Ans. 144 feet.

30. An acre contains 160 square rods; how many rods in a farm containing 769 acres? Ans. 123040 rods

31. A gentleman bought a house for three thousand forty-seven dollars, and a carriage and span of horses for five hundred seven dollars. He paid at one time two thousand seventeen dollars, and at another time nine hundred seven dollars. How much remains due? Ans. 630 dollars.

32. The erection of a factory cost 68,255 dollars; supposing this sum to be divided into 365 shares, what is the value of each? Ans. 187 dollars.

33. Bought two lots of wild land; the first contained 144 acres, for which I paid 12 dollars per acre; the second contained 108 acres, which cost 15 dollars per acre. I sold both lots at 18 dollars per acre; what was the amount of gain?
Ans. 1188 dollars.

34. Sold 17 cords of oak wood at 6 dollars per cord, 36 cords of maple at 3 dollars per cord, and 29 cords of walnut at 7 dollars per cord. What was the amount received?
Ans. 413 dollars.

35. Daniel Bailey has a fine farm of 300 acres, which cost him 73 dollars per acre. He sold 83 acres of this farm to Minot Thayer, for 97 dollars per acre; 42 acres to J. Russel, for 87 dollars per acre; 75 acres to J. Dana, at 75 dollars per acre; and the remainder to J. Webster, at 100 dollars per acre. What was his net gain? Ans. 5430 dollars.

36. J. Gale purchased 17 sheep for 3 dollars each, 19 cows at 27 dollars each, and 47 oxen at 57 dollars each. He sold his purchase for 3700 dollars. What did he gain?
Ans. 457 dollars.

37. Purchased 17 tons of copperas at 32 dollars per ton. I sold 7 tons at 29 dollars per ton, 8 tons at 36 dollars per ton,

and the remainder at 25 dollars per ton. Did I gain or lose, and how much? Ans. 3 dollars, loss.

38. John Smith bought 28 yards of broadcloth at 5 dollars per yard; and, having lost 10 yards, he sold the remainder at 9 dollars per yard. Did he gain or lose, and how much?

Ans. 22 dollars, gain.

39. Which is of the greater value, 386 acres of land at 76 dollars per acre, or 968 hogsheads of molasses at 25 dollars per hogshead? Ans. The land, by 5136 dollars.

40. Bought of J. Low 37 tons of hay at 18 dollars per ton. I paid him 75 dollars, and 12 yards of broadcloth at 4 dollars per yard. How much remains due to Low?

Ans. 543 dollars.

41. A purchased of B 40 cords of wood at 5 dollars per cord, 9 tons of hay at 17 dollars per ton, 19 grindstones at 2 dollars apiece, 37 yards of broadcloth at 4 dollars per yard, and 16 barrels of flour at 6 dollars per barrel; what is the amount of A's bill? Ans. 635 dollars.

42. John Smith, Jr., bought of R. S. Davis 18 dozen of National Arithmetics at 6 dollars per dozen, 23 dozen of Mental Arithmetics at 1 dollar per dozen, 17 dozen Family Bibles at 3 dollars per copy; what is the amount of the bill?

Ans. 743 dollars.

43. R. Hasseltine sold to John James 169 tons of timber at 7 dollars per ton, 116 cords of oak wood at 6 dollars per cord, and 37 cords of maple wood at 5 dollars per cord; James has paid Hasseltine 144 dollars in cash, and 23 yards of cloth at 4 dollars per yard; what remains due to Hasseltine?

Ans. 1828 dollars.

44. J. Frost owes me on account 375 dollars, and he has paid me 6 cords of wood at 5 dollars per cord, 15 tons of hay at 12 dollars per ton, and 32 bushels of rye at 1 dollar per bushel. How much remains due to me? Ans. 133 dollars.

45. Gave 169 dollars for a chaise, 87 dollars for a harness, and 176 dollars for a horse. I sold the chaise for 187 dollars, the harness for 107 dollars, and the horse for 165 dollars. What sum have I gained? Ans. 27 dollars.

46. Bought a farm of J. C. Bradbury for 1728 dollars, for which I paid him 75 barrels of flour at 6 dollars per barrel, 9 cords of wood at 5 dollars a cord, 17 tons of hay at 25 dollars a ton, 40 bushels of wheat at 2 dollars a bushel, and 65 bushels of beans at 3 dollars a bushel; how many dollars remain due to Bradbury? Ans. 533 dollars.

§ IX. UNITED STATES MONEY.

ART. 68. UNITED STATES MONEY, established by Congress in 1796, is the legal currency of the United States.

TABLE.

10 Mills	make	1 Cent,	marked	c.
10 Cents	"	1 Dime,	"	d.
10 Dimes	"	1 Dollar,	"	$.
10 Dollars	"	1 Eagle,	"	E.

Eagle.		Dollars.		Dimes.		Cents.		Mills.
						1	=	10
				1	=	10	=	100
		1	=	10	=	100	=	1000
1	=	10	=	100	=	1000	=	10000

SIMPLE NUMBERS, that is, numbers whose units are all of a single denomination, have thus far, in this work, been made use of alone in the operations.

But as the units or denominations of United States money in crease from right to left, and decrease from left to right, in the same manner as do the units of the several orders in simple numbers, they may, therefore, be added, subtracted, multiplied, and divided, according to the same rules.

Dollars are separated from cents by a point, called a separatrix or decimal point; the first two places at the right of the point being cents; and the third place, mills. Thus, $ 16.253 is read, sixteen dollars, twenty-five cents, three mills.

Since cents occupy two places, the place of dimes and of cents, when the number of cents is less than 10, a cipher must be written before them in the place of dimes; thus, .03, .07, &c.

The *coins* of the United States consist of the double-eagle, eagle, half-eagle, quarter-eagle, three dollars, and dollar, made of *gold;* the dollar, half-dollar, quarter-dollar, dime, half-dime, and three-cent piece, made of *silver;* the cent and half-cent, made of *copper*.

NOTE 1. — The word MILL is from the Latin word *mille* (one thousand); the word CENT, from the Latin *centum* (one hundred); the word DIME, from a French word signifying a *tithe* or *tenth;* and the reason of these

QUESTIONS. — Art. 68. What is United States money? Repeat the Table of United States Money. What is a simple number? What are the denominations of United States money? How do they increase from right to left? How are they added, subtracted, multiplied, and divided? How are dollars, cents, and mills, separated? Why must a cipher be placed before cents, when the number is less than 10? Why are two places allowed for cents, while only one is allowed for mills? Name the coins of the United States.

names, as applied to our coins, is found in the proportion which they respectively bear to the dollar.

The term DOLLAR is said to be derived from the Danish word *Daler* and this from *Dale*, the name of a town, where it was first coined.

The symbol $ represents, probably, the letter U written upon an S, denoting U. S. (United States).

NOTE 2. — All the gold and silver coins of the United States are now made of one purity, nine parts of pure metal, and one part alloy. The alloy for the silver is pure copper; and that for the gold, one part copper and one part silver. The cent is now made of pure copper and nickel. The standard weight, as fixed by present laws, of the eagle, is 258 grains, Troy; the silver dollar, $412\frac{1}{2}$ grains; half-dollar, 192 grains; quarter-dollar, 96 grains; dime, $38\frac{2}{5}$ grains; half-dime, $19\frac{1}{5}$ grains; three-cent piece, $11\frac{52}{100}$ grains; and the cent, new coinage, 72 grains.

REDUCTION OF UNITED STATES MONEY.

ART. **69.** REDUCTION of United States Money is changing the units of one of its denominations to the units of another, either of a higher or lower denomination, without altering their value.

ART. **70.** To reduce units from a higher denomination to a lower.

Ex. 1. Reduce 25 dollars to cents and mills.

Ans. 2500 cents, 25000 mills.

OPERATION.

$$\begin{array}{rl} 25 & \text{dollars.} \\ 100 & \\ \hline 2500 & \text{cents.} \\ 10 & \\ \hline 25000 & \text{mills.} \end{array}$$

We multiply the 25 by 100, because 100 cents make 1 dollar; and multiply the 2500 by 10, because 10 mills make 1 cent.

Or thus, 25000 mills.

RULE. — *To reduce dollars to cents, annex* TWO *ciphers; to reduce dollars to mills, annex* THREE *ciphers; and to reduce cents to mills, annex* ONE *cipher.*

NOTE. — Dollars, cents, and mills, expressed by a single number, are reduced to mills by merely removing the separating point; and dollars and cents, by annexing one cipher and removing the separatrix.

ART. **71.** To reduce units from a lower denomination to a higher.

Ex. 1. Reduce 25000 mills to cents and dollars.

Ans. 2500 cents, $25.

QUESTIONS. — Art. 69. What is reduction of United States Money? — Art. 70. What is the rule for reducing dollars to cents and mills? Give the reason for the rule. How do you reduce dollars and cents to cents, or dollars, cents, and mills, to mills? What is the reason for this rule?

OPERATION.

```
 10 ) 25000 mills.
      -----
100 ) 2500 cents.
      ----
      25 dollars.
```

We divide the 25000 by 10, because 10 mills make 1 cent; and divide the 2500 by 100, because 100 cents make 1 dollar.

Or thus, 2 5|0 0|0 mills.

RULE. — *To reduce mills to cents, cut off* ONE *figure on the right; to reduce cents to dollars, point off* TWO *figures; and to reduce mills to dollars, point off* THREE *figures.*

EXAMPLES FOR PRACTICE.

1. Reduce $125 to cents. Ans. 12500 cents.
2. Reduce $345 to mills. Ans. 345000 mills.
3. Reduce 297 mills to cents. Ans. $0.297.
4. Reduce 2682 mills to dollars. Ans. $2.682.
5. Reduce 4123 cents to dollars. Ans. $41.23.
6. Reduce $156.29 to cents. Ans. 15629 cents.
7. Reduce $16.428 to mills. Ans. 16428 mills.
8. Reduce $9.87 to mills. Ans. 9870 mills.

ART. 72. ADDITION OF UNITED STATES MONEY.

RULE. — *Write dollars, cents, and mills, so that units of the same denomination shall stand in the same column.*

Add as in addition of simple numbers, and place the separating point directly under that above.

Proof. — The proof is the same as in addition of simple numbers.

EXAMPLES FOR PRACTICE.

	1.	2.	3.	4.
	$. cts. m.	$. cts. m.	$. cts. m.	$. cts.
	45.243	75.643	16.705	147.86
	13.896	16.897	14.003	789.58
	93.516	43.816	18.719	496.37
	52.343	58.313	97.009	911.34
Ans.	204.998	194.669	146.436	2345.15

QUESTIONS. — Art. 71. What is the rule for reducing mills to cents? For reducing cents to dollars? For reducing mills to dollars? Give the reason for each. — Art. 72. How must the numbers be written down in addition of United States money? How added? How pointed off? Repeat the rule.

5.	6.	7.	8.
$. cts. m.	$. cts. m.	$. cts. m.	$. cts. m.
7 8 6.7 1 3	8 7.0 5 9	9 1.7 6 3	7 8 6.7 1 3
1 7 6.0 7 1	3 7.8 1 0	8 4.1 6 1	3 4 5.6 7 8
5 6 7.8 1 9	8 1.4 7 5	1 0.0 7 0	9 0 7.0 1 7
1 2 3.4 5 6	4 0.0 7 8	5 3.6 1 5	8 6 1.0 9 0
7 8 9.0 1 2	2 1.1 5 6	8 1.1 7 6	1 2 3.4 7 6
3 4 5.6 7 8	8 1.1 7 7	3 2.8 1 7	9 8 7.0 1 6
9 0 1.2 3 4	3 3.6 2 1	5 3.1 9 6	3 4 5.7 0 5
7 1 8.9 0 5	2 8.0 9 3	4 1.5 7 0	3 5 7.0 9 1

9. Bought a coat for $17.81, a vest for $3.75, a pair of pantaloons for $2.87, and a pair of boots for $7.18; what was the amount? Ans. $31.61.

10. Sold a load of wood for seven dollars six cents, five bushels of corn for four dollars seventy-five cents, and seven bushels of potatoes for two dollars six cents; what was received for the whole? Ans. $13.87.

11. Bought a barrel of flour for $6.50, a box of sugar for $9.87, a ton of coal for $12.77, and a box of raisins for $2.50; what was paid for the various articles?

Ans. $31.64.

12. Paid $4.62 for a hat, $9.75 for a coat, $5.75 for a pair of boots, and $1.50 for an umbrella; what was paid for the whole? Ans. $21.62.

13. A grocer sold a pound of tea for $0.625; 4 pounds of butter for $0.75; 4 dozen of lemons for $0.875; 9 pounds of sugar for $0.80; and 3 pounds of dates for $0.375. What was the amount of the bill? Ans. $3.425.

14. A student purchased a Latin grammar for $0.75, a Virgil for $3.75, a Greek lexicon for $4.75, a Homer for $1.25, an English dictionary for $3.75, and a Greek Testament for $0.75; what was the amount of the bill? Ans. $15.

15. Bought of J. H. Carleton a China tea-set for ten dollars eighty-two cents, a dining-set for nine dollars sixty-two cents five mills, a solar lamp for ten dollars fifty cents, a pair of vases for four dollars sixty-two cents five mills, and a set of silver spoons for twelve dollars seventy-five cents; what did the whole cost? Ans. $48.32.

16. Bought three hundred weight of beef at seven dollars seven cents per hundred weight, four cords of wood at six dollars four cents per cord, and a cheese for three dollars nine cents; what was the amount of the bill? Ans. $48.46.

ART. 73. SUBTRACTION OF UNITED STATES MONEY.

RULE. — *Write the several denominations of the subtrahend under the corresponding ones of the minuend.*

Subtract as in subtraction of simple numbers, and place the separatrix directly under that above.

Proof. — The proof is the same as in subtraction of simple numbers.

EXAMPLES FOR PRACTICE.

	1.	2.	3.	4.
	$. cts. m.	$. cts.	$. cts. m.	$. cts.
Min.	61.585	471.81	156.003	141.70
Sub.	19.197	158.19	19.009	90.91
Rem.	42.388	313.62	136.994	50.79

	5.	6.	7.	8.
	$. cts. m.	$. cts. m.	$. cts. m.	$. cts. m.
From	71.861	91.071	815.701	10781.303
Take	19.197	19.095	90.803	9999.097

9. From $71.07 take $5.09. Ans. $65.98.

10. From $100 take $17.17. Ans. $82.83.

11. From one hundred dollars there were paid to one man seventeen dollars nine cents, to another twenty-three dollars eight cents, and to another thirty-three dollars twenty-five cents, how much cash remained? Ans. $26.58.

12. From ten dollars take nine mills. Ans. $9.991.

13. A lady went "a shopping," her mother having given her fifty dollars. She purchased a dress for fifteen dollars seven cents; a shawl for eleven dollars ten cents; a bonnet for seven dollars nine cents; and a pair of shoes for two dollars. How much money had she remaining? Ans. $14.74.

14. From one hundred dollars there were taken at one time thirty-one dollars fifteen cents seven mills; at another time, seven dollars nine cents five mills; at another time, five dollars five cents; and at another time, twenty-two dollars two cents seven mills. How much cash remained of the hundred dollars? Ans. $34.671.

QUESTIONS. — Art. 73. How do you write down the numbers in subtraction of United States money? How subtract? How pointed off? Repeat the rule.

ART. 74. MULTIPLICATION OF UNITED STATES MONEY.

RULE. — *Multiply as in multiplication of simple numbers.*

The product will be in the lowest denomination in the question which must be pointed off as in reduction of United States money (Art. 71.)

Proof. — The proof is the same as in multiplication of simple numbers.

EXAMPLES FOR PRACTICE.

1. What will 143 barrels of flour cost at $7.25 per barrel? Ans. $1036.75.

OPERATION.

Multiplicand	$7.2 5
Multiplier	1 4 3
	2 1 7 5
	2 9 0 0
	7 2 5
Product	$1 0 3 6.7 5

2. What will 144 gallons of oil cost at $1.625 a gallon? Ans. $234.

OPERATION.

Multiplicand	$1.6 2,5
Multiplier	1 4 4
	6 5 0 0
	6 5 0 0
	1 6 2 5
Product	$2 3 4.0 0,0

3. What will 165 gallons of molasses cost at $0.27 a gallon? Ans. $44.55.

4. Sold 73 tons of timber at $5.68 a ton; what was the amount? Ans. $414.64.

5. What will 43 rakes cost at $0.17 apiece? Ans. $7.31.

6. What will 19 bushels of salt cost at $1.625 per bushel? Ans. $30.875.

7. What will 47 acres of land cost at $37.75 per acre? Ans. $1774.25.

8. What will 19 dozen penknives cost at $0.375 apiece? Ans. $85.50.

9. What is the value of 17 chests of souchong tea, each weighing 59 pounds, at $0.67 per pound? Ans. $672.01.

10. When 19 cords of wood are sold at $5.63 per cord, what is the amount? Ans. $106.97.

11. A merchant sold 18 barrels of pork, each weighing 200 pounds, at 12 cents 5 mills a pound; what did he receive? Ans. $450.

QUESTIONS. — Art. 74. How do you arrange the multiplicand and multiplier in multiplication of United States money? How multiply? Of what denomination is the product? How must it be pointed off? Repeat the rule.

12. What cost 132 tons of hay at $12.125 per ton?

Ans. $1600.50.

13. A farmer sold one lot of land, containing 187 acres, at $37.50 per acre; another lot, containing 89 acres, at $137.37 per acre; and another lot, containing 57 acres, at $89.29 per acre; what was the amount received for the whole?

Ans. $24 327.96.

ART. 75. DIVISION OF UNITED STATES MONEY.

RULE. — *Divide as in division of simple numbers.*

The quotient will be in the lowest denomination of the dividend, which must be pointed off as in reduction of United States money. (Art. 71.)

NOTE. — When the dividend consists of dollars only, and is either smaller than the divisor or not divisible by it without a remainder, reduce it to a lower denomination by annexing two or three ciphers, as the case may require, and the quotient will be cents or mills accordingly.

Proof. — The proof is the same as in division of simple numbers.

EXAMPLES FOR PRACTICE.

1. If 59 yards of cloth cost $90.27, what will 1 yard cost?

Ans. $1.53.

```
                OPERATION.
               Dividend.   $.
Divisor 5 9 ) 9 0.2 7 ( 1.5 3 Quotient.
              5 9
              -----
              3 1 2
              2 9 5
              -----
                1 7 7
                1 7 7
                -----
```

2. Purchased 68 ounces of indigo for $17. What did I give per ounce?

Ans. $0.25.

```
                OPERATION.
               Dividend.   $.
Divisor 6 8 ) 1 7.0 0 ( 0.2 5 Quotient.
              1 3 6
              -----
                3 4 0
                3 4 0
                -----
```

3. If 89 acres of land cost $12225.93, what is the value of 1 acre?

Ans. $137.37.

4. When 19 yards of cloth are sold for $106.97, what should be paid for 1 yard?

Ans. $5.63.

QUESTIONS. — Art. 75. How do you arrange the dividend and divisor in division of United States money? How divide? Of what denomination is the quotient? How pointed off? How do you proceed when the dividend is dollars only, and is either smaller than the divisor or not divisible by it without a remainder? Repeat the rule.

5. Gave $22.50 for 18 barrels of apples; what was paid for 1 barrel? For 5 barrels? For 10 barrels?
Ans. $20 for all.

6. Bought 153 pounds of tea for $90.27; what was it per pound? Ans. $0.59.

7. A merchant purchased a bale of cloth, containing 73 yards, for $414.64; what was the cost of 1 yard?
Ans. $5.68.

8. If 126 pounds of butter cost $16.38, what will 1 pound cost? Ans. $0.13.

9. If 63 pounds of tea cost $58.59, what will 1 pound cost?
Ans. $0.93.

10. If 76 cwt. of beef cost $249.28, what will 1 cwt. cost?
Ans. $3.28.

11. If 96,000 feet of boards cost $1120.32, what will a thousand feet cost? Ans. $11.67.

12. Sold 169 tons of timber for $790.92; what was received for 1 ton? Ans. $4.68.

13. When 369 tons of potash are sold for $48910.95, what is received for 1 ton? Ans. $132.55.

14. For 19 cords of wood I paid $109.25; what was paid for 1 cord? Ans. $5.75.

PRACTICAL QUESTIONS BY ANALYSIS.

ART. 76. ANALYSIS is an examination of a question by resolving it into its parts, in order to consider them separately, and thus render each step in the solution plain and intelligible.

ART. 77. The price of one pound, yard, bushel, &c., being given, to find the price of any quantity.

RULE. — *Multiply the price by the quantity.*

Ex. 1. If 1 ton of hay cost $12, what will 29 tons cost?
Ans. $348.

ILLUSTRATION — Since 1 ton costs $12, 29 tons will cost 29 times as much: $12 × 29 = $348.

2. If 1 bushel of salt cost 93 cents, what will 40 bushels cost? What will 97 bushels cost? Ans. $90.21.

QUESTIONS. — Art. 77. The price of 1 pound, &c., being given, how do you find the price of any quantity? Give the reason for this rule.

3. If 1 bushel of apples cost $1.65, what will 5 bushels cost? What will 18 bushels cost? Ans. $29.70.

4. If 1 ton of clay cost $0.67, what will 7 tons cost? What will 63 tons cost? Ans. $42.21.

5. When $7.83 are paid for 1 cwt. of sugar, what will 12 cwt. cost? What will 93 cwt. cost? Ans. $728.19.

6. When $0.09 are paid for 1 lb. of beef, what will 12 lb. cost? What will 760 lb. cost? Ans. $68.40.

7. A gentleman paid $38.37 for 1 acre of land; what was the cost of 20 acres. What would 144 acres cost?

Ans. $5525.28.

8. Paid $6.83 for 1 barrel of flour; what was the value of 9 barrels? What must be paid for 108 barrels?

Ans. $737.64.

ART. 78. The price of any quantity, and the quantity being given, to find the price of a unit of that quantity.

RULE. — *Divide the price by the quantity.*

9. If 15 bushels of corn cost $10.35, what will 1 bushel cost? Ans. $0.69.

ILLUSTRATION. — Since 15 bushels cost $10.35, 1 bushel will cost as many cents as 15 is contained times in $10.35 : $10.35 ÷ 15 = $0.69.

10. Bought 65 barrels of flour for $422.50; what cost one barrel? What cost 15 barrels? Ans. $97.50.

11. For 45 acres of land a farmer paid $2025; what cost 1 acre? What 180 acres? Ans. $8100.

12. For 5 pairs of gloves a lady paid $3.45; what cost 1 pair? What cost 11 pairs? Ans. $7.59.

13. If 11 tons of hay cost $214.50, what will 1 ton cost? What will 87 tons cost? Ans. $1696.50.

14. When $60 are paid for 8 dozen of arithmetics, what will 1 dozen cost? What will 87 dozen cost? Ans. $652.50.

15. Gave $5.58 for 9 bushels of potatoes; what will 1 bushel cost? What will 43 bushels cost? Ans. $26.66.

16. Bought 5 tons of hay for $85; what would 1 ton cost? What would 97 tons cost? Ans. $1649.

QUESTIONS. — Art. 78. How do you find the price of 1 pound, &c., the price of any quantity and the quantity being given? What is the reason for this rule?

17. If J. Ladd will sell 20 lb. of butter for $3.80, what should he charge for 59 lb.? Ans. $11.21.

18. Sold 27 acres of land for $472.50; what was the price of 1 acre? What should be given for 12 acres? Ans. $210.

19. Paid $39.69 for 7 cords of wood; what will 1 cord cost? What will 57 cords cost? Ans. $323.19.

20. Paid $10.08 for 144 lb. of pepper; what was the price of 1 pound? What cost 359 lb.? Ans. $25.13.

21. Paid $77.13 for 857 lb. of rice; what cost 1 lb.? What cost 359 lb.? Ans. $32.31.

22. J. Johnson paid $187.53 for 987 gal. of molasses; what cost 1 gal.? What cost 329 gal.? Ans. $62.51.

23. For 47 bushels of salt J. Ingersoll paid $26.32; what cost 1 bushel? What cost 39 bushels? Ans. $21.84.

ART. **79.** The price of any quantity and the price of a unit of that quantity being given, to find the quantity.

RULE. — *Divide the whole price by the price of a unit of the quantity required.*

24. If I expend $150 for coal at $6 per ton, how many tons can I purchase? Ans. 25 tons.

ILLUSTRATION. — Since I pay $6 for 1 ton, I can purchase as many tons with $150 as $6 is contained times in $150: $150 ÷ $6 = 25; therefore I can purchase 25 tons.

25. At $5 per ream, how many reams of paper can be bought for $175? Ans. 35 reams.

26. At $7.50 per barrel, how many barrels of flour can be obtained for $217.50? Ans. 29 barrels.

27. At $75 per ton, how many tons of iron can be purchased for $4875? Ans. 65 tons.

28. At $4 per yard, how many yards of cloth can be bought for $1728? Ans. 432 yards.

29. How many hundred weight of hay can be bought for $9.66, if $0.69 are paid for 1 hundred weight?

Ans. 14 hundred weight.

30. If $66.51 are paid for flour at $7.39 per barrel, how many barrels can be bought? Ans. 9 barrels.

31. Paid $136.50 for wood, at $3.25 per cord; how many cords did I buy? Ans. 42 cords.

QUESTIONS. — Art. 79. How do you find the quantity, the price of 1 pound, &c., being given? Give the reason for the rule.

BILLS.

ART. 80. A BILL is a paper, given by merchants, containing a statement of goods sold, and their prices.

An *invoice* is a bill of merchandise shipped or forwarded to a purchaser, or selling agent.

The *date* of a bill is the time and place of the transaction.

The bill is *against* the party owing, and in *favor* of the party who is to receive the amount due.

A bill is receipted, when the receiving of the amount due is acknowledged by the party in whose favor it is. A clerk, or any other authorized person, may, in his stead, receipt for him, as in bill 2.

When the items of a bill have been rendered at different dates, the several times may be given at the left hand, as in bill 5.

When the bill is in the form of an account, containing items of debt and credit in its settlement, it is required to find the difference due, or balance, as in bill 5.

What is the cost of each article in, and the amount due of, each of the following bills?

(1.) *New York, May* 20, 1856.

Dr. JOHN SMITH,

Bought of SOMES & GRIDLEY,

82 *gals. Temperance Wine,*	*at*	$0.75	
89 " *Port do.*	"	.92	
24 *pairs Silk Gloves,*	"	.50	
			$155.38.

Received payment,

SOMES & GRIDLEY.

(2.) *Philadelphia, March* 7, 1857.

Mr. LEVI WEBSTER,

Bought of JAMES FRANKLAND,

6 *lbs. Chocolate,*	*at*	$0.18	
12 " *Flour,*	"	.20	
6 *pairs Shoes,*	"	1.80	
30 *lbs. Candles,*	"	.26	
			$22.08.

Received payment,

JAMES FRANKLAND,
by ENOCH OSGOOD.

QUESTIONS. — Art. 80. What is a bill, in mercantile transactions? What is an invoice? When is a bill against, and when in favor of a party? How is a bill receipted?

(3.) *St. Louis, March* 19, 1856

Mr. WILLIAM GREENLEAF,

Bought of MOSES ATWOOD.

86 *Shovels,*	*at*	$0.50	
90 *Spades,*	"	.86	
18 *Ploughs,*	"	11.00	
23 *Handsaws,*	"	3.50	
14 *Hammers,*	"	.62	
12 *Mill-saws,*	"	12.12	
46 *cwt. Iron,*	"	12.00	
		———	$1105.02.

(4.) *Boston, June* 5, 1856.

Mr. AMOS DOW,

Bought of LORD & GREENLEAF,

37 *Chests Green Tea,*	*at*	$23.75	
42 " *Black do.,*	"	17.50	
43 *Casks Wine,*	"	99.00	
12 *Crates Liverpool Ware,*	"	175.00	
19 *bbl. Genesee Flour,*	"	7.00	
23 *bu. Rye,*	"	1.52	
		———	$8138.71.

(5.) *San Francisco, May* 13, 1856.

Mr. JOHN WADE,

1855.	*To* AYER, FITTS & CO., *Dr.*			
Apr. 5.	*To* 80 *pairs Hose,*	*at*	$1.20	
Aug. 7.	" 17 " *Boots,*	"	3.00	
	" 19 " *Shoes,*	"	1.08	
Nov. 1.	" 23 " *Gloves,*	"	.75	
			———	$184.77
1856.	*Cr.*			
Jan. 1.	27 *Young Readers,*	*at*	$0.20	
" "	10 *Greek Lexicons,*	"	3.90	
Feb. 10.	7 *Webster's Dictionaries,*	"	4.75	
Apr. 3.	19 *Folio Bibles,*	"	2.93	
" "	20 *Testaments,*	"	.37	
			———	$140.72.
	Balance due A., F. & Co.			$44.05.

Received payment,

AYER, FITTS & CO.

LEDGER ACCOUNTS.

ART. 81. The principal book of accounts among merchants is called a ledger. In it are brought together scattered items of account, often making long columns. As a rapid way of finding the amount of each, accountants generally add more than one column at a single operation. (Art. 24.) The examples below may be added both by the usual method and by that which is more rapid.

1.	2.	3.	4.
$. cts.	$. cts.	$. cts.	$. cts.
5.75	1.05	71.10	100.88
3.15	7.08	35.60	320.12
6.37	6.38	21.40	280.47
10.13	5.50	100.50	151.53
5.05	3.25	62.75	.92
12.50	8.19	13.13	11.08
8.00	1.13	1.37	49.13
.63	10.10	16.02	44.22
1.37	15.25	19.28	60.81
22.00	13.45	163.35	52.75
16.05	6.17	620.50	35.15
1.19	.09	75.00	70.06
.31	1.13	25.20	1050.00
10.00	8.07	53.81	3120.12
11.88	11.06	33.19	200.50
.12	35.15	17.00	16.09
9.17	18.91	10.38	900.11
.33	10.03	40.12	1825.50
6.22	30.00	15.68	105.10
2.31	1.88	71.12	35.46
7.17	2.75	13.19	67.63
15.50	1.25	10.00	81.17
11.25	5.00	18.20	10.14
.09	25.50	13.15	75.00
21.17	12.02	25.00	120.00
32.00	19.17	102.55	114.09
14.06	32.43	111.10	212.63
20.50	46.37	235.83	10300.48

QUESTIONS. — Art. 81. What is a ledger? How may ledger columns be added rapidly?

§ X. REDUCTION.

ART. 82. A SIMPLE number is a unit or a collection of units, either abstract, or concrete of a single kind or denomination; thus, 1 dollar, 9 apples, 12, are simple numbers.

A COMPOUND number is a collection of concrete units of several kinds or denominations, taken collectively; thus, 12£. 18s. 9d., is a compound number.

ART. 83. Reduction is changing numbers, either simple or compound, from one denomination to another, without altering their values.

It is of two kinds, Reduction Descending, and Reduction Ascending.

Reduction Descending is changing numbers of a higher denomination to a lower denomination; as pounds to shillings, &c. It is performed by multiplication.

Reduction Ascending is changing numbers of a lower denomination to a higher denomination; as farthings to pence, &c. It is the reverse of Reduction Descending, and is performed by division.

ENGLISH MONEY.

ART. 84. English or Sterling Money is the Currency of England.

TABLE.

4 Farthings (qr. or far.)	make	1 Penny,	d.
12 Pence	"	1 Shilling,	s.
20 Shillings	"	1 Pound,	£.
21 Shillings sterling	"	1 Guinea,	G.
20 Shillings "	"	1 Sovereign,	sov.

£.		s.		d.		far.
				1	=	4
		1	=	12	=	48
1	=	20	=	240	=	960

NOTE 1.—The symbol £. stands for the Latin word *libra*, signifying a pound; *s.* for *solidus*, a shilling; *d.* for *denarius*, a penny; *qr.* for *quadrans*, a quarter.

QUESTIONS.—Art. 82. What is a simple number? What is a compound number?—Art. 83. What is reduction? How many kinds of reduction? What are they? What is reduction descending? What is reduction ascending?—Art. 84. What is English money? Repeat the table.

NOTE 2. — Farthings are sometimes expressed in a fraction of a penny; thus, 1 *far.* = $\frac{1}{4}$ d.; 2 far. = $\frac{1}{2}$ d.; 3 far. = $\frac{3}{4}$ d.

NOTE 3. — The *Pound Sterling* is represented by a gold coin called a sovereign. Its legal value in United States money is $4.84.

NOTE 4. — The term sterling is probably from Easterling, the popular name of certain early German traders in England, whose money was noted for the purity of its quality.

MENTAL EXERCISES.

1. How many farthings in 3 pence? In 9 pence?
2. How many pence in 2 shillings? In 6 shillings?
3. How many shillings in 7 pounds? In 10 pounds?
4. How many pence in 8 farthings? In 24 farthings?
5. How many shillings in 24 pence? In 60 pence?
6. How many pounds in 40 shillings? In 80 shillings?

EXERCISES FOR THE SLATE.

ART. 85. To reduce units of a higher denomination to a lower.

1. How many farthings in 17£. 8s. 9d. 3far.?

OPERATION.

```
       17£. 8s. 9d. 3far.
       20
      ----
      348 shillings.
       12
     -----
     4185 pence.
        4
    ------
Ans. 16743 farthings.
```

We multiply the 17 by 20, because 20 shillings make 1 pound, and to this product we add the 8 shillings in the question. We then multiply by 12, because 12 pence make 1 shilling, and to the product we add the 9d. Again, we multiply by 4, because 4 farthings make 1 penny, and to this product we add the 3 far.; and we find the answer to be 16743 farthings.

RULE. — *Multiply the highest denomination given by the number required of the next lower denomination to make one in the denomination multiplied. To this product add the corresponding denomination of the multiplicand, if there be any. Proceed in this way, till the reduction is brought to the denomination required.*

QUESTIONS. — Art. 85. How do you reduce pounds to shillings? Why multiply by 20? How do you reduce shillings to pence? Why? Pence to farthings? Why? Guineas to shillings? What is the general rule for reduction descending?

ART. 86. To reduce the unit of a lower denomination to a higher.

Ex. 2. How many pounds in 16743 farthings?

OPERATION.

```
4 ) 16743 far.
   ---------
12 ) 4185 d. 3far.
    ---------
  20 ) 348 s. 9d.
      --------
        17 £. 8s.
Ans. 17£. 8s. 9d. 3far.
```

We divide by 4, because 4 farthings make 1 penny, and the result is 4185 pence, and the remainder, 3, is farthings. We divide by 12, because 12 pence make 1 shilling, and the result is 348 shillings, and the 9 remaining is pence. Lastly, we divide by 20, because 20 shillings make 1 pound, and the result is 17£. 8s. Therefore, by annexing all the remainders to the last quotient, we find the answer to be 17£. 8s. 9d. 3far.

RULE. — *Divide the lower denomination given by the number, which it takes of that denomination to make one of the next higher. The quotient thus obtained divide as before, and so proceed until it is brought to the denomination required. The last quotient, with the remainders connected, will be the answer.*

3. In 9£. 18s. 7d. how many pence?
4. In 2383d. how many pounds, &c.?
5. How many farthings in 14£. 11s. 5d. 2far.?
6. How many pounds in 13990far.?

TROY WEIGHT.

ART. 87. Troy Weight is the weight used in weighing gold, silver, and jewels.

TABLE.

24 Grains (gr.)	make	1 Pennyweight,	pwt.
20 Pennyweights	"	1 Ounce,	oz.
12 Ounces	"	1 Pound,	lb.

lb.		oz.		pwt.		gr.
				1	=	24
		1	=	20	=	480
1	=	12	=	240	=	5760

QUESTIONS. — Art. 86. How do you reduce farthings to pence? Why divide by 4? How do you reduce pence to shillings? Why? Shillings to pounds? Why? Shillings to guineas? What is the general rule for reduction ascending? What is Troy Weight used for? Repeat the Table.

NOTE 1. — The *oz.* stands for *onza*, the Spanish for ounce.

NOTE 2. — A grain or corn of wheat, gathered out of the middle of the ear, was the origin of all the weights used in England. Of these grains, 32, well dried, were to make one pennyweight. But in later times it was thought sufficient to divide the same pennyweight into 24 equal parts, still called grains, being the least weight now in use, from which the rest are computed.

NOTE 3. — Diamonds and other precious stones are weighed by what is called *Diamond Weight*, of which 16 *parts* make 1 *grain;* 4 grains, 1 *carat.* 1 grain Diamond Weight is equal to $\frac{4}{5}$ grains Troy, and 1 carat to $3\frac{1}{5}$ grains Troy.

NOTE 4. — The Troy pound is the *standard unit of weight* adopted by the United States Mint, and is the same as the Imperial Troy pound of Great Britain.

MENTAL EXERCISES.

1. How many gr. in 2pwt.? In 10pwt.?
2. How many pennyweights in 4oz.? In 20oz.?
3. How many ounces in 2lb.? In 5lb.? In 10lb.?
4. How many pennyweights in 48gr.? In 96gr.?
5. How many ounces in 40pwt.? In 120pwt.?
6. How many pounds in 24oz.? In 60oz.? In 120oz.?

EXERCISES FOR THE SLATE.

1. How many grains in 72lb. 10oz. 15pwt. 7gr.?

OPERATION.

```
 72 lb. 10oz. 15pwt. 7gr.
 12
 ---
 874 ounces.
  20
 -----
 17495 pennyweights.
    24
 -----
 69987
34990
------
Ans. 419887 grains.
```

2. In 419887 grains, how many pounds?

OPERATION.

```
24)419887 gr.
  20)17495 pwt. 7gr.
     12)874 oz. 15pwt.
         72 lb. 10oz.
Ans. 72lb. 10oz. 15pwt. 7gr.
```

QUESTIONS. — What was the original of all weights in England? How many of these grains did it take to make a pennyweight? How many grains in a pennyweight now? By what weight are diamonds weighed? What is the standard at the mint? How do you reduce pounds to grains? Give the reason of the operation. How do you reduce grains to pounds?

3. How many grains in 76pwt. 12gr.?
4. How many pennyweights in 1836gr.?
5. In 76lb. 5oz. how many grains?
6. In 440160 grains how many pounds?
7. How many pennyweights in 144lb. 9oz.?
8. How many pounds in 34740pwt.?
9. How many pounds in 17895gr.?
10. In 3lb. 1oz. 5pwt. 15gr. how many grains?
11. A valuable gem weighing 2oz. 18pwt. 12gr. was sold for $1.37 per grain; what was the sum paid? Ans. $1923.48.

APOTHECARIES' WEIGHT.

ART. 88. Apothecaries' Weight is used in mixing medicines.

TABLE.

20 Grains (gr.)	make	1 Scruple,	sc. or ℈
3 Scruples	"	1 Dram,	dr. or ʒ
8 Drams	"	1 Ounce,	oz. or ℥
12 Ounces	"	1 Pound,	lb. or ℔

℔.		oz.		dr.		sc.		gr.
						1	=	20
				1	=	3	=	60
		1	=	8	=	24	=	480
1	=	12	=	96	=	288	=	5760

NOTE 1. — In this weight the pound, ounce, and grain are the same as in Troy Weight.

NOTE 2. — Medicines are usually bought and sold by Avoirdupois Weight.

NOTE 3. — In estimating the weight of fluids, 45 drops, or a common tea-spoonful, make about 1 fluid dram; 2 common table-spoonfuls, about 1 fluid ounce.

MENTAL EXERCISES.

1. In 40 grains how many scruples? In 60gr.? In 120gr.?
2. In 5 scruples how many grains? In 10sc.? In 40sc.?
3. In 3 drams how many scruples? In 10dr.? In 17dr.?
4. How many pounds in 48 ounces? In 96oz.? In 144oz.?
5. How many ounces in 24 drams? In 64dr.? In 96dr.?

QUESTIONS. — Art 88. For what is Apothecaries' Weight used? What denominations of this weight are the same as those of Troy Weight? By what weight are medicines usually bought and sold? Repeat the table.

EXERCISES FOR THE SLATE.

1. In 40lb. 8oz. 5dr. 1sc. 7gr. how many grains?

OPERATION.

```
 4 0 lb. 8oz. 5dr. 1sc. 7gr.
   1 2
 -----
 4 8 8 ounces.
     8
 -------
 3 9 0 9 drams.
       3
 ---------
 1 1 7 2 8 scruples.
       2 0
 ---------
Ans. 2 3 4 5 6 7 grains.
```

2. How many pounds in 234567 grains?

OPERATION.

```
2 0 ) 2 3 4 5 6 7 gr.
      -----------
      3 ) 1 1 7 2 8 sc. 7gr.
          ---------
          8 ) 3 9 0 9 dr. 1sc.
              -------
            1 2 ) 4 8 8 oz. 5dr.
                  -----
                  4 0 lb. 8oz.
```

Ans. 40lb. 8oz. 5dr. 1sc. 7gr.

3. How many scruples in 76lb.?
4. How many pounds in 21888℈?
5. How many grains in 144lb.?
6. How many pounds in 829440gr.?
7. In 12℔ 8℥ 3ʒ 1℈ 18gr. how many grains?
8. In 73178 grains how many pounds?
9. How many doses are there in 7℥ 6ʒ 2℈ of tartar emetic, admitting 20 grains for each dose? Ans. 188.

AVOIRDUPOIS WEIGHT.

ART. 89. Avoirdupois Weight is used in weighing almost every kind of goods, and all metals except gold and silver.

TABLE.

16 Drams (dr.)	make	1 Ounce,	oz.
16 Ounces	"	1 Pound,	lb.
25 Pounds	"	1 Quarter,	qr.
4 Quarters	"	1 Hundred Weight,	cwt.
20 Hundred Weight	"	1 Ton,	T.

T.		cwt.		qr.		lb.		oz.		dr.
								1	=	16
						1	=	16	=	256
				1	=	25	=	400	=	6400
		1	=	4	=	100	=	1600	=	25600
1	=	20	=	80	=	20000	=	32000	=	51200

QUESTIONS. — How do you reduce pounds to grains? What is the reason for the operation? How do you reduce grains to pounds? Give the reason of the operation. — Art 89. For what is Avoirdupois Weight used? Recite the table.

NOTE 1. — In *cwt.* the *c* stands for *centum*, the Latin for *one hundred* and *wt* for *weight.*

NOTE 2. — The laws of most of the States, and common practice at the present time, make 25 pounds a quarter, as given in the table. But formerly, 28 pounds were allowed to make a quarter, 112 pounds a hundred, and 2240 pounds a ton, as is still the standard of the United States government in collecting duties at the custom-houses.

NOTE 3. — The term avoirdupois is from the French *avoir du poid*, signifying to have weight.

NOTE 4. — 1 pound Avoirdupois = 7000 gr. Troy = 1lb. 2oz. 11 pwt. 16 gr. Troy ; 1lb Troy, or Apothecary = 5760gr. Troy = 13oz. $2\frac{114}{175}$ dr. Avoirdupois ; 1oz. Troy, or Apoth. = 480gr. Troy = 1oz. $1\frac{97}{175}$ dr. Av. ; 1oz. Av. = $437\frac{1}{2}$gr. Troy = 18pwt. $5\frac{1}{2}$gr. Troy ; 1dr. Apoth. = 60gr. Troy = $2\frac{34}{175}$dr. Av. ; 1dr Av. = $27\frac{11}{32}$gr. Troy = 1pwt. $3\frac{11}{32}$gr. Troy ; 1 pwt. Troy = 24gr. Troy = $\frac{768}{875}$ of a dr. Av. ; 1sc. Apoth. = 20gr. Troy = $\frac{128}{175}$ of a dr. Av.

MENTAL EXERCISES.

1. How many drams in 3oz. ? In 7oz. ? In 10oz. ? In 12oz. ?
2. How many ounces in 10lb. ? In 15lb. ? In 12lb. ? In 100lb. ?
3. How many pounds in 2 quarters ? In 3qr. ? In 20qr. ?
4. How many quarters in 10cwt. ? In 16cwt. ? In 17cwt. ?
5. How many tons in 80cwt. ? In 100cwt. ? In 600cwt. ?
6. How many hundred weight in 16qr. ? In 48qr. ? In 96qr. ?

EXERCISES FOR THE SLATE.

1. How many pounds in 176T. 17cwt. 3qr. 15lb. ?

OPERATION.

```
 1 7 6 T. 17cwt. 3qr. 15lb.
     2 0
 -------
 3 5 3 7 hundred weight.
       4
 -------
1 4 1 5 1 quarters.
      2 5
 ---------
 7 0 7 7 0
2 8 3 0 2
---------
Ans. 3 5 3 7 9 0 pounds.
```

2. In 353790lb. how many tons ?

OPERATION.

```
 2 5 ) 3 5 3 7 9 0 lb.
      4 ) 1 4 1 5 1 qr. 15lb
      2 0 ) 3 5 3 7 cwt. 3qr.
              1 7 6 T. 17cwt
Ans. 176T. 17cwt. 3qr. 15lb.
```

QUESTIONS. — How many pounds are now allowed for a cwt., and how many for a quarter of a cwt., in most of the United States, in buying and selling articles by weight ? How many at the custom-houses ? How do you reduce tons to drams ? Give the reason for the operation. How do you reduce drams to tons ? What is the reason for the operation ?

3. In 16T. 19cwt. 0qr. 10lb. 11oz. 5dr. how many drams?

4. In 8681141 drams how many tons?

5. In 679cwt. how many pounds?

6. In 67900lb. how many cwt.?

7. What cost 17cwt. 3qr. 18lb. of beef, at 7 cents per pound? Ans. $125.51.

8. What cost 48T. 17cwt. of lead, at 8 cents per pound? Ans. $7816.00.

CLOTH MEASURE.

ART. **90.** Cloth Measure is used in measuring cloth, ribbons, lace, and other articles sold by the yard or ell.

TABLE.

$2\frac{1}{4}$ Inches (in.)	make	1 Nail,	na.
4 Nails	"	1 Quarter of a yard,	qr.
4 Quarters	"	1 Yard,	yd.
3 Quarters	"	1 Ell Flemish,	E. F.
5 Quarters	"	1 Ell English,	E. E.

E. E.		yd.		E. F.		qr.		na.		in.
								1	=	$2\frac{1}{4}$
						1	=	4	=	9
				1	=	3	=	12	=	27
		1	=	$1\frac{1}{3}$	=	4	=	16	=	36
1	=	$1\frac{1}{4}$	=	$1\frac{2}{3}$	=	5	=	20	=	45

NOTE.—The Ell French is 6 quarters; the Ell Scotch, 4qr. $1\frac{1}{5}$in.

MENTAL EXERCISES.

1. In 2 quarters how many nails? In 5qr.? In 8qr.? In 20qr.? In 25qr.? In 30qr.? In 40qr.?

2. In 3 yards how many quarters? In 7yd.? In 8yd.? In 14yd.? In 19yd.? In 100yd.? In 200yd.?

3. How many quarters in 8 nails? In 20na.? In 48na.?

4. How many yards in 20 quarters? In 40qr.? In 100qr.?

EXERCISES FOR THE SLATE.

1. How many nails in 47yd. 3qr. 1na.?

2. In 765 nails how many yards?

QUESTIONS.—Art. 90. For what is cloth measure used? Repeat the table. Is the ell French longer or shorter than the ell English? What makes an ell Scotch?

OPERATION.

```
4 7 yd. 3qr. 1na.
    4
-------
1 9 1 quarters.
    4
-------
Ans. 7 6 5 nails.
```

OPERATION.

```
4 ) 7 6 5 na.
    -----
4 ) 1 9 1 qr. 1na.
    -----
Ans. 4 7 yd. 3qr. 1na
```

3. In 144yd. 3qr. how many quarters?
4. In 579 quarters how many yards?
5. In 17E. E. 4qr. 3na. how many nails?
6. In 359 nails how many ells English?
7. In 126yd. 0qr. 3na. how many nails?
8. In 2019 nails how many yards?
9. What cost 49yd. 3qr. of cloth, at $2.17 per quarter of a yard? Ans. $431.83.
10. What cost 144yd. 1qr. 3na. of cloth, at 25 cents per nail? Ans. $577.75.

LONG MEASURE.

ART. 91. Long Measure is used in measuring distances in any direction.

TABLE.

12 Inches (in.)	make	1 Foot,	ft.
3 Feet	"	1 Yard,	yd.
$5\frac{1}{2}$ Yards, or $16\frac{1}{2}$ Feet,	"	1 Rod, or Pole,	rd.
40 Rods	"	1 Furlong,	fur
8 Furlongs, or 320 Rods,	"	1 Mile,	m
3 Miles	"	1 League,	lea.
$69\frac{1}{6}$ Miles (nearly)	"	1 Degree,	deg. or °.
360 Degrees	"	1 Circle of the Earth.	

m.		fur.		rd.		yd.		ft.		in.
								1	=	12
						1	=	3	=	36
				1	=	$5\frac{1}{2}$	=	$16\frac{1}{2}$	=	198
		1	=	40	=	220	=	660	=	7920
1	=	8	=	320	=	1760	=	5280	=	63360

QUESTIONS.—How do you reduce yards to nails? How do you reduce nails to yards? What is the reason for the operation? How is long measure used? Repeat the table.

NOTE. — 1. 12 lines make 1 inch ; 4 inches, 1 hand ; 6 feet, 1 fathom , $\frac{1}{60}$ of a degree of the circumference of the earth, 1 knot, or geographical mile, equal to $1\frac{11}{72}$ statute miles.

NOTE. — 2. The *yard* is the *standard unit* of linear measure adopted by the United States government, and it is the same as the imperial yard of Great Britain. A *metre*, the unit of linear measure, as established by the French government, is equal to about $39\frac{37}{100}$ English inches.

NOTE. — 3. The English statute mile is the same as that of the United States, but that of other countries differs in value from it ; as the German short mile is equal to 6857 yards, or about $3\frac{9}{10}$ English miles ; the German long mile, to 10125 yards, or about $5\frac{3}{4}$ English miles ; the Prussian mile, to 8237 yards, or about $4\frac{7}{10}$ English miles ; the Spanish common league, to 7416 yards, or about $4\frac{1}{5}$ English miles ; the Spanish judicial league, to 4635 yards, or about $2\frac{3}{5}$ English miles.

NOTE. — 4. A degree of longitude is $\frac{1}{360}$ of any circle of latitude. As the circles of latitude diminish in length, the degrees of longitude vary in length under different parallels of latitude. Thus, under the equator, the length of a degree of longitude is about $69\frac{1}{6}$ statute miles ; at 25° of latitude, $62\frac{7}{10}$ miles ; at 40° of latitude, 53 miles ; at 42° of latitude, $51\frac{1}{2}$ miles ; at 49° of latitude, $45\frac{1}{2}$ miles ; at 60°, $34\frac{7}{12}$ miles.

MENTAL EXERCISES.

1. How many inches in 4 feet? In 10ft.? In 12ft.? In 20ft.?
2. How many feet in 2 yards? In 5yd.? In 20yd.? In 18yd.?
3. How many rods in 2 furlongs? In 8fur.? In 1fur.? In 30 fur.? In 100fur.? In 200fur.? In 400fur.?
4. How many leagues in 9 miles? In 21m.? In 81m.? In 144m.? In 40m.? In 50m.? In 80m.?
5. How many furlongs in 120 rods? In 360rd.? In 1440rd.?
6. How many yards in 99 feet? In 66ft.? In 144ft.?
7. How many feet in 108 inches? In 144in.? In 1728in.?

EXERCISES FOR THE SLATE.

1. In 66deg. 56m. 7fur. 37rd. 12ft. 9in. how many inches?

QUESTIONS. — How many lines make 1 inch? How many inches 1 hand? How many feet 1 fathom? What is the standard unit of linear measure adopted by the United States? Is the value of the mile the same in all countries? How much is a degree of longitude under the equator? At 40° of latitude? At 42° of latitude? At 60° of latitude?

OPERATION.

```
6 6deg. 56m. 7fur. 37rd. 12ft. 9in.
 6 9 1/6
 -------
  6 0 0
4 0 1
  1 1
 -------
4 6 2 1 miles.
      8
 -------
3 6 9 7 5 fur.
      4 0
 -------------
1 4 7 9 0 3 7 rods.
          1 6 1/2
 -------------
 8 8 7 4 2 2 4
1 4 7 9 0 3 8
  7 3 9 5 1 8 1/2
 -------------
2 4 4 0 4 1 2 2 1/2 ft.
              1 2
 ---------------
2 9 2 8 4 9 4 7 9 in. Ans.
```

2. In 292849479 inches how many degrees?

OPERATION.

```
12 ) 2 9 2 8 4 9 4 7 9
     -----------------
16 1/2 ) 2 4 4 0 4 1 2 3 ft. 3in.
 2                     2
 --       ---------------
3 3    ) 4 8 8 0 8 2 4 6    [12ft. 6in
         ---------------
       4 0 ) 1 4 7 9 0 3 7 rd. 25÷2=
             -------------
           8 ) 3 6 9 7 5 fur. 37rd.
               ---------
     6 9 1/6 ) 4 6 2 1 miles 7fur
       6             6
     -----   ---------
     4 1 5 ) 2 7 7 2 6
             ---------
                    6 6 deg. 336 ÷
                            [6 = 56m.
66deg. 56m. 7fur. 37rd. 12 1/2ft. 3in.
                            1/2 = 6in.
-------------------------------------
66      56    7    37     12      9 Ans.
```

NOTE. — To multiply by $\frac{1}{2}$, we take $\frac{1}{2}$ of the multiplicand. — To divide by $16\frac{1}{2}$, we first reduce both the divisor and dividend to halves, and then divide; and the remainder being 25 half-feet, we take half of it for the true remainder = 12ft. 6in. We adopt the same principle in dividing by $69\frac{1}{6}$; the remainder being 336 sixths of miles, we divide them by 6, and the quotient is 56 miles. By adding the 6 inches to the 3 inches, we obtain the true answer.

3. In 47 miles how many feet?

4. In 248160 feet how many miles?

5. In 78deg. 50m. 7fur. 30rd. 5yd. 2ft. 10in. how many inches?

6. How many degrees in 345056794 inches?

QUESTIONS. — How do you reduce degrees to inches? Give the reason of the operation. How do you reduce inches to degrees? What is the reason for the operation? How do you multiply by $\frac{1}{2}$? How do you divide by $16\frac{1}{2}$ and find the true remainder? How do you obtain the true answer in examples of this kind?

SURVEYORS' MEASURE.

ART. **92.** This measure is used by surveyors in measuring land, roads, &c.

TABLE.

$7\frac{92}{100}$	Inches (in.)	make	1 Link,	l.
25	Links	"	1 Pole,	p.
100	Links, 4 Poles, or 66 Feet.	"	1 Chain,	cha.
10	Chains	"	1 Furlong,	fur.
8	Furlongs, or 80 Chains,	"	1 Mile,	m.

m.		fur.		cha.		p.		l.		in.
								1	=	$7\frac{92}{100}$
						1	=	25	=	198
				1	=	4	=	100	=	792
		1	=	10	=	40	=	1000	=	7920
1	=	8	=	80	=	320	=	8000	=	63360

NOTE. — Gunter's chain, in length 4 poles, or 66 feet, and divided into 100 links, is that mostly used in ordinary land surveys; but in locating roads, and like public works, an engineer's chain is usually 100 feet in length, containing 120 links, each 10 inches long.

MENTAL EXERCISES.

1. In 2 poles how many links? In 4 poles? In 7 poles?
2. In 5 chains how many links? In 8cha.? In 10cha.?
3. How many poles in 50 links? In 75l.? In 125l.?

EXERCISES FOR THE SLATE.

1. How many links in 7m. 5fur. 6cha. 30l.?

OPERATION.

```
7 m. 5fur. 6cha. 30l.
8
6 1 furlongs.
1 0
6 1 6 chains.
1 0 0
6 1 6 3 0 links. Ans.
```

2. In 61630 links how many miles?

OPERATION.

```
100)61630 l.
   10)616 cha 30l.
      8)61 fur. 6cha.
        7 m. 5fur.
Ans. 7m. 5fur. 6cha. 30l.
```

3. How many miles in 4386 chains?

QUESTIONS. — Art. 92. For what is surveyors' measure used? Recite the table. How do you reduce miles to links? What is the reason for the operation? How do you reduce inches to chains? To miles? Give the reason of the operation.

4. In 54m. 66cha. how many chains?
5. In 75m. 49cha. how many poles?
6. How many miles in 24196 poles?
7. How many links in 7m. 4fur. 30rd.?
8. How many miles in 60750 links?

SQUARE MEASURE.

ART. **93.** Square Measure is used in measuring surfaces of all kinds.

TABLE.

144	Square inches	make	1 Square foot,	ft.
9	Square feet	"	1 Square yard,	yd.
$30\frac{1}{4}$ / $272\frac{1}{4}$	Square yards, or / Square feet,	"	1 Square rod or pole,	p.
40	Square rods or poles	"	1 Rood,	R.
4	Roods, or 160 Poles,	"	1 Acre,	A.
640	Acres	"	1 Square mile,	S. M.

S. M.	A.	R.	p.	yd.	ft.	in.
					1 =	144
				1 =	9 =	1296
			1 =	$30\frac{1}{4}$ =	$272\frac{1}{4}$ =	39204
		1 =	40 =	1210 =	10890 =	1568160
	1 =	4 =	160 =	4840 =	43560 =	6272640
1 =	640 =	2560 =	102400 =	3097600 =	27878400 =	4014489600

NOTE. — A *square* is a figure bounded by four equal lines, perpendicular to each other.

When the four lines are each 1 foot in length, the space enclosed is 1 *square foot;* when 1 yard in length, 1 *square yard;* when 1 rod in length, 1 *square rod;* and so for any other dimension.

3ft. = 1yd.

Square foot.

3ft. = 1yd.

In this diagram the *large* square represents a square *yard*, and each of the *smaller* squares within it represents one square *foot.* Now, since there are *three* rows of small squares, and *three* square feet in each row, there will be 3 sq. ft. × 3 = 9 sq. ft. in the large square. But the large square is 3 ft. in length, and 3 ft. in breadth; hence,

To find the contents of a square, multiply the numbers denoting its length and breadth together.

QUESTIONS. — Art. 93. For what is square measure used? Repeat the table. What is a square? What is a square foot? How may the contents of a square be found? Explain by the diagram the reason of the operation.

Mental Exercises.

1. In 2 square feet how many square inches?
2 In 3 square yards how many square feet? In 10 sq. yd.?
3. In 5 roods how many poles? In 20 roods? In 30 roods?
4. In 7 acres how many roods? In 24 acres? In 40 acres?

Exercises for the Slate.

1. How many square inches in 12A. 3R. 24p. 144ft. 72in.?

OPERATION.

```
      12 A. 3R. 24p. 144ft. 72in.
       4
      ---
      51 roods.
      40
     ----
     2064 poles.
     272¼
    -----
     4132
   14452
   4129
     516
   ------
   562068 feet.
      144
  -------
  2248274
 2248279
 562068
 --------
Ans. 80937864 inches.
```

NOTE. — To multiply by ¼, we take ¼ of the multiplicand

2. In 80937864 square inches how many inches?

OPERATION.

```
    144)80937864 inches.
  272¼)562068 ft. 72in.
     4      4
1089)2248272 fourths of a foot.
         40)2064 poles.   576 ÷ 4 = 144ft.
            4)51 R. 24p.
        Ans. 12 A. 3R. 24p. 144ft. 72in.
```

NOTE. — To divide by the 272¼, we first reduce the divisor and dividend to *fourths*, and then divide. The remainder obtained, being *fourths*, is reduced to whole numbers by dividing by 4.

QUESTIONS. — How do you reduce acres to square inches? Give the reason for the operation. How do you reduce square inches to acres? What is the reason for the operation? How do you multiply by ¼?

3. In 49A. 3R. 16p. how many square feet?

4. In 2171466 square feet how many acres?

5. What is the value of 365A. 3R. 17p. at $1.75 per square rod or pole? Ans. $102,439.75.

6. Sold a valuable piece of land, containing 3A. 1R. 30p., at $1.25 per square foot; what was received for the land?
Ans. $187,171.87,5.

7. In a tract of land 12 miles square, how many square miles? How many acres? Ans. 92160 acres.

8. In 18A. 0R. 16p. how many square feet?
Ans. 788436 square feet.

9. Purchased 48A. 3R. 14p. of land for $2.25 per square rod, and sold the same for $3.15 per square rod; what did I gain by my bargain? Ans. $7032.60.

CUBIC OR SOLID MEASURE.

ART. **94.** Cubic or Solid Measure is used in measuring such bodies or things as have length, breadth, and thickness; as timber, stone, &c.

TABLE.

1728	Cubic inches (cu. in.)	make	1 Cubic foot,	cu. ft.
27	" feet	"	1 " yard,	cu. yd.
40	" feet	"	1 Ton,	T.
16	" feet	"	1 Cord foot,	c. ft.
8	Cord feet, or }	"	1 Cord of wood,	C.
128	Cubic feet, }			

C.		T.		yd.		ft.		in.
						1	=	1728
				1	=	27	=	46656
		1	=	$1\frac{13}{27}$	=	40	=	69120
1	=	$3\frac{1}{5}$	=	$4\frac{20}{27}$	=	128	=	221184

NOTE 1.—A pile of wood 8ft. in length, 4ft. in breadth, and 4ft. in height, contains a cord.

Also, one ton of timber, as usually surveyed, contains $50\frac{92}{100}$ cubic or solid feet.

QUESTIONS.—How do you divide by 272¼? Of what denomination is the remainder? How is the true remainder found?—Art. 94. For what is cubic measure used? Recite the table. What are the dimensions of a pile of wood containing 1 cord? How many solid feet does a ton of timber contain, as usually surveyed?

NOTE 2.—A *cube* is a solid bounded by six square and equal sides.

If the cube is 1 foot long, 1 foot wide, and 1 foot high, it is called a *cubic* or *solid foot*. If the cube is 3 feet long, 3 feet wide, and 3 feet thick, it is called a *cubic* or *solid yard*.

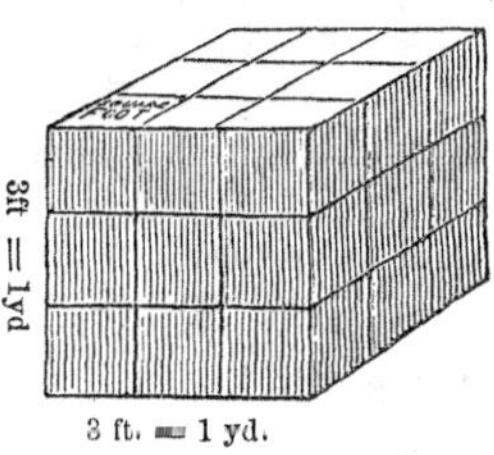

Now, since each side of a cubic yard, as represented in the diagram, contains 9 sq. ft. of surface (Art. 93), it is plain, if a block be cut off from one side, *one* foot thick, it can be divided into 9 solid blocks, with sides 1 foot in length, breadth, and thickness, and therefore will contain 9 solid feet; and since the whole block or cube is *three* feet thick, it must contain 9 solid feet × 3 = 27 solid feet; or 3 solid feet × 3 × 3 = 27 solid feet. Hence,

To find the contents of a cubic body, multiply together the numbers denoting the length, breadth, and thickness.

MENTAL EXERCISES.

1. In 2 cubic feet how many cubic inches? In 4 cu. ft.?
2. In 3 cubic yards how many cubic feet? In 10 cu. yd.?
3. How many cords of wood in 64 cord feet? In 96 c. ft.?
4. How many tons in 80 cu. ft. of timber? In 160 cu. ft.?

EXERCISES FOR THE SLATE.

1. In 48 cu. yd. and 15 cu. ft. how many cubic inches?

OPERATION.

```
        48 yd. 15ft.
        27
       ---
       341
       97
      ----
      1311 feet.
      1728
     -----
     10488
     2622
    9177
   1311
   -------
Ans. 2265408 inches.
```

2. In 2265408 cubic inches how many cubic yards?

OPERATION.

```
1728)2265408 cu. in.
     27)1311 cu. ft.
     Ans. 48 yd. 15ft.
```

QUESTIONS.—What is a cube? How do you find the contents of a cube? Give the reason for the operation. Describe a cubic foot. How do you reduce a ton to cubic inches? Give the reason for the operation. How do you reduce cubic inches to cubic yards? Give the reason for the operation.

3. In 45 cords of wood how many cubic inches?

4. In 9953280 cubic inches how many cords of wood?

5. How many cubic feet in a pile of wood 15ft. long, 4ft wide, and 6½ft. high? How many cords? Ans. 3C. 6 cu. ft.

6. How many cubic inches in a block of marble 4ft. long 3¼ft. wide, and 2ft. thick? Ans. 44928.

7. In a room 14ft. long, 12ft. wide, and 8ft. high, how many cubic feet? Ans. 1344.

8. What will 9080 cubic feet of ship-timber cost, at $11.50 per ton? Ans. $2610.50

WINE OR LIQUID MEASURE.

ART. **95.** Wine or Liquid Measure is used in measuring all kinds of liquids, except, in some places, beer, ale, porter, and milk.

TABLE.

4 Gills (gi.)	make	1 Pint,	pt.
2 Pints	"	1 Quart,	qt.
4 Quarts	"	1 Gallon,	gal.
63 Gallons	"	1 Hogshead,	hhd.
2 Hogsheads	"	1 Pipe, or Butt,	pi.
2 Pipes	"	1 Tun,	tun.

tun.		pi.		hhd.		gal.		qt.		pt.		gi.
										1	=	4
								1	=	2	=	8
						1	=	4	=	8	=	32
				1	=	63	=	252	=	504	=	2016
		1	=	2	=	126	=	504	=	1008	=	4032
1	=	2	=	4	=	252	=	1008	=	2016	=	8064

NOTE 1.—By laws of Massachusetts, 32 gallons make 1 *barrel*. In some states 31½ gallons, and in others from 28 to 32 gallons, make 1 barrel. 42 gallons make 1 *tierce*, and 2 tierces, 1 *puncheon*.

NOTE 2.—The term hogshead is often applied to any large cask that may contain from 50 to 120 gallons, or more.

NOTE 3.—The *Standard Unit* of *Liquid Measure* adopted by the government of the United States is the *Winchester Wine Gallon*, which contains 231 cubic inches. It has the name Winchester, from its standard having been formerly kept at Winchester, England. The *Imperial Gallon*, now adopted in Great Britain, contains $277\frac{274}{1000}$ cubic inches; so that 6 Winchester gallons make about 5 Imperial gallons.

QUESTIONS.—Art. 95. For what is wine or liquid measure used? Repeat the table. How many gallons make a barrel? How many a tierce? How many a puncheon? How is the term hogshead often applied? What is the standard unit of liquid measure?

MENTAL EXERCISES

1. In 3 pints how many gills? In 5 pints? In 9 pints?
2. In 4 quarts how many pints? In 6 quarts? In 8 quarts?
3. In 5 gallons how many quarts? In 7 gallons?
4. How many gallons in 12 quarts? In 18 quarts?

EXERCISES FOR THE SLATE.

1. In 47 tuns of wine how many gills?

OPERATION.

```
   47 tuns.
    4
  ---
  188 hogsheads.
   63
  ---
  564
 1128
 ----
11844 gallons.
    4
 ----
47376 quarts.
    2
 ----
94752 pints.
    4
 ----
Ans. 379008 gills.
```

2. In 379008 gills how many tuns?

OPERATION.

```
 4)379008 gi.
   ------
  2)94752 pt.
   ------
  4)47376 qt.
   ------
 63)11844 gal.
   ------
    4)188 hhd.
   ------
Ans. 47 tuns.
```

3. Reduce 197 tuns 3hhd. 60gal. 3qu. 1pt. to gills.
4. In 1596604 gills how many tuns?
5. What will 7 hogsheads of wine cost, at 5 cents a pint?
6. What cost 18 tuns 1hhd. 47gal. of oil, at $1.25 per gallon? Ans. $5807.50.

BEER MEASURE.

ART. 96. Beer Measure is used in measuring beer, ale, porter, and milk.

TABLE.

2 Pints (pt.)	make	1 Quart,	qt.
4 Quarts	"	1 Gallon,	gal.
54 Gallons	"	1 Hogshead,	hhd.

hhd.		gal.		qt.		pt.
				1	=	2
		1	=	4	=	8
1	=	54	=	216	=	432

QUESTION. — Art. 96. Repeat the table of beer measure.

NOTE 1. — The gallon of beer measure contains 282 cubic inches ; and as has been usually reckoned, 36 gallons equal 1 barrel ; 2 hogsheads, or 108 gallons, 1 butt ; 2 butts, or 216 gallons, 1 tun. 1 gallon beer measure = 1gall. 1pt. $3\frac{5}{77}$ gi. wine measure.

NOTE 2. — Beer Measure is becoming obsolete. Milk and malt liquors, at the present time, are bought and sold, very generally, by wine or liquid measure.

EXERCISES FOR THE SLATE.

1. How many quarts in 76 hogsheads?

OPERATION.

```
   7 6 hhd.
   5 4
  ------
   3 0 4
 3 8 0
 -------
 4 1 0 4 gallons.
       4
 -------
Ans. 1 6 4 1 6 quarts.
```

2. In 16416 quarts how many hogsheads?

OPERATION.

```
  4 ) 1 6 4 1 6 qt.
 5 4 ) 4 1 0 4 gal.
       Ans. 7 6 hhd.
```

3. In 4 tuns 1hhd. 17gal. 1pt. how many pints?

4. How many tuns in 7481 pints?

5. What cost 7hhd. 18gal. of beer at 4 cents a quart? Ans. $63.36.

6. At 15 cents per gallon, what will 18hhds. of ale cost? Ans. $145.80.

DRY MEASURE.

ART. 97. This measure is used in measuring grain, fruit salt, coal, &c.

TABLE.

2 Pints (pt.)	make	1 Quart,	qt.
8 Quarts	"	1 Peck,	pk.
4 Pecks	"	1 Bushel,	bu.
8 Bushels	"	1 Quarter,	qr.
36 Bushels	"	1 Chaldron,	ch.

ch.		bu.		pk.		qt.		pt.
						1	=	2
				1	=	8	=	16
		1	=	4	=	32	=	64
1	=	36	=	144	=	1152	=	2304

QUESTIONS. — How many cubic inches does the beer gallon contain? How do you reduce hogsheads to quarts? How do you reduce quarts to hogsheads? -- Art. 97. For what is dry measure used? Repeat the table.

NOTE 1. — The *Standard Unit* of *Dry Measure* adopted by the United States government is the *Winchester* bushel, which is in form a cylinder, $18\frac{1}{2}$ inches in diameter, and 8 inches deep, containing $2150\frac{42}{100}$ cubic inches. The *Standard Imperial* bushel of Great Britain contains $2218\frac{192}{1000}$ cubic inches, so that 32 Imperial bushels equal about 33 Winchester bushels. The gallon in Dry Measure contains $268\frac{4}{5}$ cubic inches.

NOTE 2. — 1gal. Dry Measure = $268\frac{4}{5}$cu. in. = 1gal. 1pt. $1\frac{1}{5}\frac{3}{5}$gi. Wine Measure = 3qt. $1\frac{1}{2}\frac{4}{3}\frac{7}{5}$pt. Beer Measure; 1gal. W. M. = 231cu. in. = 3 qt. $\frac{7}{8}$pt. D. M. = 3qt. $\frac{2}{4}\frac{6}{7}$ pt. B. M.; 1gal. B. M. = 282cu. in. = 1gal. 1pt. $3\frac{5}{7}\frac{}{7}$gi. W. M. = 1gal. $\frac{1}{2}\frac{1}{8}$pt. D. M.; 1qt. D. M. = $67\frac{1}{5}$cu. in. = 1qt. $1\frac{1}{5}\frac{7}{5}$gi. W. M.; 1qt. W. M. = $57\frac{3}{4}$cu. in. = $1\frac{2}{3}\frac{3}{2}$pt. D. M.; 1pt. D. M. = $33\frac{3}{5}$cu. in. = 1pt. $\frac{1}{5}\frac{7}{5}$ gi. W. M.; 1pt. W. M. = $28\frac{7}{8}$cu. in. = $\frac{5}{6}\frac{5}{4}$pt. D. M.

MENTAL EXERCISES.

1. In 2 quarts how many pints? In 5 quarts? In 7 quarts?
2. In 3 pecks how many quarts? In 6 pecks? In 9 pecks?
3. In 5 bushels how many pecks? In 10 bushels?
4. How many pecks in 16 quarts? In 25 quarts?

EXERCISES FOR THE SLATE.

1. How many quarts in 49ch. 8bu. 3pk. and 3qt.?

OPERATION.

```
   4 9 ch. 8bu. 3pk. 3qt.
     3 6
   -----
   3 0 2
  1 4 7
  -------
  1 7 7 2 bushels.
        4
  -------
  7 0 9 1 pecks.
        8
  -------
Ans. 5 6 7 3 1 quarts.
```

2. In 56731 quarts how many chaldrons?

OPERATION.

```
 8 ) 5 6 7 3 1 qt.
     ---------
   4 ) 7 0 9 1 pk. 3qt.
       -------
 3 6 ) 1 7 7 2 bu. 3pk.
       -------
           4 9 ch. 8bu.
```

Ans. 49ch. 8bu. 3pk. 3qt.

3. Reduce 97ch. 30bu. 2pk. to quarts.
4. In 112720 quarts how many chaldrons?
5. How many pints in 35bu. 1pt.?
6. Reduce 2241 pints to bushels.
7. Reduce 18qr. 3pk. 5qt. to quarts.
8. How many quarters in 4637 quarts?
9. In 19bu. 3pk. 7qt. 1pt. how many pints?
10. In 1279 pints how many bushels?

QUESTION. — What is the standard unit of Dry Measure?

MEASURE OF TIME.

ART. 98. This measure is applied to the various divisions and subdivisions into which time is divided.

TABLE.

60 Seconds (sec.)	make	1 Minute,	m.
60 Minutes	"	1 Hour,	h
24 Hours	"	1 Day,	da.
7 Days	"	1 Week,	w
$365\frac{1}{4}$ Days, or 52 weeks $1\frac{1}{4}$ days,	"	1 Julian Year,	y.
12 Calendar Months (mo.)	"	1 Year,	y.

y.		w.		d.		h.		m.		sec
								1	=	60
						1	=	60	=	3600
				1	=	24	=	1440	=	86400
		1	=	7	=	168	=	10080	=	604800
1	=	$52\frac{5}{28}$	=	$365\frac{1}{4}$	=	8766	=	525960	=	31557600

NOTE 1. — The true *Solar* or *Tropical Year* is the time measured from the sun's leaving either equinox or solstice to its return to the same again, and is 365d. 5h. 48m. 49sec. nearly.

The *Julian Year*, so called from the calendar instituted by Julius Cæsar, contains $365\frac{1}{4}$ days, as a medium ; three years in succession containing 365 days, and the fourth year 366 days ; which, as compared with the true solar year, produces an average yearly error of 11m. $10\frac{3}{10}$ sec., or a difference that would amount to 1 whole day in about 120 years.

The *Gregorian Year*, or that instituted by Pope Gregory XIII., in the year 1582, and which is now the *Civil* or *Legal Year* in use among the different nations of the earth, contains, like the Julian year, 365 days for three years in succession, and 366 days for the fourth, *excepting centennial years whose number cannot be exactly divided by* 400. The Gregorian year is so nearly correct as to err only 1 day in 3866 years, a difference so little as hardly to be worth taking into account.

A *Common Year* is one of 365 days, and a *Leap* or *Bissextile Year* is one of 366 days. Any year is Leap Year whose number can be divided by 4 without a remainder, except years whose number can be divided without a remainder by 100, but not by 400.

A *Sidereal Year* is the time in which the earth revolves round the sun, and is 365d. 6h. 9m. $9\frac{6}{10}$sec.

NOTE 2. — The names of the 12 calendar months, composing the civil year, are January, February, March, April, May, June, July, August, September, October, November, December, and the number of days in each may be readily remembered by the following lines :

"Thirty days hath September,
April, June, and November ;

QUESTIONS. — Art. 98. To what is the measure of time applied? Repeat the table. How is the true solar year measured? How long is it? Why is the Julian year so called? Who instituted the Gregorian year? What is a Common year? What is a Sidereal year?

And all the rest have thirty-one,
Save February, which alone
Hath twenty-eight ; and this, in fine,
One year in four hath twenty-nine."

TABLE.

SHOWING THE NUMBER OF DAYS FROM ANY DAY OF ONE MONTH TO THE SAME DAY OF ANY OTHER MONTH IN THE SAME YEAR.

FROM ANY DAY OF	TO THE SAME DAY OF											
	Jan.	Feb.	Mar.	Apr.	May.	June.	July.	Aug.	Sept.	Oct.	Nov.	Dec.
January	365	31	59	90	120	151	181	212	243	273	304	334
February	334	365	28	59	89	120	150	181	212	242	273	303
March	306	337	365	31	61	92	122	153	184	214	245	275
April	275	306	334	365	30	61	91	122	153	183	214	244
May	245	267	304	335	365	31	61	92	123	153	184	214
June	214	245	273	304	334	365	30	61	92	122	153	183
July	184	215	243	274	304	335	365	31	62	92	123	153
August	153	184	212	243	273	304	334	365	31	61	92	122
September	122	153	181	212	242	273	303	334	365	30	61	91
October	92	123	151	182	212	243	273	304	335	365	31	61
November	61	92	120	151	181	212	242	273	304	334	365	30
December	31	62	90	121	151	182	212	243	274	304	335	365

For example, suppose we wish to find the number of days from April 4th to November 4th, we look for April in the left-hand vertical column, and November at the top, and, where the lines intersect, is 214, the number sought. Again, if we wish the number of days from June 10th to September 16th, we find the difference between June 10th and September 10th to be 92 days, and add 6 days for the excess of the 16th over the 10th of September, so we have 98 days as the exact difference.

If the end of February be included between the points of time, a day must be added in leap year.

When the time includes more than one year, there must be added 365 days for each year.

MENTAL EXERCISES

1. In 3 minutes how many seconds? In 5 minutes?
2. In 2 hours how many minutes? In 4 hours?
3. In 4 weeks how many days? In 6 weeks? In 9 weeks?
4. In 2 days how many hours? In 3 days? In 7 days?
5. How many weeks in 21 days? In 30 days? In 50 days?
6. How many calendar months in 2 years? In 8 years? In 10 years? In 12 years? In 20 years?

QUESTIONS. — Name the months in their order. How many days has each month? How do you find by the table the number of days from April 4th to November 4th? When the time sought for is more than one year, how many days must be added?

EXERCISES FOR THE SLATE.

1. How many seconds in 365da. 5h. 48m. 49sec., or one solar year?

OPERATION.

```
 3 6 5 da. 5h. 48m. 49sec.
     2 4
 1 4 6 5
 7 3 0
 8 7 6 5 hours.
     6 0
5 2 5 9 4 8 minutes.
        6 0
Ans. 3 1 5 5 6 9 2 9 seconds.
```

2. In 31556929 seconds how many days?

OPERATION.

```
60) 3 1 5 5 6 9 2 9
  60) 5 2 5 9 4 8 m. 49sec
      24) 8 7 6 5 h. 48m.
            3 6 5 da. 5h.
Ans. 365da. 5h. 48m 49sec
```

3. Reduce 296da. 18h. 32m. to minutes.
4. In 427352 minutes how many days?
5. How many seconds in 30 solar years 262da. 17h. 28m 42sec.?
6. In 969407592 seconds how many solar years?
7. How many weeks in 684592 minutes?
8. In 67w. 6d. 9h. 52m. how many minutes?
9. How many days from June 5th to Dec. 11th?
10. How many days from March 17th, 1856, to May 16th 1857? Ans. 425 days.
11. How many days from December 18th, 1856, to January 30th, 1857?
12. How many days from August 30th, 1857, to June 1st, 1858?
13. How many days from July 4th, 1859, to July 4th, 1860?
14. How many days from April 25th, 1855, to August 20th, 1858? Ans. 1213 days.

NOTE. — The last six examples are to be performed by aid of the table on page 103.

QUESTIONS. — How do you reduce years to seconds? Give the reason for the operation. How do you reduce seconds to days? To years? Give the reason for the operation.

CIRCULAR MEASURE.

ART. **99.** Circular Measure is applied to the measurement of circles and angles, and is used in reckoning latitude and longitude, and the revolutions of the planets round the sun.

TABLE.

60 Seconds (″)	make	1 Minute,	′.
60 Minutes	"	1 Degree,	°.
30 Degrees	"	1 Sign,	S.
12 Signs, or 360 Degrees,	"	The Circle of the Zodiac,	C.

C.		S.		°		′		″
						1	=	60
				1	=	60	=	3600
		1	=	30	=	1800	=	108000
1	=	12	=	360	=	21600	=	1296000

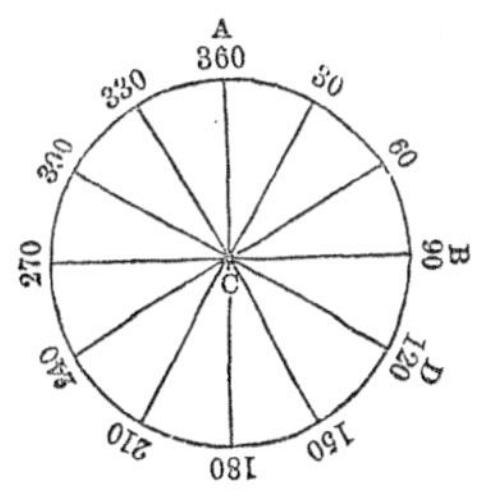

NOTE. — 1. A *Circle* is a plane figure bounded by a curve line, every part of which is equally distant from a point called its centre.

The *Circumference* of a circle is the line which bounds it, as shown by the diagram.

An *Arc* of a circle is any part of its circumference; as AB.

A *Radius* of a circle is a straight line drawn from its centre to its circumference; as CA, CB, or CD.

Every circle, large or small, is supposed to be divided into 360 equal parts, called degrees.

A *Quadrant* is one fourth of a circle, or an arc of 90°; as AB.

An *Angle*, as ACB, is the inclination or opening of two lines which meet at a point, as C. The point is the *vertex* of the angle. If a circle be drawn around the vertex of an angle as a centre, the two sides of the angle, as *radii* of the circle, will include an arc, which is the *measure* of the angle; as the arc AD = 120° is the measure of the angle ACD, and AB = 90°, the measure of the angle ACB; hence the one is called an angle of 120°, and the other an angle of 90°.

NOTE. — 2. As the earth turns on its axis from west to east every 24 hours, the sun appears to pass from east to west $\frac{1}{24}$ of 360° of longitude every hour, or over 15° of longitude in 1 hour's time, or 1° in 4 minutes of time, and 1′ in 4 seconds of time; so that, for instance, at any place, when it is noon, it is 1 hour earlier for every 15° of longitude westward, or 1 hour later for every 15° of longitude eastward. Thus, Boston being 71° 4′ west of Greenwich, and San Francisco 51° 17′ west of Boston, when it is noon at Boston, it is 4h. 44m. 16sec. past noon at Greenwich, and wanting 3h. 25m. 8sec. of noon at San Francisco.

QUESTIONS. — Art. 99. To what is circular measure applied? Recite the table. What is a circle? What is an angle?

EXERCISES FOR THE SLATE.

1. How many minutes in 11S. 18° 57′?

OPERATION.

```
11S. 18° 57′
 30
———
348 degrees.
 60
———
Ans. 20937 minutes.
```

2. In 20937 minutes how many signs?

OPERATION.

```
60)20937′
   ————————
30)348° 57′
   ————————
   11S. 18°.
```

Ans. 11S. 18° 57′.

3. In 27S. 19° 51′ 28″ how many seconds?
4. How many signs in 2987488 seconds?

MISCELLANEOUS TABLE.

ART. **100.** This table embraces a variety of things in business important to be known.

12 units	make	1 dozen.
12 dozen	"	1 gross.
12 gross	"	1 great gross.
20 units	"	1 score.
14 pounds of Iron or Lead	"	1 stone.
60 pounds of Wheat	"	1 bushel.
60 pounds of Clover-seed	"	1 bushel.
60 pounds of Beans	"	1 bushel.
60 pounds of Potatoes	"	1 bushel.
52 pounds of Onions	"	1 bushel.
70 pounds of Corn on the Cob	"	1 bushel.
56 pounds of Shelled Corn	"	1 bushel.
56 pounds of Rye	"	1 bushel.
56 pounds of Flax-seed	"	1 bushel.
45 pounds of Timothy-seed	"	1 bushel.
20 pounds of Bran	"	1 bushel.
48 pounds of Barley	"	1 bushel.
52 pounds of Buckwheat	"	1 bushel in Ky.
48 pounds of Buckwheat	"	1 bushel in Mass. and Pa.
32 pounds of Oats	"	1 bushel in Ms., Ill., O., &c.
30 pounds of Oats	"	1 bushel in Me., N.H., Pa., &c.

QUESTIONS. — How do you reduce signs to seconds? Give the reason of the operation. How do you reduce seconds to degrees? To signs? Give the reason for the operation. How many degrees in a circle? — Art. 100. What is embraced in the miscellaneous table?

196 pounds of Flour	make	1 barrel.
200 pounds of Beef	"	1 barrel.
200 pounds of Pork	"	1 barrel.
100 pounds of Fish	"	1 quintal.
200 pounds of Shad or Salmon	"	1 barrel in N. Y., Ct.
220 pounds of Fish	"	1 barrel in Md.
30 gallons of Fish	"	1 barrel in Mass.
5 bushels of Corn	"	1 barrel in Md., Tenn., &c.

OF BOOKS.

A sheet	folded in	2 leaves	forms	a folio.
A sheet	"	4 leaves	"	a quarto.
A sheet	"	8 leaves	"	an octavo.
A sheet	"	12 leaves	"	a 12mo.
A sheet	"	18 leaves	"	an 18mo.
A sheet	"	24 leaves	"	a 24mo.

MISCELLANEOUS EXERCISES IN REDUCTION.

1. In $345.18 how many mills?
2. How many dollars in 345180 mills?
3. In 46£ 18s. 5d. how many farthings?
4. How many pounds in 45044 farthings?
5. Reduce 61lb. 0oz. 17pwt. 17gr. troy to grains.
6. In 351785 grains troy how many pounds?
7. How many scruples in 27℔ 3℥ 1ʒ 1℈?
8. In 7852 scruples how many pounds?
9. In 83T. 11cwt. 3qr. 18lb. how many ounces?
10. How many tons in 2675088 ounces?
11. How many nails in 97yd. 3qr. 3na.?
12. In 1567 nails how many yards?
13. In 57 ells English how many yards?
14. How many ells English in 71yd. 1qr.?
15. How many inches in 15m. 7fur. 18rd. 10ft. 6in.?
16. In 1009530 inches how many miles?
17. In 95,000,000 of miles how many inches?
18. How many miles in 6,019,200,000,000 inches?
19. In 48deg. 18m. 7fur. 18rd. how many feet?
20. In 17629557 feet how many degrees?
21. How many square feet in 7A. 3R. 16p. 218ft.?
22. In 342164 square feet how many acres?
23. How many square inches in 25 square miles?

QUESTION.—What gives name to the size or form of books?

24. In 100362240000 square inches how many square miles?
25. How many cubic inches in 15 tons of timber?
26. In 1036800 cubic inches how many tons?
27. How many gills of wine in 5hhd. 17gal. 3qt.?
28. In 10648 gills how many hogsheads of wine?
29. How many quarts of beer in 29hhd. 30gal. 3qt.?
30. In 6387 quarts of beer how many hogsheads?
31. How many pints in 15ch. 16bu. 3pk. of wheat?
32. In 35632 pints of wheat how many chaldrons?
33. How many seconds of time in 365 days 6 hours?
34. In 31557600 seconds how many days?
35. How many hours in 1842 years (of 365da. 6h. each)?
36. In 16146972 hours how many years?
37. How many seconds in 8S. 14° 18′ 17″?
38. In 915497″ how many signs?
39. What will be the cost of 13 gross of steel pens, at $2\frac{1}{2}$ cents per pen? Ans. $46.80.
40. Bought 12 reams of paper at 20 cents per quire; how much did it cost? Ans. $48.
41. I wish to put 2 hogsheads of wine into bottles that will contain 3 quarts each; how many bottles are required? Ans. 168 bottles.
42. When $1480 are paid for 25 acres of land, what costs 1 acre? What costs 1 rood? What cost 37A. 2R. 18p.? Ans. $2226.66.
43. John Webster bought 5cwt. 3qr. 18lb. of sugar at 9 cents per lb., for which he paid 25 barrels of apples at $1.75 per barrel; how much remains due? Ans. $9.62.
44. Bought a silver tankard weighing 2lb. 7oz. for $46.50; what did it cost per oz.? How much per lb.? Ans. $18.
45. Bought 3T. 1cwt. 18lb. of leather at 12 cents per lb., and sold it at 9 cents per lb.; what did I lose? Ans. $183.54.
46. Phineas Bailey has agreed to grade a certain railroad at $5.75 per rod; what will he receive for grading a road between two cities, whose distance from each other is 37m. 7fur. 29rd.? Ans. $69856.75.
47. If it cost $17.29 per rod to grade a certain piece of railroad, what will be the expense of grading 15m. 6fur. 37rd.? Ans. $87,781.33.
48. What is the value of a house-lot, containing 40 square rods and 200 square feet, at $1.50 per square foot? Ans. $16635.

49. How many yards of carpeting, that is one yard in width, will be required to carpet a room 18ft. long and 15ft. wide?
Ans. 30 yards.

50. A certain machine will cut 120 shingle-nails in a minute how many will it cut in 47 days 7 hours, admitting the machine to be in operation 10 hours per day? Ans. 3434400 nails.

51. In a field 80 rods long and 50 rods wide, how many square rods? How many acres? Ans. 25 acres.

52. How long will it take to count 18 millions, counting at the rate of 90 a minute? Ans. 138da. 21h. 20m.

53. A merchant purchased 9 bales of cloth, each containing 15 pieces, each piece 23 yards, at 8 cents per yard; what was the amount paid? Ans. $248.40.

54. Suppose a certain township is 6 miles long and 4½ miles wide, how many lots of land of 90 acres each does it contain?
Ans. 192 lots.

55. The pendulum of a certain clock vibrates 47 times in 1 minute; how many times will it vibrate in 196 days 49m.?
Ans. 13267583 times.

56. How many shingles will it take to cover a roof, each of whose equal sides is 36 feet long, with rafters 16 feet in length, supposing 1 shingle to cover 27 square inches?
Ans. 6144 shingles.

57. How many times will the large wheels of an engine turn round in going from Boston to Portland, a distance of 110 miles, supposing the wheels to be 12 feet and 6 inches in circumference? Ans. 46464 times.

58. In a certain house there are 25 rooms, in each room 7 bureaus, in each bureau 5 drawers, in each drawer 12 boxes, in each box 15 purses, in each purse 178 sovereigns, each sovereign valued at $4.84; what is the amount of the money?
Ans. $135689400.

59. In 18rd. 5yd. 2ft. 11in. how many inches?
Ans. 3779 inches.

60. In 3779 inches how many rods?
Ans. 18rd. 5yd. 2ft. 11in.

61. Sold 5T. 17cwt. 3qr. 18lb. of potash for 3 cents per pound; what was the amount? Ans. $353.79.

62. A gentleman purchased a house-lot that was 25 rods long and 16 rods wide for $100,000, and sold the same for $1.25 per square foot; what did he gain by his purchase?
Ans. $36,125.

10

§ XI. ADDITION OF COMPOUND NUMBERS.

ART. **101.** ADDITION of Compound Numbers is the process of finding the amount of two or more compound numbers.

ENGLISH MONEY.

Ex. 1. Paid a London tailor 7£. 13s. 6d. 2far. for a coat; 2£. 17s. 9d. 1far. for a vest; 3£. 8s. 3d. 3far. for pantaloons; 9£. 11s. 8d. 3far. for a surtout; what was the amount of the bill? Ans. 23£. 11s. 4d. 1far.

OPERATION.

	£.	s.	d.	far.
	7	13	6	2
	2	17	9	1
	3	8	3	3
	9	11	8	3
Ans.	23	11	4	1

Having written units of the same denomination in the same column, we find the sum of farthings in the right-hand column to be 9 farthings, equal to 2d. and 1far. We write the 1far. under the column of farthings, and carry the 2d. to the column of pence; the sum of which is 28d., equal to 2s. 4d. We write the 4d. under the column of pence, and carry the 2s. to the column of shillings; the sum of which is 51s., equal to 2£. 11s. Having written the 11s. under the column of shillings, we carry the 2£. to the column of pounds, and find the whole amount to be 23£. 11s. 4d. 1far.

The same result can be arrived at by *reducing the numbers as they are added* in their respective columns. Thus, in working the example, we can, beginning with farthings, add in this way: 3far. and 3far. are 6far., equal to 1d. 2far., and 1far. are 1d. 3far., and 2far. are 1d. 5far., equal 2d. 1far. Writing the 1far. under the column of farthings, carry the 2d. to the column of pence; add 2d. (carried) and 8d. are 10d., and 3d. are 13d., equal to 1s. 1d., and 9d. are 1s. 10d., and 6d. are 1s. 16d., equal to 2s. 4d. Writing the 4d. under the column of pence, carry the 2s. to the column of shillings; add 2s. (carried) and 11s. are 13s., and 8s. are 21s., equal to 1£. 1s., and 17s. are 1£. 18s., and 13s. are 1£. 31s., equal to 2£. 11s. Writing 11s. under the column of shillings, carry the 2£. to the column of pounds, and so find the whole amount to be, as before, 23£. 11s. 4d. 1far.

Thus the adding of compound numbers is like that of simple numbers, except in carrying; which difference holds also in subtracting, multiplying, and dividing of compound numbers.

QUESTIONS. — Art. 101. What is addition of compound numbers? How do you arrange compound numbers for addition? Why? What is the difference between addition of compound and addition of simple numbers?

RULE. — *Write all the given numbers so that units of the same denomination may stand in the same column.*

Add as in addition of simple numbers; and carry, from column to column, one for as many units as it takes of the denomination added to make a unit of the denomination next higher.

PROOF. — The proof is the same as in addition of simple numbers.

EXAMPLES FOR PRACTICE.

TROY WEIGHT.

2.

lb.	oz.	pwt.	gr.
15	11	19	22
71	10	13	17
65	9	17	14
73	11	13	13
14	8	9	9
242	4	14	3

3.

lb.	oz.	pwt.	gr.
10	10	10	10
81	11	19	23
47	7	8	19
16	9	10	14
33	10	9	21

APOTHECARIES' WEIGHT.

4.

℔	℥	ʒ	℈	gr.
81	11	6	1	19
75	10	7	2	13
14	9	7	1	12
37	8	1	1	11
61	11	3	2	3
272	4	3	0	18

5.

℔	℥	ʒ	℈	gr.
35	9	6	2	19
71	1	1	1	11
37	3	3	2	12
14	4	7	1	13
75	5	6	1	17

AVOIRDUPOIS WEIGHT.

6.

T.	cwt.	qr.	lb.	oz.	dr.
71	19	3	17	14	13
14	13	1	11	13	12
39	9	3	13	9	9
15	17	3	16	10	14
61	16	3	13	7	8
203	17	3	23	8	8

7.

T.	cwt.	qr.	lb.	oz.	dr.
14	13	2	15	15	15
13	17	3	13	11	13
46	16	3	11	13	10
14	15	2	7	6	9
11	17	3	10	15	11

QUESTIONS. — What is the rule? The proof?

CLOTH MEASURE.

8.

yd.	qr.	na.	in.
5	3	3	2
7	1	1	2
8	3	3	1
9	1	2	2
4	3	3	2
36	3	0	0

9.

E. E.	qr.	na.	in
16	3	2	1
71	1	1	2
13	3	2	1
47	3	2	2
39	2	3	2

LONG MEASURE.

10.

deg.	m.	fur.	rd.	ft.	in.
18	19	7	15	11	1
61	47	6	39	10	11
78	32	5	14	9	9
17	59	7	36	16	10
28	56	1	30	16	1
205	$8\frac{1}{2}$	5	17	$14\frac{1}{2}$	8
	$\frac{1}{2}=4$				$\frac{1}{2}=6$
205	9	1	17	15	2

11.

m.	fur.	rd.	yd.	ft.	in.
12	7	35	5	2	11
13	6	15	3	1	10
16	1	17	1	2	5
13	4	13	2	1	9
17	7	36	5	2	7

SURVEYORS' MEASURE.

12.

m.	fur.	ch.	p.	l.
17	5	8	3	24
16	3	7	1	21
47	7	9	3	19
19	6	6	1	16
31	7	1	0	20
133	7	4	0	0

13.

m.	fur.	ch.	p.	l.
14	7	9	3	21
37	1	0	3	16
17	7	8	3	17
61	6	5	3	16
47	1	1	0	23

SQUARE MEASURE.

14.

A.	R.	p.	ft.	in.
67	3	39	272	143
78	3	14	260	116
14	2	31	167	135
67	1	17	176	131
49	3	31	69	117
278	3	15	$131\frac{1}{4}$	66
				$\frac{1}{4}=36$
278	3	15	131	102

15.

A.	R.	p.	yd.	ft.	in.
43	1	15	30	8	17
16	3	39	19	7	141
47	1	16	27	5	79
38	3	17	18	8	17
15	1	32	11	1	117

SOLID MEASURE.

16.

Tun.	ft.	in.
17	39	1371
61	17	1711
47	16	1666
71	38	1711
47	17	1617
246	11	1164

17.

Cord.	ft.	in.
14	116	1169
67	113	1711
96	127	969
19	98	1376
14	37	1414

WINE MEASURE.

18.

Tun.	hhd.	gal.	qt.	pt.
61	1	62	3	1
71	3	14	1	1
60	0	17	3	0
14	1	51	1	1
57	3	14	3	1
265	2	35	1	0

19.

Tun.	hhd.	gal.	qt.	pt.
14	3	18	3	0
81	1	60	3	1
17	3	61	3	0
61	3	57	3	1
17	1	17	1	0

BEER MEASURE.

20.

Tun.	hhd.	gal	qt.	pt.
15	3	50	3	1
67	3	17	3	1
17	1	44	1	0
71	3	12	3	1
81	1	18	1	0
254	1	36	0	1

21.

Tun.	hhd.	gal.	qt.	pt.
67	1	51	1	0
15	3	16	3	1
44	1	45	1	1
15	2	12	2	1
67	3	35	1	0

DRY MEASURE.

22.

ch.	bu.	pk.	qt.	pt.
15	35	3	7	1
61	16	3	6	1
51	30	1	5	0
42	17	2	2	1
14	14	1	4	1
186	7	1	2	0

23.

ch.	bu.	pk.	qt.	pt.
71	17	1	1	1
16	31	3	3	0
41	14	3	1	1
71	17	1	0	1
10	10	2	3	0

TIME

24.

y.	da.	h.	m.	s.
57	300	23	59	17
47	169	15	17	38
29	364	23	42	17
18	178	16	38	47
49	317	20	52	57
03	236	10	30	56

25.

w.	da.	h.	m.	s.
15	6	23	15	17
61	5	15	27	18
71	6	21	57	58
18	5	19	39	49
87	6	19	18	57

CIRCULAR MEASURE.

26.

S.	°	′	″
11	28	56	58
10	21	51	37
8	13	39	57
8	19	38	49
7	17	47	48
11	11	55	09

27.

S.	°	′	″
6	17	17	18
7	09	19	51
8	18	57	45
4	17	16	39
7	27	38	48

NOTE.—The sum of the signs, in circular motion, must always be divided by 12, and the remainder only be written down, as in Ex. 26.

---

§ XII. SUBTRACTION OF COMPOUND NUMBERS.

ART. **102.** SUBTRACTION of Compound Numbers is the process of finding the difference between two compound numbers.

ENGLISH MONEY.

Ex. 1. From 87£. 9s. 6d. 3far., take 52£. 11s. 7d. 1far.
Ans. 34£. 17s. 11d. 2far.

OPERATION.

	£.	s.	d.	far.
Min.	87	9	6	3
Sub.	52	11	7	1
Rem.	34	17	11	2

Having placed the less number under the greater, farthings under farthings, pence under pence, &c., we begin with the farthings, thus: 1 far. from 3 far. leaves 2 far., which we set under the column of farthings. As we cannot

---

QUESTIONS.—Art. 102. What is subtraction of compound numbers? How do you arrange the numbers for subtraction?

take 7d. from 6d., we add 12d. = 1s. to the 6d., making 18d., and then subtract the 7d. from it, and set the remainder, 11d., under the column of pence. We then add 1s. = 12d. to the 11s. in the subtrahend, making 12s., to compensate for the 12d. we added to the 6d. in the minuend. (Art. 30.) Again, since we cannot take 12s. from 9s., we add 20s. = 1£. to the 9s., making 29s., from which we take the 12s., and set the remainder, 17s., under the column of shillings. Having added 1£. = 20s. to the 52£., to compensate for the 20s. added to the 9s. in the minuend, we subtract the pounds as in subtraction of simple numbers, and obtain 34£. for the remainder, and as the result complete, 34£. 17s. 11d. 2far.

RULE. — *Write the less compound number under the greater, so that units of the same denomination shall stand in the same column.*

Subtract as in subtraction of simple numbers.

If any number in the subtrahend is larger than that above it, add to the upper number as many units as make one of the next higher denomination before subtracting, and carry one to the next lower number before subtracting it.

PROOF. — The proof is the same as in simple subtraction.

EXAMPLES FOR PRACTICE.

2.

£.	s.	d.	far.
78	11	5	2
41	13	3	3
36	18	1	3

3.

£.	s	d.	far
765	16	10	1
713	17	11	3

TROY WEIGHT.

4.

lb	oz.	pwt.	gr.
15	3	12	14
9	11	17	21
5	3	14	17

5.

lb.	oz.	pwt.	gr.
711	1	3	17
19	3	18	19

APOTHECARIES' WEIGHT.

6.

℔	℥	ʒ	℈	gr.
15	7	1	2	15
11	9	7	1	19
3	9	2	0	16

7.

℔	℥	ʒ	℈	gr.
161	6	3	1	17
97	7	1	2	18

QUESTIONS. — What do you do when the upper number is smaller than the lower? How many do you carry to the next denomination? What is the rule for subtraction? The proof?

AVOIRDUPOIS WEIGHT.

8.

T.	cwt.	qr.	lb.	oz.	dr.
117	16	1	5	0	14
19	17	3	17	1	15
97	18	1	12	14	15

9.

T.	cwt.	qr.	lb.	oz.	dr
11	1	0	1	1	13
9	18	3	1	13	15

CLOTH MEASURE.

10.

yd.	qr.	na.	in.
15	1	1	2
9	3	3	1
5	1	2	1

11.

E. E.	qr.	na.	in.
171	2	2	1
19	3	0	2

LONG MEASURE.

12.

deg.	m.	fur.	rd.	yd.	ft.	in.
97	3	7	31	1	1	3
19	17	1	39	1	2	7
77	55$\frac{1}{6}$	5	31	4$\frac{1}{2}$	1	8
	$\frac{1}{6}$=	1	13	1	2	6
77	55	7	5	1	1	2

13.

deg.	m.	fur.	rd.	ft.	in.
18	19	1	1	3	7
9	28	7	1	16	9

SURVEYORS' MEASURE.

14.

m.	fur.	cha.	p.	l.
21	3	5	2	17
9	5	8	1	20
11	5	7	0	22

15.

m.	fur.	cha.	p.	l.
31	7	1	1	19
18	1	7	3	23

SQUARE MEASURE.

16.

A.	R.	p.	ft.	in.
116	1	13	100	113
87	3	17	200	117
28	1	35	171$\frac{1}{4}$	140
				$\frac{1}{4}$=36
28	1	35	172	32

17.

A.	R.	p.	yd.	ft.	in.
139	1	17	18	1	30
97	3	18	30	1	31

SOLID MEASURE.

18.

T.	ft.	in.
171	30	1000
98	37	1234
72	32	1494

19.

Cords.	ft.	in.
571	18	1234
199	19	1279

WINE MEASURE.

20.

T.	hhd.	gal.	qt.	pt.	gi.
171	3	8	1	1	1
99	1	19	3	1	3
72	1	51	1	1	2

21.

T.	hhd.	gal.	qt.	pt.	gi.
71	1	1	1	1	1
9	3	3	3	1	3

BEER MEASURE.

22.

T.	hhd.	gal.	qt.	pt.
15	1	17	1	0
9	3	19	3	1
5	1	51	1	1

23.

T.	hhd.	gal.	qt.	pt.
79	2	2	2	0
19	3	13	3	1

DRY MEASURE.

24.

ch.	bu.	pk.	qt.	pt.
716	1	2	1	0
19	9	3	1	1
696	27	2	7	1

25.

ch.	bu.	pk.	qt.	pt.
73	13	3	0	1
19	18	1	3	1

TIME.

26.

y.	da.	h.	m.	sec.
375	15	13	17	5
199	137	15	1	39
175	243	4	15	26

27.

w.	da.	h.	m.	sec.
14	1	3	4	15
9	6	17	37	48

CIRCULAR MEASURE.

28.

S.	°	′	″
11	7	13	15
9	29	17	36
1	7	55	39

29.

S.	°	′	″
1	23	37	39
9	15	38	47
4	7	58	52

NOTE. — In Circular Measure, the minuend is sometimes less than the subtrahend, as in Ex. 29, in which case it must be increased by 12 signs.

ART. 103. To find the time between two different dates.

Ex. 1. What is the difference of time between October 16th 1852, and August 9th, 1854? Ans. 1y. 9mo. 23da.

FIRST OPERATION.

	y.	mo.	da.
Min.	1854	7	9
Sub.	1852	9	16
Rem.	1	9	23

SECOND OPERATION.

Min.	1854	8	9
Sub.	1852	10	16
Rem.	1	9	23

Commencing with January, the first month in the year, and counting the months and days in the later date up to August 9th, we find that 7mo. and 9 da. have elapsed; and counting the months and days in the earlier date, up to October 16th, we find that 9mo. and 16da. have elapsed. We, therefore, write the numbers for subtraction as in the first operation. The same result, however, could be obtained, as some prefer, by reckoning the *number* of the given months instead of the *number of months* that have elapsed since the beginning of the year, and writing the numbers as in the second operation; — written either way,

The earlier date being placed under the later, is subtracted, as by the preceding rule.

NOTE. — In finding the difference between two dates, and in computing interest for less than a month, 30 days are considered a month. In *legal* transactions, a month is reckoned from any day in one month to the corresponding day of the following month, if it has a corresponding day, otherwise to its end.

EXAMPLES FOR PRACTICE.

2. What is the time from March 21st, 1853, to Jan. 6th, 1857? Ans. 3y. 9m. 15da.

3. A note was given Nov. 15th, 1852, and paid April 25th, 1857; how long was it on interest? Ans. 4y. 5mo. 10da.

4. John Quincy Adams was born at Braintree, Mass., July 11th, 1767, and died at Washington, D. C., Feb. 23, 1848; to what age did he live? Ans. 80y. 7mo. 12da.

5. Andrew Jackson was born at Waxaw, S. C., March 15th, 1767, and died at Nashville, Tenn., June 8th, 1845; at what age did he die? Ans. 78y. 2mo. 23da.

QUESTIONS. — Art. 103. From what period do you count the months and days in preparing dates for subtraction? How do you arrange the dates for subtraction? How subtract? How many days are considered a month in business transactions? What is the second method of preparing dates for subtraction?

§ XIII. MISCELLANEOUS EXERCISES IN ADDITION AND SUBTRACTION OF COMPOUND NUMBERS.

1. WHAT is the amount of the following quantities of gold 4lb. 8oz. 13pwt. 8gr., 5lb. 11oz. 19pwt. 23gr., 8lb. 0oz. 17pwt. 15gr., and 18lb. 9oz. 14pwt. 10gr.?
Ans. 37lb. 7oz. 5pwt. 8gr.

2. An apothecary would mix 7℔ 3℥ 2ʒ 2℈ 1gr. of rhubarb, 2℔ 10℥ 0ʒ 1℈ 13gr. of cantharides, and 2℔ 3℥ 7ʒ 2℈ 17gr. of opium; what is the weight of the compound?
Ans. 12℔ 5℥ 3ʒ 0℈ 11gr.

3. Add together 17T. 11cwt. 3qr. 11lb. 12oz., 11T. 17cwt. 1qr. 19lb. 11oz., 53T. 19cwt. 1qr. 17lb. 8oz., 27T. 19cwt. 3qr. 18lb. 9oz., and 16T. 3cwt. 3qr. 0lb. 13oz.
Ans. 127T. 12cwt. 1qr. 18lb. 5oz.

4. A merchant owes a debt in London amounting to 7671£.; what remains due after he has paid 1728£. 17s. 9d.?
Ans. 5942£. 2s. 3d.

5. From 73lb. of silver there were made 26lb. 11oz. 13pwt. 14gr. of plate; what quantity remained?
Ans. 46lb. 0oz. 6pwt. 10gr.

6. From 71℔ 8℥ 1ʒ 1℈ 14gr. take 7℔ 9℥ 1ʒ 1℈ 17 gr.
Ans. 63℔ 10℥ 7ʒ 2℈ 17gr.

7. From 28T. 13cwt. take 10T. 17cwt. 19lb. 14oz.
Ans. 17T. 15cwt. 3qr. 5lb. 2oz.

8. A merchant has 3 pieces of cloth; the first contains 37yd. 3qr. 3na., the second 18yd. 1qr. 3na., and the third 31yd. 1qr. 2na.; what is the whole quantity? Ans. 87yd. 3qr. 0na.

9. Sold 3 loads of hay; the first weighed 2T. 13cwt. 1qr. 17lb., the second 3T. 17lb., and the third 1T. 3qr. 11lb.; what did they all weigh? Ans. 6T. 14cwt. 1qr. 20lb.

10. What is the sum of the following distances: 16m. 7fur. 18rd. 14ft. 11in., 19m. 1fur. 13rd. 16ft. 9in., 97m. 3fur. 27rd. 13ft. 3in., and 47m. 5fur. 37rd. 13ft. 10in.?
Ans. 181m. 2fur. 18rd. 9ft. 3in.

11. From 76yd. take 18yd. 3qr. 2na. Ans. 57yd. 0qr. 2na.

12. From 20m. take 3m. 4fur. 18rd. 13ft. 8in.
Ans. 16m. 3fur. 21rd. 2ft. 10in.

13. From 144A. 3R. take 18A. 1R. 17p. 200ft. 100in.
Ans. 126A. 1R. 22p. 71ft. 80in.

14. From 18 cords take 3 cords 100ft. 1000in.

Ans. 14 cords 27ft. 728in.

15. A gentleman has three farms; the first contains 169A. 3R. 15p. 227ft., the second 187A. 1R. 15p. 165ft., and the third 217A. 2R. 28p. 165ft.; what is the whole quantity?

Ans. 574A. 3R. 20p. $12\frac{1}{2}$ft.

16. There are 3 piles of wood; the first contains 18 cords 116ft. 1000in., the second 17 cords 111ft. 1600in., and the third 21 cords 109ft. 1716in.; how much in all?

Ans. 58 cords 82ft. 860in.

17. From 17T. take 5T. 18ft. 765 in.

Ans. 11T. 21ft. 963in.

18. From 169gal. take 76gal. 3qt. 1pt.

Ans. 92gal. 0qt. 1pt.

19. From 17ch. 18bu. take 5ch. 20bu. 1pk. 7qt.

Ans. 11ch. 33bu. 2pk. 1qt.

20. From 83y. take 47y. 10mo. 27d. 18h. 50m. 14s.

Ans. 35y. 1mo. 2d. 5h. 9m. 46s.

21. From 11S. 15° 36′ 15″ take 5S. 18° 50′ 18″.

Ans. 5S. 26° 45′ 57″.

22. John Thomson has 4 casks of molasses; the first contains 167gal. 3qt. 1pt., the second 186gal. 1qt. 1pt., the third 108gal. 2qt. 1pt., and the fourth 123gal. 3qt. 0pt.; how much is the whole quantity? Ans. 586gal. 2qt. 1pt.

23. Add together 17bu. 1pk. 7qt. 1pt., 18bu. 3pk. 2qt. 19bu. 1pk. 3qt. 1pt., and 51bu. 3pk. 0qt. 1pt.

Ans. 107bu. 1pk. 5qt. 1pt.

24. James is 13y. 4mo. 13d. old, Samuel is 12y. 11mo. 23d., and Daniel is 18y. 9mo. 29d.; what is the sum of their united ages? Ans. 45y. 2mo. 5d.

25. Add together 18y. 345d. 13h. 37m. 15s., 87y. 169d. 12h. 16m. 28s., 316y. 144d. 20h. 53m. 18s., and 13y. 360d. 21h. 57m. 15s. Ans. 436y. 290d. 8h. 44m. 16s.

26. A carpenter sent two of his apprentices to ascertain the length of a certain fence. The first stated it was 17rd. 16ft. 11in., the second said it was 18rd. 5in. The carpenter, finding a discrepancy in their statements, and fearing they might both be wrong, ascertained the true length himself, which was 17rd. 5yd. 1ft. 11in.; how much did each differ from the other?

27. From a mass of silver weighing 106lb., a goldsmith made 36 spoons, weighing 5lb. 11oz. 12pwt. 15gr.; a tankard, 3lb. 0oz. 13pwt. 14gr.; a vase, 7lb. 11oz. 14pwt. 23gr.; how much unwrought silver remains?

Ans. 88lb. 11oz. 18pwt. 20gr

28. From a piece of cloth, containing 17yd. 3qr., there were taken two garments, the first measuring 3yd. 3qr. 2na., the second 4yd. 1qr. 3na.; how much remained?

Ans. 9yd. 1qr. 3na.

29. Venus is 3S. 18° 45′ 15″ east of the Sun, Mars is 7S. 15° 36′ 18″ east of Venus, and Jupiter is 5S. 21° 38′ 27″ east of Mars; how far is Jupiter east of the Sun? Ans. 4S. 26°.

30. The longitude of a certain star is 3S. 18° 14′ 35″, and the longitude of Jupiter is 11S. 25° 30′ 50″; how far will Jupiter have to move in his orbit to be in the same longitude with the star? Ans. 3S. 22° 43′ 45″.

---

§ XIV. MULTIPLICATION OF COMPOUND NUMBERS.

ART. **104.** MULTIPLICATION of Compound Numbers is the process of taking a compound number any proposed number of times.

ART. **105.** To multiply when the multiplier is not more than 12.

Ex. 1. If an acre of land cost 14£. 5s. 8d. 2far., what will 9 acres cost? Ans. 128£. 11s. 4d. 2far.

OPERATION.

	£.	s.	d.	far.
Multiplicand	14	5	8	2
Multiplier				9
Product	128	11	4	2

We write the multiplier under the lowest denomination of the multiplicand, and then say 9 times 2far. are 18far., equal to 4d. and 2far. We set down the 2far. under the number multiplied, reserving the 4d. to be added to the next product. We then say 9 times 8d. are 72d., and the 4d. make 76d., equal to 6s. and 4d., and set the 4d. under the column of pence, reserving the 6s. to be added to the next product. Then, 9 times 5s. are 45s., and 6s. make 51s., equal to 2£. and 11s. We place the 11s. under the column of shillings, reserving the 2£. to be added to the next product. Again, 9 times 14£. are 126£., and 2£. make 128£. This, placed under the column of pounds, gives us 128£. 11s. 4d. 2far. for the answer.

QUESTIONS. — Art. 104. What is multiplication of compound numbers? — Art. 105. Explain the operation. By what do you divide the product of each denomination? What do you do with the quotient and remainders thus obtained?

RULE. — *Multiply each denomination of the compound number as in multiplication of simple numbers, and carry as in addition of compouna numbers.*

NOTE. — Going a second time carefully over the work is a good way of testing its accuracy. On learning Division of Compound Numbers, the pupil will find that rule a better method of proving multiplication of compound numbers.

EXAMPLES FOR PRACTICE.

2.

£.	s.	d.
5	6	8
		2
10	13	4

3.

£.	s.	d.
19	11	7
		3
58	14	9

4.

£.	s.	d.
25	17	11
		5
129	9	7

5.

£.	s.	d.
18	15	$8\frac{3}{4}$
		6
112	14	$4\frac{1}{2}$

6.

cwt.	qr.	lb.	oz.
18	3	17	10
			6
113	2	5	12

7.

Ton.	cwt.	qr.	lb.
14	15	3	12
			7
103	11	0	9

8

cwt.	qr.	lb.	oz.
19	1	8	15
			8
154	2	21	8

9.

lb.	oz.	dr.
15	14	13
		9
143	5	5

10.

m.	fur.	rd.	ft.
97	7	14	13
			6
587	4	8	12

11.

deg.	m.	fur.	rd.
18	12	6	18
			8
145	33	2	$10\frac{2}{3}$

12.

rd.	yd.	ft.	in.
23	3	2	9
			9
213	2	0	9

13.

fur.	rd.	ft.	in
9	31	16	11
			10
98	0	4	2

NOTE. — The answers to the following questions are found in the corresponding questions in Division of Compound Numbers, p. 128.

14. What cost 7 yards of cloth at 18s. 9d. per yard?

15. If a man travel 12m. 3fur. 29rd. in one day, how far will he travel in 9 days?

16. If 1 acre produce 2 tons 13cwt. 19lb. of hay, what will 8 acres produce?

QUESTIONS. — What is the rule? How may the work be tested?

17. If a family consume 49gal. 3qt. 1pt. of molasses in one month, what quantity will be sufficient for one year?

18. John Smith has 12 silver spoons, each weighing 3oz. 17pwt. 14gr.; what is the weight of all?

19. Samuel Johnson bought 7 loads of timber, each measuring 7 tons 37ft.; what was the whole quantity?

20. If the moon move in her orbit 13° 11′ 35″ in 1 day, how far will she move in 10 days?

21. If 1 dollar will purchase 2℔ 8℥ 7ʒ 1℈ 10gr. of ipecacuanha, what quantity would 9 dollars buy?

22. If 1 dollar will buy 2A. 3R. 15p. 30yd. 8ft. 100in. of wild land, what quantity may be purchased for 12 dollars?

23. Joseph Doe will cut 2 cords 97ft. of wood in 1 day; how much will he cut in 9 days?

24. If 1 acre of land produce 3ch. 6bu. 2pk. 7qt. 1pt. of corn, what will 8 acres produce?

ART. 106. When the multiplier is a composite number, and none of its factors exceed 12.

Ex. 1. What cost 24 yards of broadcloth at 2£. 7s. 11d. per yard? Ans. 57£. 10s. 0d.

OPERATION.

£.	s.	d.	
2	7	11	= price of 1 yard.
		4	
9	11	8	= price of 4 yards.
		6	
57	10	0	= price of 24 yards.

We find the number 24 equal to the product of 4 and 6; we therefore multiply the price first by 4, and then that product by 6, and the last product is the answer.

Ex. 2. What cost 360 tons of iron at 17£. 16s. 1d. per ton? Ans. 6409£. 10s. 0d.

OPERATION.

£.	s.	d.	
17	16	1	= price of 1 ton.
		6	
106	16	6	= price of 6 tons.
		6	
640	19	0	= price of 36 tons.
		10	
6409	10	0	= price of 360 tons.

We find the factors of 360 to be 6, 6, and 10. We first multiply by 6, and then that product by 6, and then again the last product by 10.

RULE. — *Multiply by the factors of the composite number in succession.*

EXAMPLES FOR PRACTICE.

3. If a man travel 3m. 7fur. 18rd. in one day, how far would he travel in 30 days?

4. If a load of hay weigh 2 tons 7cwt. 3qr. 18lb., what would be the weight of 84 similar loads?

5. When it requires 7yd. 3qr. 2na. of silk to make a lady's dress, what quantity would be sufficient to make 72 similar dresses?

6. A tailor has an order from the navy agent to make 132 garments for seamen; how much cloth will it take, supposing each garment to require 3yd. 2qr. 1na.?

ART. **107.** When the multiplier is not a composite number, and exceeds 12, or, if a composite number, and any of its factors exceed 12.

Ex. 1. What cost 379cwt. of iron at 3£. 16s. 8d. per cwt.?
Ans. 1452£. 16s. 8d.

OPERATION.

£.	s.	d.	
3	16	8	× 9 units.
		10	
38	6	8	× 7 tens.
		10	
383	6	8	
		3	hundreds.
1150	0	0	cost of 300cwt.
268	6	8	cost of 70cwt.
34	10	0	cost of 9cwt.
1452	16	8	cost of 379cwt.

Since 379 is not a composite number, we cannot resolve it into factors; but we may separate it into parts, and find the value of each part separately; thus, 379 = 300 + 70 + 9. In the operation, we first multiply by 10, and then this product by 10, to get the cost of 100cwt. To find the cost of 300cwt., we multiply the last product by 3; and to find the cost of 70cwt., we multiply the cost of 10cwt. by 7; and then, to find the cost of 9cwt., we multiply the cost of 1cwt. by 9. Adding the several products, we obtain 1452£. 16s. 8d. for the answer.

RULE. — *Having resolved the multiplier into any convenient parts, as of units, tens, &c., multiply by these several parts, adding together the products thus obtained for the required result.*

QUESTIONS. — Art. 106. What is the rule for multiplying by a composite number? Give the reason for the rule. — Art. 107. How do you find the cost of 300cwt. in the example? Of 70cwt.? Of 9cwt.? What is the rule when the multiplier is large, and is not a composite number?

EXAMPLES FOR PRACTICE.

2. If 1 dollar will buy 17lb. 10oz. 13dr. of beef, how much may be bought for 62 dollars?

3. What cost 97 tons of lead at 2£. 17s. 9½d. per ton?

4. If a man travel 17m. 3fur. 19rd. 3yd. 2ft. 7in. in one day, how far would he travel in 38 days?

5. If 1 acre will produce 27bu. 3pk. 6qt. 1pt. of corn, what will 98 acres produce?

6. If it require 7yd. 3qr. 2na. to make 1 cloak, what quantity would it require to make 347 cloaks?

7. One ton of iron will buy 13A. 3R. 14p. 18yd. 7ft. 76in. of land; how many acres will 19 tons buy?

8. If 1 ton of copper ore will purchase 17T. 14cwt. 3qr. 18lb 14oz. of iron ore, how much can be purchased for 451 tons?

Ans. 8003T. 17cwt. 1qr. 12lb. 10oz.

---

§ XV. DIVISION OF COMPOUND NUMBERS.

ART. **108.** DIVISION of Compound Numbers is the process of dividing compound numbers into any proposed number of equal parts.

ART. **109.** To divide when the divisor does not exceed 12.

Ex. 1. If 9 acres of land cost 128£. 11s. 4d. 2far., what is the value of 1 acre? Ans. 14£. 5s. 8d. 2far.

OPERATION.

	£.	s.	d.	far.
9)	128	11	4	2
	14	5	8	2

Having divided the 128£. by 9, we find the quotient to be 14£. and 2£. remaining. We place the quotient 14£. under the 128£., and to the remainder 2£., equal to 40s., we add the 11s. in the question, and divide the amount, 51s., by 9. We write the quotient 5s. under the 11s., and to the remainder 6s., equal to 72d., we add the 4d., making 76d., which we divide by 9, and write the quotient 8d. under the 4d. To the remainder 4d., equal to 16far., we add

---

QUESTIONS. — Art. 108. What is division of compound numbers? — Art. 109. Where do you begin to divide? Why? When there is a remainder after dividing any one denomination, what must be done with it?

the 2far., and divide the amount, 18far., by 9, and obtain 2far. for a quotient, which we place under the 2far. in the dividend. Thus we find the answer to be 14£ 5s. 8d. 2far.

RULE. — *Divide as in division of simple numbers, each denomination in its order, beginning with the highest.*

If there be a remainder, reduce it to the next lower denomination adding in the number already of this denomination, if any, and divide as before.

PROOF. — The same as in simple numbers.

NOTE. — When the divisor and dividend are both compound numbers, they must be reduced to the same denomination, and the division then is that of simple numbers.

2.

	£.	s.	d.
2)	10	13	4
	5	6	8

3.

	£.	s.	d.
3)	58	14	9
	19	11	7

4.

	£.	s.	d.
5)	129	9	7
	25	17	11

5.

	£.	s.	d.	far.
6)	112	14	4	2
	18	15	8	3

6.

	cwt.	qr.	lb.	oz.
6)	113	2	5	12
	18	3	17	10

7.

	ton.	cwt.	qr.	lb.
7)	103	11	0	9
	14	15	3	12

8.

	cwt.	qr.	lb.	oz.
8)	154	2	21	8
	19	1	8	15

9.

	lb.	oz.	dr.
9)	143	5	5
	15	14	13

10.

	m.	fur.	rd.	ft.
6)	587	4	8	12

11.

	deg.	m.	fur.	rd.
8)	145	33	2	10$\frac{2}{3}$

12.

	rd.	yd.	ft.	in.
9)	213	2	0	9

13.

	fur.	rd.	ft.	in.
10)	98	0	4	2

NOTE. — The answers to the following questions are found in the corresponding numbers in Multiplication of Compound Numbers.

14. What costs 1 yard of cloth, when 7yd. can be bought for 6£. 11s. 3d.?

15. If a man, in 9 days, travel 112m. 1fur. 21rd., how far will he travel in 1 day?

16. If 8 acres produce 21T. 5cwt. 2qr. 2lb. of hay, what will 1 acre produce?

QUESTION. — What is the rule for division of compound numbers?

17. If a family consume in 1 year 598 gal. 2qt. of molasses, how much will be necessary for 1 month?

18. John Smith has 12 silver spoons, weighing 3lb. 10oz. 11pwt.; what is the weight of each spoon?

19. Samuel Johnson bought 7 loads of timber, measuring 55T. 19ft.; what was the quantity in each load?

20. If the moon, in 10 days, move in her orbit 4S. 11° 55′ 50″, how far does she move in 1 day?

21. If $9 will buy 24℔ 8℥ 3ʒ 1℈ 10gr. of ipecacuanha, how large a quantity will $1 purchase?

22. When $12 will buy 34A. 0R. 32p. 8yd. 5ft. 48in. of wild land, how much will $1 buy?

23. Joseph Doe will cut 24 cords 105 feet of wood in 9 days; how much will he cut in 1 day?

24. When 8 acres of land produce 25ch. 17bu. 3pk. 4qt. of grain, what will 1 acre produce?

ART. **110.** When the divisor is a composite number, and none of its factors exceed 12.

Ex. 1. When 24 yards of broadcloth are sold for 57£. 10s. 0d., what is the price of 1 yard? Ans. 2£. 7s. 11d.

OPERATION.

	£.	s.	d.	
6)	57	10	0	= price of 24 yards.
4)	9	11	8	= price of 4 yards.
	2	7	11	= price of 1 yard.

We find the component parts, or factors, of 24, are 6 and 4. We therefore divide the price by one of these numbers, and the quotient by the other.

RULE. — *Divide by the factors of the composite number in succession.*

EXAMPLES FOR PRACTICE.

2. If 360 tons of iron cost 6409£. 10s. 0d., what is the cost of 1 ton?

3. If a man travel 117m. 7fur. 20rd. in 30 days, how far will he travel in 1 day?

4. If 84 loads of hay weigh 201 tons 6cwt. 0qr. 12lb., what will 1 load weigh?

5. When 72 ladies require 567yd. 0qr. 0na. for their dresses, how many yards will be necessary for one lady?

QUESTIONS. — Art. 110. How does it appear that dividing by 6 in Ex. 1 gives the price of 4 yards? How do you divide by a composite number?

6. When 132 sailors require 470yd. 1qr. of cloth to make their garments, how many yards will be necessary for 1 sailor?

ART. **111.** When the divisor is not a composite number, and exceeds 12, or, if a composite number, and any of its factors exceed 12, *the whole operation can be written down*, as in the following example:

Ex. 1. If 23cwt. of iron cost 171£. 1s. 3d., what cost 1cwt.?
Ans. 7£. 8s. 9d.

OPERATION.

```
       £.  s. d.
23)171  1  3(7£.
   161
   ---
    10
    20
   ---
 23)201(8s.
    184
    ---
     17
     12
    ---
 23)207(9d.
    207
    ---
```

We divide the pounds by 23, and obtain 7 for the quotient, and 10£. remaining, which we reduce to shillings, and add the 1s., and again divide by 23, and obtain 8s. for the quotient. The remainder, 17s., we reduce to pence, and add the 3d., and again divide by 23, and obtain 9d. for the quotient. Thus, the method of operation is the same as by the general rule (Art. 109), excepting more of the work is written down; and, by uniting the several quotients, we find the answer to be 7£. 8s. 9d.

2. If $62 will buy 1095lb. 14oz. 6dr. of beef, how much may be obtained for $1?

3. Paid 280£. 5s. 9½d. for 97 tons of lead; what did it cost per ton?

4. If a man travel 662m. 4fur. 28rd. 3yd. 2ft. 2in. in 38 days, how far will he travel in 1 day?

5. When 98 acres produce 2739 bu. 1pk. 5qt. of grain, what will 1 acre produce?

6. A tailor made 347 garments from 2732yd. 2qr. 2na. of cloth; what quantity did it take to make 1 garment?

7. When 19 tons of iron will purchase 262A. 3R. 37p. 25yd. 1ft. 40in. of land, how much may be obtained for 1 ton?

8. If 451 tons of copper ore will purchase 8003T. 17cwt. 1qr 12lb. 10oz. of iron ore, how much will 1 ton purchase?
Ans. 17T. 14cwt. 3qr. 18lb. 14oz.

QUESTION. — Art. 111. When the divisor is large, and not a composite number, how is the division performed?

§ XVI. MISCELLANEOUS EXAMPLES IN MULTIPLICATION AND DIVISION OF COMPOUND NUMBERS.

1. BOUGHT 30 boxes of sugar, each containing 8cwt. 3qr. 20lb., but having lost 68cwt. 2qr. 0lb., I sold the remainder for 1£. 17s. 6d. per cwt.; what sum did I receive? Ans. 375£.

2. A company of 144 persons purchased a tract of land containing 11067A. 1R. 8p. John Smith, who was one of the company and owned an equal share with the others, sold his part of the land for 1s. 9½d. per square rod; what sum did he receive? Ans. 1101£. 12s. 1½d

3. The exact distance from Boston to the mouth of the Columbia River is 2644m. 3fur. 12rd. A man, starting from Boston, travelled 100 days, going 18m. 7fur. 32rd. each day; required his distance from the mouth of the Columbia at the end of that time. Ans. 746m. 7fur. 12rd.

4. James Bent was born July 4, 1798, at 3h. 17m. A. M.; how long had he lived Sept. 9, 1807, at 11h. 19m. P. M., reckoning 365 days for each year, excepting the leap year 1804, which has 366 days? Ans. 3353da. 20h. 2m.

5. The distance from Vera Cruz, in a straight line, to the city of Mexico, is 121m. 5fur. If a man set out from Vera Cruz to travel this distance, on the first day of January, 1848, which was Saturday, and travelled 3124rd. per day until the eleventh day of January, omitting, however, as in duty bound, to travel on the Lord's day, how far would he be from the city of Mexico on the morning of that day? Ans. 43m. 4fur. 8rd.

6. Bought 16 casks of potash, each containing 7cwt. 3qr. 18lb., at 5 cents per pound. I disposed of 9 casks at 6 cents per pound, and sold the remainder at 7 cents per pound; what did I gain? Ans. $182.39.

7. A merchant purchased in London 17 bales of cloth for 17£. 18s. 10d. per bale. He disposed of the cloth at Havana for sugar at 1£. 17s. 6d. per cwt. Now, if he purchased 144cwt. of sugar, what balance did he receive? Ans. 35£. 0s. 2d.

8. A and B commenced travelling, the same way, round an island 50 miles in circumference. A travels 17m. 4fur. 30rd. a day, and B travels 12m. 3fur. 20rd. a day; required how far they are apart at the end of 10 days. Ans. 1m. 4fur. 20rd.

9. Bought 760 barrels of flour at $5.75 per barrel, which I paid for in iron at 2 cents per pound. The purchaser afterwards sold one half of the iron to an axe manufacturer; what quantity did he sell? Ans. 54T. 12cwt. 2qr

10. Bought 17 house-lots, each containing 44 perches, 200 square feet. From this purchase I sold 2A. 2R. 240ft., and the remaining quantity I disposed of at 1s. 2½d. per square foot; what amount did I receive for the last sale?
Ans. 5914£. 19s. 5½d.

11. J. Spofford's farm is 100 rods square. From this he sold H. Spaulding a fine house-lot and garden, containing 5A. 3R. 17p., and to D. Fitts a farm 50rd. square, and to R. Thornton a farm containing 3000 square rods; what is the value of the remainder, at $1.75 per square rod? Ans. $6235.25.

12. Bought 78A. 3R. 30p. of land for $7000, and, having sold 10 house-lots, each 30rd. square, for $8.50 per square rod, I dispose of the remainder for 2 cents per square foot. How much do I gain by my bargain? Ans. $89265.35

---

§ XVII. PROPERTIES AND RELATIONS OF NUMBERS.

ART. **112.** AN INTEGER is a whole number; as 1, 6, 13.

All numbers are either *odd* or *even.*

An *odd* number is a number that cannot be divided by 2 without a remainder; thus, 3, 7, 11.

An *even* number is a number that can be divided by 2 without a remainder; thus, 4, 8, 12.

Numbers are also either *prime* or *composite.*

A *prime* number is a number which can be exactly divided only by itself or 1; as 1, 3, 5, 7.

A *composite* number is a number which can be exactly divided other than by itself or 1; as 6, 9, 14.

Numbers are prime to each other when they have no factor in common; thus, 7 and 11 are prime to each other, as are, also, 4, 15, and 19.

QUESTIONS. — Art. 112. What is an integer? What are all numbers? What is an odd number? What is an even number? What other distinctions of numbers are mentioned? What is a prime number? When are numbers prime to each other? What is a composite number?

All the prime numbers not larger than 1109 are included in the following

TABLE OF PRIME NUMBERS.

1	59	139	233	337	439	557	653	769	883	1013
2	61	149	239	347	443	563	659	773	887	1019
3	67	151	241	349	449	569	661	787	907	1021
5	71	157	251	353	457	571	673	797	911	1031
7	73	163	257	359	461	577	677	809	919	1033
11	79	167	263	367	463	587	683	811	929	1039
13	83	173	269	373	467	593	691	821	937	1049
17	89	179	271	379	479	599	701	823	941	1051
19	97	181	277	383	487	601	709	827	947	1061
23	101	191	281	389	491	607	719	829	953	1063
29	103	193	283	397	499	613	727	839	967	1069
31	107	197	293	401	503	617	733	853	971	1087
37	109	199	307	409	509	619	739	857	977	1091
41	113	211	311	419	521	631	743	859	983	1093
43	127	223	313	421	523	641	751	863	991	1097
47	131	227	317	431	541	643	757	877	997	1103
53	137	229	331	433	547	647	761	881	1009	1109

ART. **113.** A *prime* factor of a number is a prime number that will exactly divide it; thus, the prime factors of 21 are the prime numbers 1, 3, and 7.

A *composite* factor of a number is a composite number that will exactly divide it; thus, the composite factors of 24 are the composite numbers 4 and 6.

NOTE 1. — Unity or 1 is not regarded as a material prime factor, since multiplying or dividing any number by 1 does not alter its value; it will be omitted when speaking of the prime factors of numbers.

NOTE 2. — There has been discovered no direct process by which prime numbers may be found. The following facts, however, if kept in mind, will aid in ascertaining whether a number is prime or not; and, if not prime, indicate one or more of its factors:

1. 2 is the only even prime number.
2. 2 is a factor of every even number.
3. 3 is a factor of every number the sum of whose digits 3 will exactly divide; thus, 15, 81, and 546, have each 3 as a factor.
4. 4 is a factor of every number whose two right-hand figures 4 will exactly divide; thus, 316, 532, and 1724, have each 4 as a factor.
5. 5 is the only prime number having 5 for a unit or right-hand figure.

QUESTIONS. — Art. 113. What is a prime factor? What is a composite factor? How is unity or 1 regarded? Is there any direct process for determining prime numbers? Which is the only even prime number? Of what numbers is 2 a factor? Of what numbers is 3 a factor? Of what numbers is 4 a factor?

6. 5 is a factor of every number whose right-hand figure is either 5 or 0; as, 15, 20, &c.

7. 6 is a factor of every even number that 3 will exactly divide; thus, 24, 108, and 360, have each 6 as a factor.

8. 7 is a factor of every number occupying four places whose two right-hand figures are contained in the left-hand figure or figures exactly 3 times; thus, 602, 2107, and 3913, have each 7 as a factor.

9. 7 is a factor of every number occupying three or four places, when the two right-hand figures contain the left-hand figure or figures exactly 5 times; thus, 840, 945, and 1155, have each 7 as a factor.

10. 8 is a factor of every number whose three right-hand figures 8 will exactly divide; thus, 5072, 11240, and 17128, have each 8 as a factor.

11. 9 is a factor of every number the sum of whose digits 9 will exactly divide; thus, 27, 432, and 20304, have each 9 as a factor.

12. 10 is a factor of every number whose right-hand figure is 0; as, 20, 30, &c.

13. 7, 11 and 13, are factors of any number occupying four places in which two like figures have two ciphers between them; as, 3003, 4004, 9009, &c.

14. Every prime number, except 2 and 5, has 1, 3, 7, or 9, for the right-hand figure.

ART. **114.** Method of finding the prime factors of numbers.

Ex. 1. It is required to find the prime factors of 24.

Ans. 2, 2, 2, 3.

OPERATION.

```
2 | 24
  ----
2 | 12
  ----
2 |  6
  ----
     3
```

$2 \times 2 \times 2 \times 3 = 24$

We divide by 2, the least prime number greater than 1, and obtain the quotient 12 And since 12 is a composite number, we divide this also by 2, and obtain a quotient 6. We divide 6 by 2, and obtain 3 for a quotient, which is a prime number. The several divisors and the last quotient, all being prime, constitute all the prime factors of 24, which, multiplied together, they equal.

RULE. — *Divide the given number by the least prime number, greater than* 1, *that will divide it, and the quotient, if a composite number, in the same manner; and continue dividing until a prime number is obtained for a quotient. The several divisors and the last quotient will be the prime factors required.*

NOTE. — The composite factors of any number may be found by multiplying together two or more of its prime factors.

QUESTIONS. — Of what numbers is 5 a factor? Of what numbers is 6 a factor? Of what numbers is 7 a factor? Of what numbers is 8 a factor? Of what numbers is 9 a factor? What is the right-hand figure of every prime number? What is the rule for finding the prime factors of numbers? How may the composite factors of numbers be found?

EXAMPLES FOR PRACTICE.

2. What are the prime factors of 36? Ans. 2, 2, 3, 3.
3. What are the prime factors of 48? Ans. 2, 2, 2, 2, 3.
4. What are the prime factors of 56? Ans. 2, 2, 2, 7.
5. What are the prime factors of 144?
Ans. 2, 2, 2, 2, 3, 3.
6. What are the prime factors of 3420?
Ans. 2, 2, 3, 3, 5, 19.
7. What are the prime factors of 18500?
Ans. 2, 2, 5, 5, 5, 37.
8. What are the prime factors of 19965?
Ans. 3, 5, 11, 11, 11.
9. What are the prime factors of 12496?
Ans. 2, 2, 2, 2, 11, 71
10. What are the prime factors of 17199?
Ans. 3, 3, 3, 7, 7, 13.
11. What are the prime factors of 7800?
Ans. 2, 2, 2, 3, 5, 5, 13.

CANCELLATION.

ART. **115.** *If the dividend and divisor are both divided by the same number, the quotient is not changed.* Thus, if the dividend is 20 and the divisor 4, the quotient will be 5. Now, if we divide the dividend and divisor by some number, as 2, their proportion is not changed, and we obtain 10 and 2 respectively; and $10 \div 2 = 5$, the same as the original quotient.

ART. **116.** *If a factor in any number is cancelled, the number is divided by that factor.* Thus, if 15 is the dividend and 5 the divisor, the quotient will be 3. Now, since the divisor and quotient are the two factors, which, being multiplied together, produce the dividend (Art. 50), it is plain, if we *cross out* or *cancel* the factor 5, the remaining 3 is the quotient, and by the operation the dividend 15 has been divided by 5.

ART. **117.** *Cancellation* is the method of shortening arithmetical operations by rejecting any factor or factors common to the divisor and dividend.

QUESTIONS. — Art. 115. What is the effect on the quotient when the dividend and divisor are divided by the same number? What is the effect of cancelling a factor of any number? What is cancellation?

Ex. 1. A man sold 25 hundred weight of iron at 5 dollars per hundred weight, and expended the money for flour at 5 dollars per barrel; how many barrels did he purchase?

Ans. 25 barrels.

OPERATION.

$$\begin{array}{l} \text{Dividend} \\ \text{Divisor} \end{array} \frac{\not{5} \times 25}{\not{5}} = 25.$$

We first indicate by their signs the multiplication and division required by the question. We then, observing 5 to be a common factor of the divisor and dividend, divide the divisor and dividend by this factor, or, which is the same thing, cancel or reject it in both, and obtain 25 for the quotient.

2. Divide the product of 12, 7, and 5, by the product of 5, 4, and 2.

Ans. $10\frac{1}{2}$.

OPERATION.

$$\begin{array}{l} \text{Dividend} \\ \text{Divisor} \end{array} \frac{\overset{3}{\not{1}\not{2}} \times 7 \times \not{5}}{\not{5} \times \not{4} \times 2} = \frac{21}{2} = 10\tfrac{1}{2} \text{ Quotient.}$$

Finding 4 in the divisor to be a factor of 12 in the dividend, we divide 12 by 4, cancelling these numbers, and use the 3 instead of 12. The factor 5, common to both dividend and divisor, having been cancelled, we divide the product of the remaining factors in the dividend by the product of those in the divisor, and obtain the quotient $10\frac{1}{2}$.

3. Divide the product of 8, 5, 16, and 21, by the product of 10, 4, 12, and 7.

OPERATION.

$$\begin{array}{l} \text{Dividend} \\ \text{Divisor} \end{array} \frac{\not{8} \times \not{5} \times \overset{4}{\not{1}\not{6}} \times \not{2}\not{1}}{\not{1}\not{0} \times \not{4} \times \underset{\not{3}}{\not{1}\not{2}} \times \not{7}} = 4, \text{ Quotient.}$$

The product of the factors 8 and 5 in the dividend is equal to the product of 10 and 4 in the divisor; therefore we cancel these factors. Finding 16 in the dividend and 12 in the divisor may be divided by 4, they are cancelled, and use made of their quotients. Again, as the product of the factors 3 and 7 of the divisor equals the 21 of the dividend, we cancel the 3, 7, and 21. The factor 4 alone remaining is the quotient.

QUESTIONS. — How do you arrange the dividend and divisor for cancellation? How do you then proceed? Is the factor 5, in Ex. 1, reduced to 0 or 1 by being cancelled? How do you proceed when a number in the dividend and another in the divisor have a common factor? How do you proceed when the products of two or more factors in the dividend and divisor are alike?

RULE. — *Cancel the factor or factors common to the dividend and divisor, and then divide the product of the factors remaining in the dividend by the product of those remaining in the divisor.*

NOTE. — 1. In arranging the numbers for cancellation, the dividend may be written above the divisor with a horizontal line between them, as in division (Art. 47); or, as some prefer, the dividend may be written on the right of the divisor, with a vertical line between them.

NOTE. — 2. Cancelling a factor does not leave 0, but the quotient 1, to take its place, since rejecting a factor is the same as dividing by that factor (Art. 116). Therefore, for every factor cancelled, either in the dividend or divisor, the factor 1 remains.

EXAMPLES FOR PRACTICE.

4. Divide 42×19 by 19. Ans. 42.

5. Divide the product of 8, 6, and 3, by the product of 6, 3, and 4. Ans. 2.

6. Divide the product of 17, 6, and 2, by the product of 6, 2, and 17. Ans. 1.

7. Sold 15 pieces of shirting, and in each piece there were 30 yards, for which I received 10 cents per yard; expended the money for 10 pieces of calico, each containing 15 yards; what was the calico per yard? Ans. 30 cents.

8. Divide the product of 12, 7, and 5, by the product of 2, 4, and 3. Ans. $17\frac{1}{2}$.

9. Divide the product of 20, 13, and 9, by the product of 13, 16, and 1. Ans. $11\frac{1}{4}$.

10. Divide the product of 9, 8, 2, and 14, by the product of 3, 4, 6, and 7. Ans. 4.

11. Divide the product of 16, 5, 10, and 18, by the product of 8, 6, 2, and 12. Ans. $12\frac{1}{2}$.

12. Divide the product of 22, 9, 12, and 5, by the product of 3, 11, 6, and 4. Ans. 15.

13. Divide the product of 25, 7, 14, and 36, by the product of 4, 10, 21, and 54. Ans. $1\frac{7}{18}$.

14. Divide the product of 26, 72, 81, and 12, by the product of 36, 13, 24, and 54. Ans. 3.

15. Divide the product of 8, 5, 3, 16, and 28, by the product of 10, 4, 12, 4, and 7. Ans. 4.

16. Divide the product of 8, 4, 9, 2, 12, 16, and 5, by the product of 4, 6, 6, 3, 8, 4, and 20. Ans. 2.

17. Divide the product of 6, 15, 16, 24, 12, 21, and 27, by the product of 2, 10, 9, 8, 36, 7, and 81. Ans. 8.

QUESTIONS. — What is the rule for cancellation? How may the numbers be arranged for cancelling? What takes the place of a cancelled factor? What remains for every factor cancelled either in the dividend or divisor?

A COMMON DIVISOR.

ART. **118.** A common divisor of two or more numbers is any number that will divide them without a remainder; thus, 2 is a common divisor of 2, 4, 6, and 8.

ART. **119.** To find a common divisor of two or more numbers.

Ex. 1. What is the common divisor of 10, 15, and 25? Ans. 5.

OPERATION.

$10 = 5 \times 2$
$15 = 5 \times 3$
$25 = 5 \times 5$

We resolve each of the given numbers into two factors, one of which is common to all of them. In the operation 5 is the common factor, and therefore must be a common divisor of the numbers.

RULE. — *Resolve each of the given numbers into two factors one of which is common to all of them, and this common factor is a common divisor.*

EXAMPLES FOR PRACTICE.

2. What is the common divisor of 3, 9, 18, 24? Ans. 3.
3. What is the common divisor of 4, 12, 16, 28? Ans. 2 or 4.

ART. **120.** A divisor of any factor of a number is a divisor of the number itself. Thus 3, a divisor of 9, a factor of 45, is a divisor of 45 itself.

ART. **121.** A common divisor of two numbers is a divisor of their *sum* and of their *difference.* Thus 4, a common divisor of 16 and 12, is a divisor of their sum, 28, and of their difference, 4.

ART. **122.** A common divisor of the *remainder* and the *divisor* is a divisor of the *dividend.* Thus, in a division having 12 for remainder, 36 for divisor, and 48 for dividend, 12, a common divisor of the 12 and the 36, is also a divisor of the 48.

THE GREATEST COMMON DIVISOR.

ART. **123.** The greatest common divisor of two or more numbers is the greatest number that will divide each of them without a remainder. Thus 6 is the greatest common divisor of 12, 18, and 24.

QUESTIONS. — Art. 118. What is a common divisor of two or more numbers? — Art. 119. What is the rule? — Art. 121. Of what is the common divisor of two numbers a divisor? — Art. 122. Of what is a common divisor of the less of two numbers and of their difference a divisor? — Art. 123. What is the greatest common divisor of two or more numbers?

ART. 124. To find the greatest common divisor of two or more numbers.

Ex. 1. What is the greatest common divisor or measure of 84 and 132? Ans. 12.

FIRST OPERATION.

$$84 = 2 \times 2 \times 3 \times 7$$
$$132 = 2 \times 2 \times 3 \times 11$$
$$2 \times 2 \times 3 = 12.$$

Resolving the numbers into their prime factors (Art. 114), thus, $84 = 2 \times 2 \times 3 \times 7$, and $132 = 2 \times 2 \times 3 \times 11$, we find the factors $2 \times 2 \times 3$ are common to both. Since only these common factors, or the product of two or more of such factors, will exactly divide both numbers, it follows *that the product of all their common prime factors must be the greatest factor that will exactly divide both of them.* Therefore $2 \times 2 \times 3 = 12$ is the greatest common divisor required.

The same result may be obtained by a sort of trial process, as by the second operation.

SECOND OPERATION.

```
84)132(1
   84
   --
   48)84(1
      48
      --
      36)48(1
         36
         --
         12)36(3
            36
```

It is evident, since 84 cannot be exactly divided by a number greater than itself, if it will also exactly divide 132, it will be the *greatest common divisor* sought. But, on trial, we find 84 will not exactly divide 132, there being a remainder, 48. Therefore 84 is not a common divisor of the two numbers.

We know a common divisor of 48 and 84 will also be a divisor of 132 (Art. 122). We next try to find that divisor. It cannot be greater than 48. But 48 will not exactly divide 84, there being a remainder, 36; therefore 48 is not the greatest common divisor.

Again, as the common divisor of 36 and 48 will also be a divisor of 84 (Art. 122), we try to find that divisor, knowing that it cannot be greater than 36. But 36 will not exactly divide 48, there being a remainder, 12; therefore 36 is not the greatest common divisor.

As before, the common divisor of 12 and 36 will be a divisor of 48 (Art. 122); we make a trial to find that divisor, knowing that it cannot be greater than 12, and find 12 will exactly divide 36. Therefore 12 is the greatest common divisor required.

RULE 1. — *Resolve the given numbers into their prime factors. The product of all the factors common to the several numbers will be the greatest common divisor.* Or,

RULE 2. — *Divide the greater number by the less, and if there be a*

QUESTION. — Art. 124. What are the rules for finding the greatest common divisor of two or more numbers?

remainder divide the preceding divisor by it, and so continue dividing until nothing remains. The last divisor will be the greatest common divisor.

NOTE. — When the greatest common divisor is required of *more* than *two* numbers, find it of two of them, and then of that common divisor and of one of the other numbers, and so on for all the given numbers The last common divisor will be the greatest common divisor required.

EXAMPLES FOR PRACTICE.

2. What is the greatest common divisor of 85 and 95? Ans. 5.

3. What is the greatest common divisor of 72 and 168? Ans. 24.

4. What is the greatest common divisor of 119 and 121? Ans. 1.

5. What is the greatest common divisor of 12, 18, 24, and 30? Ans. 6.

6. Having three rooms, the first 12 feet wide, the second 15 feet, and the third 18 feet, I wish to purchase a roll of the widest carpeting that will exactly fit each room without any cutting as to width. How wide must it be? Ans. 3 feet.

A COMMON MULTIPLE.

ART. **125.** A *multiple* of a number is a number that can be divided by it without a remainder; thus 6 is a multiple of 3.

ART. **126.** A *common multiple* of two or more numbers is a number that can be divided by each of them without a remainder; thus 12 is a common multiple of 3 and 4.

ART. **127.** The *least common multiple* of two or more numbers is the *least* number that can be divided by each of them without a remainder; thus 30 is the least common multiple of 10 and 15.

NOTE. — A multiple of a number contains all the prime factors of that number; and the common multiple of two or more numbers contains all the prime factors of each of the numbers. Therefore, the least common multiple of two or more numbers must be the *least* number that will contain all the prime factors of them, and none others. Hence it will have each prime factor taken only the greatest number of times it is found in any of the several numbers.

QUESTIONS. — Art. 125. What is a multiple of a number? — Art. 127 What is the least common multiple of a number?

ART. 128. To find the least common multiple.

EX. 1. What is the least common multiple of 6, 9, 12?

Ans. 36.

FIRST OPERATION.

$6 = 2 \times 3$
$9 = 3 \times 3$
$12 = 2 \times 2 \times 3$
$2 \times 2 \times 3 \times 3 = 36$

Resolving the numbers into their prime factors, — thus, $6 = 2 \times 3$, and $9 = 3 \times 3$, and $12 = 2 \times 2 \times 3$, — we find their *different* prime factors to be 2 and 3. The greatest number of times the 2 occurs as a factor in any of the numbers is twice, as 2×2 in 12; and the greatest number of times the 3 occurs in any of the numbers is also twice, as 3×3 in 9. Hence $2 \times 2 \times 3 \times 3$ must be all the prime factors that are necessary in composing 6, 9, and 12; and, consequently, the product of these factors must be the least number that can be exactly divided by 6, 9, and 12. Therefore $2 \times 2 \times 3 \times 3 = 36$ is the least common multiple required.

SECOND OPERATION.

3	6	9	12
2	2	3	4
	1	3	2

$3 \times 2 \times 3 \times 2 = 36$

Another method, and one usually preferred, is as by second operation. Having arranged the numbers on a horizontal line, we divide by 3, a prime number that will divide all of them without a remainder, and write the quotients in a line below. We next divide by 2, a prime number that will divide without a remainder most of them, writing down the quotients and undivided numbers as before. Then, since these numbers are prime to each other, we multiply together the divisors and the numbers on the lower line, which are all the prime factors of 6, 9, and 12, and thus obtain 36 for the least common multiple.

RULE 1. — *Resolve the given numbers into their prime factors. The product of these factors, taking each factor the greatest number of times it occurs in any of the numbers, will be the least common multiple.* Or,

RULE 2. —*Having arranged the numbers on a horizontal line, divide by such a prime number as will divide most of them without a remainder, and write the quotients and undivided numbers in a line beneath. So continue to divide until no prime number greater than* 1 *will divide two or more of them. The product of the divisors and the numbers of the line below will be the least common multiple.*

NOTE 1. — When numbers are prime to each other, their product is their least common multiple.

NOTE 2. — When one or more of the given numbers are factors of any one of the other numbers the factor or factors may be cancelled.

QUESTION. — Art. 128. What are the rules for finding the least common multiple?

EXAMPLES FOR PRACTICE.

2. What is the least common multiple of 7, 14, 21, and 15? Ans. 210.

OPERATION.

7	~~7~~ 14 21 15
	2 ~~3~~ 15

$7 \times 2 \times 15 = 210$

Since 7 is a factor of 14, another of the numbers, we cancel it; and since 3 is a factor of 15, we also cancel that (Note 2): thus the work is rendered shorter.

3. What is the least common multiple of 3, 4, 5, 6, 7, and 8? Ans. 840.

4. What is the least number that 10, 12, 16, 20, and 24, will divide without a remainder? Ans. 240.

5. What is the least common multiple of 9, 8, 12, 18, 24, 36, and 72? Ans. 72.

6. Five men start from the same place to go round a certain island. The first can go round it in 10 days; the second, in 12 days; the third, in 16 days; the fourth, in 18 days; the fifth, in 20 days. In what time will they all meet at the place from which they started? Ans. 720 days.

---

§ XVIII. FRACTIONS.

ART. **129.** A FRACTION is an expression denoting one or more equal *parts* of a unit.

The term *fraction* is derived from the Latin word *frango*, which signifies *to break;* from the idea that a number or thing is broken or separated into parts.

Fractions are of two kinds, *Common* and *Decimal.*

COMMON FRACTIONS.

ART. **130.** A COMMON FRACTION is expressed by two numbers one above the other, with a line between them.

The number *below* the line is called the *denominator;* and the number *above*, the *numerator*.

Thus, { Numerator $\frac{3}{5}$ Three / Denominator Fifths.

QUESTIONS. — Art. 129. What is a fraction? From what is the term derived, and what does it signify? How many kinds of fractions, and what are they called? — Art. 130. How is a common fraction expressed? What is the number below the line called? The number above the line?

The denominator shows into how many parts the whole number is divided, and gives a name to the fraction. The numerator shows how many of these parts are taken, or expressed by the fraction.

A *proper* fraction is one whose numerator is *less* than the denominator; as, $\frac{5}{7}$.

An *improper* fraction is one whose numerator is equal to, or greater than, the denominator; as, $\frac{8}{8}$, $\frac{9}{5}$.

NOTE. — A fraction, strictly speaking, is less than a unit; hence, if the numerator is equal to, or greater than, the denominator, it expresses a unit or more than a unit, and is therefore called an improper fraction.

A *mixed* number is a whole number with a fraction; as, $7\frac{6}{11}$, $5\frac{3}{8}$.

A *simple* or *single* fraction has but one numerator and one denominator, and may be either proper or improper; as, $\frac{7}{8}$, $\frac{5}{3}$.

A *compound* fraction is a fraction of a fraction, connected by the word *of*; as, $\frac{7}{8}$ of $\frac{5}{6}$ of $\frac{8}{9}$.

A *complex* fraction is a fraction having a fraction or a mixed number for its numerator or denominator, or both; as $\frac{\frac{3}{4}}{\frac{7}{9}}$, $\frac{7}{9\frac{7}{8}}$, $\frac{8\frac{1}{4}}{11}$, $\frac{7\frac{1}{4}}{9\frac{1}{11}}$.

ART. **131.** The *terms* of a fraction are its numerator and denominator.

The *unit of a fraction* is the unit or whole thing from which its fractional parts, or *fractional units*, are obtained.

A whole number may be expressed fractionally, by writing 1 for the denominator. Thus, 5 may be written $\frac{5}{1}$, and read 5 ones; and 9 may be written $\frac{9}{1}$, and read 9 ones.

ART. **132.** *Fractions originate from division; the numerator answers to the dividend, and the denominator to the divisor.* Thus, when we divide 479956 by 6 (Art. 49, Ex. 12), we had a remainder of 4, which could not be divided by 6, and therefore we wrote it over the divisor, with a line between them. This expression originating from division is a fraction; the number above the line being the numerator, and the one below the denominator.

QUESTIONS. — What does the denominator of a fraction show? What does the numerator show? What is a proper fraction? What is an improper fraction? What is a mixed number? What is a simple fraction? What is a compound fraction? What is a complex fraction? — Art. 131. What are the terms of a fraction? What is the unit of a fraction? How may a whole number be expressed fractionally? From what do fractions originate?

ART. 133. From what has preceded, we perceive that *the value of a fraction is the quotient arising from the division of the numerator by the denominator.* Thus, the value of $\frac{6}{2}$, or $6 \div 2$, is 3; and the value of $\frac{3}{4}$, or $3 \div 4$, is $\frac{3}{4}$.

REDUCTION OF COMMON FRACTIONS.

ART. 134. Reduction of Fractions is the process of changing their *form* of expression without altering their value.

A fraction is in its lowest terms, when its terms are prime to each other. (Art. 112.)

ART. 135. To reduce a fraction to its lowest terms.

Ex. 1. Reduce $\frac{6}{18}$ to its lowest terms. Ans. $\frac{1}{3}$.

OPERATION.

$2)\ \frac{6}{18} = \frac{3}{9}$

$3)\ \frac{3}{9} = \frac{1}{3}$

We divide the terms of the fraction by 2, a factor common to them both, and obtain $\frac{3}{9}$. We divide, again, both terms of $\frac{3}{9}$ by 3, a factor common to them, and obtain $\frac{1}{3}$. Now, as 1 and 3 are numbers prime to each other, the fraction $\frac{1}{3}$ is in its lowest terms. The same result would have been produced, if we had divided the terms by 6, the greatest common divisor.

Since the numerator and denominator of a fraction correspond to the dividend and divisor in division (Art. 132), dividing both by the same number, or cancelling equal factors in both (Art. 115), changes only the form of the fraction, while the value expressed remains the same. Therefore,

Dividing the numerator and denominator of a fraction by the same number does not alter the value of the fraction.

RULE. — *Divide the numerator and denominator by any number greater than* 1, *that will divide them both without a remainder, and thus proceed until they are prime to each other.* Or,

Divide both the numerator and denominator by their greatest common divisor.

EXAMPLES FOR PRACTICE.

2. Reduce $\frac{5}{25}$ to its lowest terms. Ans. $\frac{1}{5}$.
3. Reduce $\frac{8}{36}$ to its lowest terms. Ans. $\frac{2}{9}$.
4. Reduce $\frac{12}{96}$ to its lowest terms. Ans. $\frac{1}{8}$.
5. Reduce $\frac{96}{144}$ to its lowest terms. Ans. $\frac{2}{3}$.
6. Reduce $\frac{107}{214}$ to its lowest terms. Ans. $\frac{1}{2}$.
7. Reduce $\frac{123}{386}$ to its lowest terms. Ans. $\frac{123}{386}$.
8. Reduce $\frac{81}{567}$ to its lowest terms. Ans. $\frac{1}{7}$.

QUESTIONS. — What is the value of a fraction? — Art. 134. What is reduction of fractions? When is a fraction in its lowest terms? — Art. 135. Why does dividing both terms of a fraction by the same number not alter the value? Has $\frac{1}{3}$ the same value as $\frac{6}{18}$? Why? Repeat the rule.

9. Reduce $\frac{7891}{8116}$ to its lowest terms. Ans. $\frac{7891}{8116}$.
10. What is the lowest expression of $\frac{346}{618}$? Ans. $\frac{173}{309}$.

ART. 136. To reduce a mixed number to an improper fraction.

Ex. 1. In $7\frac{3}{5}$ how many fifths? Ans. $\frac{38}{5}$.

OPERATION.

$$\begin{array}{l} 7\frac{3}{5} \\ \underline{5} \\ 35 \text{ fifths.} \\ \underline{3} \\ 38 \text{ fifths} = \frac{38}{5} \end{array}$$

Since there are 5 fifths in 1 whole one, there will be 5 times as many *fifths* as whole ones; therefore, in 7 there are 35 fifths, and the 3 fifths being added make 38 fifths, which are expressed thus, $\frac{38}{5}$.

RULE. — *Multiply the whole number by the denominator of the fraction, and to the product add the numerator, and place the sum over the given denominator.*

NOTE. — To reduce a whole number to a fraction of the same value, having a *given* denominator, we *multiply the whole number by the given denominator, and make the product the numerator;* thus, 5, reduced to a fraction, having 3 for a denominator, becomes $\frac{15}{3}$.

EXAMPLES FOR PRACTICE.

2. In $8\frac{3}{7}$ dollars how many sevenths? Ans. $\frac{59}{7}$.
3. In $3\frac{1}{4}$ oranges how many fourths? Ans. $\frac{13}{4}$.
4. In $9\frac{4}{11}$ gallons how many elevenths? Ans. $\frac{103}{11}$.
5. Reduce $8\frac{3}{11}$ to an improper fraction. Ans. $\frac{91}{11}$.
6. Reduce $15\frac{7}{12}$ to an improper fraction. Ans. $\frac{187}{12}$.
7. In $18\frac{7}{9}$ how many ninths? Ans. $\frac{169}{9}$.
8. In $161\frac{11}{117}$ how many one hundred and seventeenths? Ans. $\frac{18848}{117}$.
9. Change $43\frac{111}{117}$ to an improper fraction. Ans. $\frac{5142}{117}$
10. What improper fraction will express $27\frac{9}{13}$? Ans. $\frac{360}{13}$.
11. Change $111\frac{111}{111}$ to an improper fraction. Ans. $\frac{12332}{111}$.
12. Change 125 to an improper fraction. Ans. $\frac{125}{1}$.
13. Change 25 to an improper fraction, having 6 for a denominator. Ans. $\frac{150}{6}$.
14. Reduce 75 to ninths. Ans. $\frac{675}{9}$.
15. Change 343 to the form of a fraction. Ans. $\frac{343}{1}$.
16. Reduce 84 to fifteenths. Ans. $\frac{1260}{15}$.

QUESTIONS. — Art. 136. What is the rule for reducing a mixed number to an improper fraction? Give the reason. How do you reduce a whole number to a fraction of the same value, having a given denominator?

ART. **137.** To reduce improper fractions to whole or mixed numbers.

Ex. 1. How many dollars in $\frac{37}{16}$ dollars? Ans. $\$2\frac{5}{16}$.

OPERATION.

```
16 ) 37 ( 2 5/16
     32
     --
      5
```

This question may be analyzed by saying, As 16 sixteenths make one dollar, there will be as many dollars in 37 sixteenths of a dollar as 37 contains times 16, which is $2\frac{5}{16}$ times. Therefore $\$ 2\frac{5}{16}$ is the answer.

RULE. — *Divide the numerator by the denominator, and the quotient will be the whole or mixed number.*

EXAMPLES FOR PRACTICE.

2. Reduce $\frac{96}{8}$ to a whole number. Ans. 12.
3. Change $\frac{178}{17}$ to a mixed number. Ans. $10\frac{8}{17}$.
4. Change $\frac{1111}{111}$ to a mixed number. Ans. $10\frac{1}{111}$.
5. Change $\frac{1735}{876}$ to a mixed number. Ans. $1\frac{859}{876}$.
6. Reduce $\frac{1000}{7}$ to a mixed number. Ans. $142\frac{6}{7}$.
7. Reduce $\frac{378}{378}$ to a whole number. Ans. 1.
8. Change $\frac{567}{1}$ to a whole number. Ans. 567.
9. Reduce $\frac{743}{79}$ to a mixed number. Ans. $9\frac{32}{79}$.
10. Reduce $\frac{1848}{459}$ to a mixed number. Ans. $4\frac{4}{153}$.

ART. **138.** To reduce a compound fraction to a simple fraction.

Ex. 1. Reduce $\frac{4}{5}$ of $\frac{7}{11}$ to a simple fraction. Ans. $\frac{28}{55}$.

OPERATION.

$\frac{4}{5} \times \frac{7}{11} = \frac{28}{55}$

To show the reason of the operation, this question may be analyzed by saying, that, if $\frac{1}{11}$ of an apple be divided into 5 equal parts, one of these parts is $\frac{1}{55}$ of an apple; and, if $\frac{1}{5}$ of $\frac{1}{11}$ be $\frac{1}{55}$, it is evident that $\frac{1}{5}$ of $\frac{7}{11}$ will be 7 times as much. 7 times $\frac{1}{55}$ is $\frac{7}{55}$; and, if $\frac{1}{5}$ of $\frac{7}{11}$ be $\frac{7}{55}$, $\frac{4}{5}$ of $\frac{7}{11}$ will be 4 times as much. 4 times $\frac{7}{55}$ are $\frac{28}{55}$.

Or, by multiplying the denominator of $\frac{7}{11}$ by 5, the denominator of $\frac{4}{5}$, it is evident we obtain $\frac{1}{5}$ of $\frac{7}{11} = \frac{7}{55}$, since the parts into which the number or thing is divided are 5 times as many, and consequently only $\frac{1}{5}$ as large as before. Again, since $\frac{1}{5}$

QUESTIONS. — Art. 137. What is the rule for reducing improper fractions to whole or mixed numbers? Give a reason for the rule. — Art. 138. How do you reduce a compound fraction to a simple one? Give the reason for the operation.

of $\frac{7}{11}=\frac{7}{55}$, $\frac{4}{5}$ of $\frac{7}{11}$ will be 4 times as much; and 4 times $\frac{7}{55}=\frac{28}{55}$. This process will be seen to be precisely like the operation.

Ex. 2. Reduce $\frac{3}{4}$ of $\frac{4}{5}$ of $\frac{5}{7}$ of $\frac{6}{9}$ of $\frac{7}{11}$ to a simple fraction. Ans. $\frac{2}{11}$.

OPERATION BY CANCELLATION.

$$\frac{\cancel{3}\times\cancel{4}\times\cancel{5}\times\overset{2}{\cancel{6}}\times\cancel{7}}{\cancel{4}\times\cancel{5}\times\cancel{7}\times\underset{\cancel{3}}{\cancel{9}}\times 11}=\frac{2}{11}$$

Since some of the numerators and denominators to be multiplied together are alike, we may cancel these common factors, according to the principles of cancellation.

RULE. — *Multiply all the numerators together for a new numerator, and all the denominators for a new denominator.*

NOTE 1. — All whole and mixed numbers in the compound fraction must be reduced to improper fractions, before multiplying the numerators and denominators.

NOTE 2. — When there are factors common to both numerator and denominator, they may be cancelled in the operation.

EXAMPLES FOR PRACTICE.

3. What is $\frac{2}{3}$ of $\frac{4}{5}$ of $\frac{6}{7}$? Ans. $\frac{48}{105}=\frac{16}{35}$.
4. What is $\frac{7}{8}$ of $\frac{9}{11}$ of 7? Ans. $5\frac{1}{88}$.
5. What is $\frac{7}{8}$ of $\frac{9}{11}$ of $\frac{3}{8}$ of $\frac{4}{7}$? Ans. $\frac{27}{176}$.
6. Change $\frac{11}{17}$ of $\frac{1}{2}$ of $\frac{3}{4}$ of $\frac{1}{20}$ of 7 to a simple fraction. Ans. $\frac{231}{2720}$.
7. Required the value of $\frac{3}{5}$ of $\frac{4}{11}$ of $\frac{11}{17}$ of $\frac{17}{23}$ of $5\frac{3}{4}$. Ans. $\frac{3}{5}$.
8. Reduce $\frac{1}{5}$ of $\frac{8}{9}$ of $\frac{9}{11}$ of $\frac{5}{8}$ of $\frac{3}{7}$ to a simple fraction. Ans. $\frac{3}{77}$.
9. Reduce $\frac{3}{7}$ of $\frac{4}{11}$ of $\frac{7}{9}$ of $\frac{9}{10}$ of $4\frac{1}{3}$ to a simple fraction. Ans. $\frac{26}{55}$.
10. Reduce $\frac{15}{16}$ of $\frac{8}{9}$ of $\frac{7}{11}$ to a simple fraction. Ans. $\frac{35}{66}$.
11. Reduce $\frac{8}{11}$ of $\frac{22}{35}$ of $\frac{15}{22}$ of $9\frac{5}{8}$ to a whole number. Ans. 3.
12. Reduce $\frac{5}{7}$ of $\frac{3}{15}$ of $\frac{4}{16}$ of $8\frac{3}{4}$ of $\frac{11}{5}$ to a simple fraction. Ans. $\frac{11}{16}$.

QUESTIONS. — When there are common factors in the numerator and denominator, how may the operation be shortened? What is the rule? What must be done with all whole and mixed numbers in the compound fraction? How may the operation be shortened by cancelling?

A Common Denominator.

Art. 139. A common denominator of two or more fractions is a common multiple of their denominators. The *least* common denominator is the least common multiple.

Note. — Fractions have a common denominator, when all their denominators are alike.

Art. 140. To reduce fractions to a common denominator.

Ex. 1. Reduce $\frac{3}{4}$, $\frac{5}{6}$, and $\frac{7}{8}$, to a common denominator.
Ans. $\frac{144}{192}$, $\frac{160}{192}$, $\frac{168}{192}$.

OPERATION.

$3 \times 6 \times 8 = 144$ new numerator for $\frac{3}{4} = \frac{144}{192}$.
$5 \times 4 \times 8 = 160$ " " " $\frac{5}{6} = \frac{160}{192}$.
$7 \times 4 \times 6 = 168$ " " " $\frac{7}{8} = \frac{168}{192}$.
$4 \times 6 \times 8 = 192$ common denominator.

We first multiply the numerator of $\frac{3}{4}$ by the denominators 6 and 8, and obtain 144 for its numerator. We next multiply the numerator of $\frac{5}{6}$ by the denominators 4 and 8, and obtain 160 for its numerator; and then we multiply the numerator of $\frac{7}{8}$ by the denominators 4 and 6, and obtain 168 for its numerator. Finally, we multiply all the denominators together for a *common denominator*, and write it under the several numerators, as in the operation.

By this process, since the numerator and denominator of each fraction are multiplied by the same numbers, only the form of the fraction is changed, while the quotient arising from dividing the numerator by the denominator, or the value of the fraction (Art. 133), remains the same. Therefore,

Multiplying the numerator and denominator of a fraction by the same number does not alter the value of the fraction.

Rule. — *Multiply each numerator by all the denominators except its own, for the new numerators; and all the denominators together for a common denominator.*

Note 1. — Compound fractions, if any, must first be reduced to simple ones, and whole or mixed numbers to improper fractions.

Note 2. — Fractions may often be reduced to lower terms, without destroying their common denominator, by dividing all their numerators and denominators by a common divisor.

Questions. — Art. 139. What is a common denominator of two or more fractions? What is the least common denominator? When have fractions a common denominator? — Art. 140. How do you find a common denominator of two or more fractions? Give the reason of the operation. What inference is drawn from it? What is the rule for finding a common denominator? How may fractions having a common denominator be reduced to lower terms?

EXAMPLES FOR PRACTICE.

2. Reduce $\frac{3}{4}$ and $\frac{5}{6}$ to common denominators.
Ans. $\frac{18}{24}$, $\frac{20}{24}$, or $\frac{9}{12}$, $\frac{10}{12}$.

3. Reduce $\frac{7}{9}$, $\frac{4}{5}$, and $\frac{1}{2}$, to a common denominator.
Ans. $\frac{70}{90}$, $\frac{72}{90}$, $\frac{45}{90}$.

4. Reduce $\frac{4}{7}$, $\frac{3}{8}$, and $\frac{5}{11}$, to a common denominator.
Ans. $\frac{352}{616}$, $\frac{231}{616}$, $\frac{280}{616}$.

5. Reduce $\frac{8}{9}$, $\frac{5}{12}$, and $\frac{2}{3}$, to a common denominator.
Ans. $\frac{288}{324}$, $\frac{135}{324}$, $\frac{216}{324}$, or $\frac{32}{36}$, $\frac{15}{36}$, $\frac{24}{36}$.

6. Reduce $\frac{1}{6}$, $\frac{2}{5}$, $\frac{7}{8}$, and $\frac{1}{4}$, to a common denominator.
Ans. $\frac{160}{960}$, $\frac{384}{960}$, $\frac{840}{960}$, $\frac{240}{960}$, or $\frac{20}{120}$, $\frac{48}{120}$, $\frac{105}{120}$, $\frac{30}{120}$.

ART. 141. To reduce fractions to their least common denominator.

Ex. 1. Reduce $\frac{2}{3}$, $\frac{5}{6}$, and $\frac{7}{12}$, to the least common denominator.

OPERATION.

```
3|3  6  12        |12 common denominator.
 |--------        |----
2|1  2   4       3| 4 × 2 =  8 numerator for 2/3 = 8/12.
 |--------       6| 2 × 5 = 10 numerator for 5/6 = 10/12.
  1  1   2      12| 1 × 7 =  7 numerator for 7/12 = 7/12.
```

$3 \times 2 \times 2 = 12$, the least common denominator.

Having first obtained a common multiple, or denominator of the given fractions, we take the part of it expressed by each of these fractions separately for their new numerators. Thus, to get a new numerator for $\frac{2}{3}$, we take $\frac{2}{3}$ of 12, the common denominator, by dividing it by 3, and multiplying the quotient 4 by 2. We proceed in this manner with each of the fractions, and write the numerators thus obtained over the common denominator.

NOTE. — The change in the terms of the fractions, in reducing them to the *least* common denominator by this process, depends upon the same principle as explained in the preceding article.

RULE. — 1. *Find the least common multiple of the denominators for the least common denominator.*

2. *Divide the least common denominator by the denominator of each of the given fractions, and multiply the quotients by their respective numerators, for the new numerators.*

NOTE. — Compound fractions must be reduced to simple ones, whole

QUESTIONS. — Art. 141. How do you find the least common denominator of two or more fractions? Upon what principle does this process depend? What is the rule for reducing fractions to their least common denominator? What must be done with compound fractions, whole numbers, and mixed numbers?

and mixed numbers to improper fractions, and all to their lowest terms before finding the least common denominator

EXAMPLES FOR PRACTICE.

2. Reduce $\frac{3}{4}$, $\frac{4}{5}$, $\frac{5}{6}$, and $\frac{7}{8}$, to the least common denominator. Ans. $\frac{90}{120}$, $\frac{96}{120}$, $\frac{100}{120}$, $\frac{105}{120}$.

3. Reduce $\frac{3}{4}$, $\frac{2}{5}$, $\frac{4}{9}$, and $\frac{2}{11}$, to the least common denominator. Ans. $\frac{1485}{1980}$, $\frac{792}{1980}$, $\frac{880}{1980}$, $\frac{360}{1980}$.

4. Reduce $\frac{7}{8}$, $\frac{9}{10}$, and $7\frac{3}{4}$, to the least common denominator. Ans. $\frac{35}{40}$, $\frac{36}{40}$, $\frac{310}{40}$.

5. Reduce $\frac{3}{7}$, $\frac{9}{14}$, $\frac{11}{28}$, and $5\frac{3}{7}$, to the least common denominator. Ans. $\frac{12}{28}$, $\frac{18}{28}$, $\frac{11}{28}$, $\frac{152}{28}$.

6. Reduce $\frac{1}{2}$, $\frac{3}{4}$, $\frac{5}{6}$, $\frac{5}{8}$, $\frac{7}{8}$, and $\frac{5}{12}$, to the least common denominator. Ans. $\frac{12}{24}$, $\frac{18}{24}$, $\frac{20}{24}$, $\frac{15}{24}$, $\frac{21}{24}$, $\frac{10}{24}$.

7. Reduce $\frac{4}{9}$, $\frac{2}{3}$, $\frac{1}{3}$, $\frac{1}{4}$, $\frac{1}{6}$, and $\frac{1}{12}$, to the least common denominator. Ans. $\frac{16}{36}$, $\frac{24}{36}$, $\frac{12}{36}$, $\frac{9}{36}$, $\frac{6}{36}$, $\frac{3}{36}$.

8. Reduce $\frac{5}{6}$, $\frac{4}{9}$, and $\frac{7}{12}$, to the least common denominator. Ans. $\frac{30}{36}$, $\frac{16}{36}$, $\frac{21}{36}$.

9. Reduce $7\frac{3}{4}$, $5\frac{6}{11}$, 7, and 8, to the least common denominator. Ans. $\frac{341}{44}$, $\frac{244}{44}$, $\frac{308}{44}$, $\frac{352}{44}$.

10. Reduce $\frac{3}{4}$, 4, 5, 7, and 9, to the least common denominator. Ans. $\frac{3}{4}$, $\frac{16}{4}$, $\frac{20}{4}$, $\frac{28}{4}$, $\frac{36}{4}$.

ADDITION OF COMMON FRACTIONS.

ART. **142.** ADDITION of Fractions is the process of finding the value of two or more fractions in one sum.

ART. **143.** To add fractions that have a common denominator.

Ex. 1. Add $\frac{1}{7}$, $\frac{2}{7}$, $\frac{4}{7}$, $\frac{5}{7}$, and $\frac{6}{7}$, together. Ans. $2\frac{4}{7}$.

OPERATION.

$$\frac{1}{7}+\frac{2}{7}+\frac{4}{7}+\frac{5}{7}+\frac{6}{7}=\frac{18}{7}=2\frac{4}{7}.$$

These fractions all being *sevenths*, that is, having 7 for a common denominator, we add their numerators together, and write their sum, 18, over the common denominator, 7. Thus we obtain $\frac{18}{7}=2\frac{4}{7}$, the sum required.

Hence, to add fractions having a common denominator,

Write their sum over the common denominator, and reduce the fraction, if necessary.

QUESTIONS. — Art. 142. What is addition of fractions? — Art. 143. How are fractions having a common denominator added? Give the reason.

EXAMPLES FOR PRACTICE.

2. Add $\frac{4}{11}$, $\frac{5}{11}$, $\frac{7}{11}$, $\frac{8}{11}$, $\frac{9}{11}$, and $\frac{10}{11}$, together. Ans. $3\frac{10}{11}$.
3. Add $\frac{4}{17}$, $\frac{3}{17}$, $\frac{8}{17}$, $\frac{9}{17}$, and $\frac{11}{17}$, together. Ans. $2\frac{1}{17}$.
4. Add $\frac{3}{25}$, $\frac{8}{25}$, $\frac{19}{25}$, and $\frac{21}{25}$, together. Ans. $2\frac{1}{25}$.
5. Add $\frac{17}{47}$, $\frac{18}{47}$, $\frac{37}{47}$, and $\frac{41}{47}$, together. Ans. $2\frac{19}{47}$.
6. Add $\frac{119}{137}$, $\frac{111}{137}$, and $\frac{19}{137}$, together. Ans. $1\frac{112}{137}$.
7. Add $\frac{1678}{1771}$, $\frac{1135}{1771}$, and $\frac{19}{1771}$, together. Ans. $1\frac{1061}{1771}$.

ART. **144.** To add fractions that have not a common denominator.

Ex. 1. What is the sum of $\frac{5}{6}$, $\frac{3}{8}$, and $\frac{7}{12}$? Ans. $1\frac{19}{24}$.

OPERATION.

2	6	8	12
3	3	4	6
2	1	4	2
	1	2	1

$2 \times 3 \times 2 \times 2 = 24$.

24 common denominator.

6	$4 \times 5 = 20$	new numerators.
8	$3 \times 3 = 9$	
12	$2 \times 7 = 14$	

Sum of numerators, 43; Com. denominator, 24: $\frac{43}{24} = 1\frac{19}{24}$.

Having found the common denominator and new numerators, as in Art. 141, we add the numerators together, and write their sum over the common denominator, and reduce the fraction.

RULE. — *Reduce the given fractions to a common denominator. Add the numerators, and write their sum over the common denominator.*

NOTE 1. — Mixed numbers must be reduced to improper fractions, and compound fractions to simple fractions, and each fraction to its lowest terms, before attempting to find their common denominator.

NOTE 2. — In adding mixed numbers, when deemed most convenient, the fractional parts may be added separately, and their sum added to the amount of the whole numbers.

EXAMPLES FOR PRACTICE.

2. What is the sum of $\frac{5}{8}$, $\frac{11}{12}$, and $\frac{13}{16}$? Ans. $2\frac{17}{48}$.
3. What is the sum of $\frac{9}{20}$, $\frac{11}{18}$, and $\frac{5}{14}$? Ans. $1\frac{527}{1260}$.
4. What is the sum of $\frac{19}{21}$ and $\frac{31}{37}$? Ans. $1\frac{577}{777}$.
5. What is the sum of $\frac{3}{4}$, $\frac{5}{6}$, $\frac{3}{8}$, and $\frac{1}{12}$? Ans. $2\frac{1}{24}$.
6. Add $\frac{4}{9}$, $\frac{8}{21}$, $\frac{11}{24}$, and $\frac{1}{2}$, together. Ans. $1\frac{395}{504}$.
7. Add $\frac{19}{72}$, $\frac{51}{84}$, and $\frac{71}{96}$, together. Ans. $1\frac{1231}{2016}$.
8. Add $\frac{3}{25}$, $\frac{49}{50}$, $\frac{74}{75}$, and $\frac{81}{100}$, together. Ans. $2\frac{269}{300}$.

QUESTIONS. — Art. 144. What is the rule for adding fractions not having a common denominator? How may mixed numbers be conveniently added?

9. Add $\frac{1}{2}$, $\frac{2}{3}$, $\frac{3}{4}$, $\frac{4}{5}$, $\frac{5}{6}$, $\frac{6}{7}$, and $\frac{7}{8}$, together. Ans. $5\frac{79}{280}$.

10. Add $\frac{8}{9}$, $\frac{9}{10}$, $\frac{10}{11}$, $\frac{11}{12}$, $\frac{12}{13}$, $\frac{13}{14}$, and $\frac{14}{15}$, together. Ans. $6\frac{144401}{360360}$.

11. Add $\frac{2}{3}$ of $\frac{3}{4}$ to $\frac{5}{6}$ of $\frac{7}{8}$. Ans. $1\frac{11}{48}$.

12. Add $\frac{3}{4}$ of $\frac{7}{8}$ to $\frac{11}{12}$ of $\frac{1}{2}$. Ans. $1\frac{11}{96}$.

13. Add $\frac{1}{9}$ of $\frac{2}{3}$ to $\frac{1}{5}$ of $\frac{7}{10}$. Ans. $\frac{289}{1350}$.

14. Add $\frac{2}{3}$ of $\frac{3}{4}$ of $\frac{4}{5}$ to $\frac{5}{6}$ of $\frac{6}{7}$ of $\frac{7}{10}$. Ans. $\frac{9}{10}$.

15. Add $\frac{1}{3}$ of $\frac{3}{11}$ of $\frac{11}{12}$ to $\frac{1}{2}$ of $\frac{2}{9}$. Ans. $\frac{7}{36}$.

16. Add $3\frac{3}{7}$ to $4\frac{11}{14}$. Ans. $8\frac{3}{14}$.

17. Add $4\frac{3}{4}$ to $5\frac{6}{7}$. Ans. $10\frac{17}{28}$.

18. Add $17\frac{3}{4}$ to $18\frac{5}{12}$. Ans. $36\frac{1}{6}$.

ART. 145. To add any two fractions whose numerators are a unit.

Ex. 1. Add $\frac{1}{4}$ to $\frac{1}{5}$. Ans. $\frac{9}{20}$

OPERATION.

Sum of the denominators, $4 + 5 = 9$
Product of the denominators, $\overline{4 \times 5 = 20}$

We first find the product of the denominators, which is 20, and then their sum, which is 9, and write the former for the denominator of the required fraction, and the latter for the numerator.

The reason of this operation will be seen, when we consider that the process reduces the fractions to a common denominator, and then adds their numerators. Hence, to add two fractions whose numerators are a unit, simply

Write the sum of the given denominators over their product.

EXAMPLES FOR PRACTICE.

2. Add $\frac{1}{4}$ to $\frac{1}{6}$, $\frac{1}{7}$ to $\frac{1}{8}$, $\frac{1}{2}$ to $\frac{1}{3}$, $\frac{1}{5}$ to $\frac{1}{6}$, $\frac{1}{2}$ to $\frac{1}{9}$.
3. Add $\frac{1}{2}$ to $\frac{1}{11}$, $\frac{1}{8}$ to $\frac{1}{9}$, $\frac{1}{7}$ to $\frac{1}{6}$, $\frac{1}{5}$ to $\frac{1}{12}$, $\frac{1}{3}$ to $\frac{1}{10}$, $\frac{1}{8}$ to $\frac{1}{4}$.
4. Add $\frac{1}{5}$ to $\frac{1}{7}$, $\frac{1}{7}$ to $\frac{1}{12}$, $\frac{1}{6}$ to $\frac{1}{3}$, $\frac{1}{8}$ to $\frac{1}{12}$, $\frac{1}{3}$ to $\frac{1}{5}$, $\frac{1}{11}$ to $\frac{1}{12}$.
5. Add $\frac{1}{3}$ to $\frac{1}{11}$, $\frac{1}{3}$ to $\frac{1}{12}$, $\frac{1}{3}$ to $\frac{1}{10}$, $\frac{1}{6}$ to $\frac{1}{2}$, $\frac{1}{6}$ to $\frac{1}{5}$, $\frac{1}{6}$ to $\frac{1}{6}$.
6. Add $\frac{1}{7}$ to $\frac{1}{3}$, $\frac{1}{7}$ to $\frac{1}{4}$, $\frac{1}{7}$ to $\frac{1}{5}$, $\frac{1}{7}$ to $\frac{1}{6}$, $\frac{1}{7}$ to $\frac{1}{7}$, $\frac{1}{7}$ to $\frac{1}{111}$.
7. Add $\frac{1}{8}$ to $\frac{1}{7}$, $\frac{1}{8}$ to $\frac{1}{8}$, $\frac{1}{8}$ to $\frac{1}{9}$, $\frac{1}{8}$ to $\frac{1}{10}$, $\frac{1}{8}$ to $\frac{1}{11}$, $\frac{1}{8}$ to $\frac{1}{12}$.

SUBTRACTION OF COMMON FRACTIONS.

ART. 146. SUBTRACTION of Fractions is the process of finding the difference between two fractions.

QUESTIONS. — Art. 145. How can you add two fractions when the numerators are a unit? What is the reason for this? — Art. 146. What is subtraction of fractions?

ART. **147.** To subtract fractions that have a common denominator.

Ex. 1. From $\frac{7}{9}$ take $\frac{2}{9}$. Ans. $\frac{5}{9}$.

OPERATION. $\frac{7}{9}-\frac{2}{9}=\frac{5}{9}$ The fractions both being *ninths*, having 9 for a common denominator, we subtract the less numerator from the greater, and write the difference, 5, over the common denominator, 9. Thus we have $\frac{5}{9}$ as the difference required. Hence, to subtract fractions having a common denominator,

Write the difference of their numerators over the common denominators, and reduce the fraction, if necessary.

EXAMPLES FOR PRACTICE.

2. From $\frac{7}{11}$ take $\frac{2}{11}$. Ans. $\frac{5}{11}$.
3. From $\frac{11}{19}$ take $\frac{7}{19}$. Ans. $\frac{4}{19}$.
4. From $\frac{27}{37}$ take $\frac{4}{37}$. Ans. $\frac{23}{37}$.
5. From $\frac{167}{711}$ take $\frac{19}{711}$. Ans. $\frac{148}{711}$.
6. From $\frac{628}{1728}$ take $\frac{150}{1728}$. Ans. $\frac{239}{864}$.
7. From $\frac{7}{20}$ take $\frac{5}{20}$. Ans. $\frac{1}{10}$.
8. From $\frac{97}{100}$ take $\frac{17}{100}$. Ans. $\frac{4}{5}$.

ART. **148.** To subtract fractions that have not a common denominator.

Ex. 1. From $\frac{13}{16}$ take $\frac{7}{12}$. Ans. $\frac{11}{48}$.

OPERATION.

$$
\begin{array}{r|rr}
4 & 16 & 12 \\
\hline
 & 4 & 3
\end{array}
\qquad
\begin{array}{r|l}
 & 48 \text{ common denominator.} \\
\hline
16 & 3\times13=39 \\
12 & 4\times\ \ 7=28
\end{array}
\Big\}\text{ new numerators.}
$$

$4\times4\times3=48.$

$$
\begin{array}{l}
11 \text{ difference of numerators.} \\
\hline
48 \text{ common denominator.}
\end{array}
$$

Having found the common denominator and new numerators as in Art. 141, we subtract the less numerator from the greater, and place the difference over the common denominator.

RULE. — *Reduce the fractions to a common denominator, then write the difference of the numerators over the common denominator.*

NOTE. — If the minuend or subtrahend, or both, are compound fractions, they must be reduced to simple ones.

QUESTIONS. — Art. 147. How do you subtract fractions having a common denominator? — Art. 148. What is the rule for subtracting fractions not having a common denominator? If the minuend or subtrahend is a compound fraction, what must be done?

EXAMPLES FOR PRACTICE.

2. From $\frac{7}{18}$ take $\frac{4}{21}$. Ans. $\frac{25}{126}$.
3. From $\frac{19}{20}$ take $\frac{11}{16}$. Ans. $\frac{21}{80}$.
4. From $\frac{17}{24}$ take $\frac{7}{20}$. Ans. $\frac{43}{120}$.
5. From $\frac{11}{34}$ take $\frac{1}{10}$. Ans. $\frac{19}{85}$.
6. From $\frac{31}{36}$ take $\frac{9}{16}$. Ans. $\frac{43}{144}$.
7. From $\frac{18}{37}$ take $\frac{3}{11}$. Ans. $\frac{87}{407}$.
8. From $\frac{111}{200}$ take $\frac{1}{19}$. Ans. $\frac{1909}{3800}$.
9. From $\frac{1}{10}$ take $\frac{1}{1000}$. Ans. $\frac{99}{1000}$.
10. From $\frac{2}{3}$ of $\frac{9}{11}$ take $\frac{1}{4}$ of $\frac{2}{7}$. Ans. $\frac{73}{154}$.
11. From $\frac{1}{9}$ of $\frac{9}{10}$ take $\frac{1}{12}$ of $\frac{12}{13}$. Ans. $\frac{3}{130}$.
12. From $\frac{3}{8}$ of $12\frac{5}{6}$ take $\frac{2}{5}$ of $9\frac{7}{12}$. Ans. $\frac{47}{48}$.

ART. **149.** To subtract a proper fraction or a mixed number from a whole number.

Ex. 1. From 16 take $2\frac{1}{4}$. Ans. $13\frac{3}{4}$.

OPERATION.	
From	16
Take	$2\frac{1}{4}$
Rem.	$13\frac{3}{4}$

Since we have no fraction from which to subtract the $\frac{1}{4}$, we must add 1, equal to $\frac{4}{4}$, to the minuend, and say $\frac{1}{4}$ from $\frac{4}{4}$ leaves $\frac{3}{4}$. We write the $\frac{3}{4}$ below the line, and carry 1 to the 2 in the subtrahend, and subtract as in subtraction of simple numbers.

The same result will be obtained, if we

Subtract the numerator from the denominator of the fraction, and under the remainder write the denominator, and carry one to the subtrahend to be subtracted from the minuend.

NOTE. — When the subtrahend is a mixed number, we may, if we choose, reduce it to an improper fraction, and change the whole number in the minuend to a fraction having the same denominator, and then proceed as in Art. 148.

EXAMPLES FOR PRACTICE.

	2.	3.	4.	5.	6.
From	12	19	13	14	17
Take	$4\frac{3}{4}$	$3\frac{3}{7}$	$9\frac{1}{11}$	$8\frac{3}{7}$	$6\frac{11}{12}$
Ans.	$7\frac{1}{4}$	$15\frac{4}{7}$	$3\frac{10}{11}$	$5\frac{4}{7}$	$10\frac{1}{12}$

7. From 23 take $13\frac{1}{3}$. Ans. $9\frac{2}{3}$.
8. From 47 take $\frac{13}{20}$. Ans. $46\frac{7}{20}$.
9. From 139 take $75\frac{11}{25}$. Ans. $63\frac{14}{25}$.

QUESTIONS. — Art. 149. How do you subtract a proper fraction or mixed number from a whole number? Give the reason for this rule.

ART. **150.** To subtract a mixed number from a mixed number.

Ex. 1. From $9\frac{2}{7}$ take $3\frac{3}{5}$. Ans. $5\frac{24}{35}$.

FIRST OPERATION.

From $9\frac{2}{7} = 9\frac{10}{35}$
Take $3\frac{3}{5} = 3\frac{21}{35}$
Rem. $5\frac{24}{35}$

We first reduce the fractional parts to a common denominator by multiplying the terms of the fraction $\frac{2}{7}$ by 5, the denominator of the other, thus: $\frac{2\times 5 = 10}{7\times 5 = 35}$; and then the terms of the fraction $\frac{3}{5}$ by 7, the denominator of the first, thus: $\frac{3\times 7 = 21}{5\times 7 = 35}$. Now, since we cannot take $\frac{21}{35}$ from $\frac{10}{35}$, we add 1, equal to $\frac{35}{35}$, to the $\frac{10}{35}$ in the minuend, and obtain $\frac{45}{35}$. We next subtract $\frac{21}{35}$ from $\frac{45}{35}$, and write the remainder, $\frac{24}{35}$, below the line, and carry 1 to the 3 in the subtrahend, and subtract as in simple numbers.

SECOND OPERATION.

From $9\frac{2}{7} = \frac{65}{7} = \frac{325}{35}$
Take $3\frac{3}{5} = \frac{18}{5} = \frac{126}{35}$
Rem. $\frac{199}{35} = 5\frac{24}{35}$

In this operation, we reduce the mixed numbers to improper fractions, and these fractions to a common denominator, as in the first operation. We then subtract the less fraction from the greater, and, reducing the remainder to a mixed number, obtain $5\frac{24}{35}$, as before. Hence, in performing like examples, we may

Reduce the fractional parts, if necessary, to a common denominator, and subtract the fractional part of the subtrahend from that of the minuend, as in Art. 147; *remembering to increase the fractional part of the minuend, when otherwise it would be less than that of the subtrahend, before subtracting, by as many fractional units as it takes to make a unit of the fraction* (Art. 131), *and carry* 1 *to the whole number of the subtrahend before subtracting it from the whole number of the minuend.* Or,

Reduce the mixed numbers to improper fractions, then to a common denominator, and subtract the less fraction from the greater.

EXAMPLES FOR PRACTICE.

	2.	3.	4.	5.	6.
From	$9\frac{7}{8}$	$7\frac{1}{4}$	$8\frac{3}{7}$	$9\frac{1}{4}$	$10\frac{3}{4}$
Take	$5\frac{11}{12}$	$3\frac{7}{9}$	$4\frac{4}{5}$	$3\frac{7}{8}$	$10\frac{1}{19}$
Ans.	$3\frac{23}{24}$	$3\frac{17}{36}$	$3\frac{22}{35}$	$5\frac{3}{8}$	$\frac{53}{76}$

QUESTIONS. — Art. 150. How do you reduce the fractions of mixed numbers to a common denominator? How does it appear that this process reduces them to a common denominator? How do you then proceed? What other method of subtracting mixed numbers? How may all like examples be performed?

	7.	8.	9.	10.	11.
From	$12\frac{3}{7}$	$16\frac{3}{11}$	$19\frac{3}{5}$	$97\frac{1}{4}$	$87\frac{11}{23}$
Take	$9\frac{1}{2}$	$5\frac{6}{7}$	$15\frac{6}{7}$	$18\frac{3}{11}$	$19\frac{3}{4}$
Ans.	$2\frac{13}{14}$	$10\frac{32}{77}$	$3\frac{26}{35}$	$78\frac{43}{44}$	$67\frac{67}{92}$

12. From $19\frac{1}{6}$ take $7\frac{3}{11}$. Ans. $11\frac{59}{66}$.
13. From $15\frac{1}{7}$ take $8\frac{1}{4}$. Ans. $6\frac{25}{28}$.
14. From $9\frac{1}{13}$ take $3\frac{18}{19}$. Ans. $5\frac{32}{247}$.
15. From $71\frac{1}{15}$ take $13\frac{7}{12}$. Ans. $57\frac{29}{60}$.
16. From $61\frac{11}{17}$ take $33\frac{13}{14}$. Ans. $27\frac{171}{238}$.
17. From a hogshead of wine there leaked out $12\frac{3}{8}$ gallons; how much remained? Ans. $50\frac{5}{8}$ gallons.
18. From \$10, \$$2\frac{1}{8}$ were given to Benjamin, \$$3\frac{1}{4}$ to Lydia \$$1\frac{1}{2}$ to Emily, and the remainder to Betsey; what did she receive? Ans. \$$3\frac{1}{8}$.

ART. 151. To subtract one fraction from another, when both fractions have a unit for a numerator.

Ex. 1. What is the difference between $\frac{1}{3}$ and $\frac{1}{7}$? Ans. $\frac{4}{21}$.

OPERATION.

Difference of the denominators, $7 - 3 = 4$

Product of the denominators, $7 \times 3 = 21$

We first find the product of the denominators, which is 21, and then their difference, which is 4, and write the former for the denominator of the required fraction, and the latter for the numerator. By this process the fractions are reduced to a common denominator, and their difference found. Hence, to subtract such fractions, we may simply

Write the difference of the denominators over their product.

EXAMPLES FOR PRACTICE.

2. Take $\frac{1}{6}$ from $\frac{1}{3}$, $\frac{1}{7}$ from $\frac{1}{2}$, $\frac{1}{9}$ from $\frac{1}{4}$, $\frac{1}{7}$ from $\frac{1}{3}$.
3. Take $\frac{1}{9}$ from $\frac{1}{4}$, $\frac{1}{8}$ from $\frac{1}{7}$, $\frac{1}{4}$ from $\frac{1}{3}$, $\frac{1}{5}$ from $\frac{1}{4}$.
4. Take $\frac{1}{8}$ from $\frac{1}{3}$, $\frac{1}{7}$ from $\frac{1}{2}$, $\frac{1}{12}$ from $\frac{1}{7}$, $\frac{1}{11}$ from $\frac{1}{3}$.
5. Take $\frac{1}{6}$ from $\frac{1}{5}$, $\frac{1}{8}$ from $\frac{1}{7}$, $\frac{1}{10}$ from $\frac{1}{9}$, $\frac{1}{7}$ from $\frac{1}{6}$.
6. Take $\frac{1}{7}$ from $\frac{1}{6}$, $\frac{1}{9}$ from $\frac{1}{8}$, $\frac{1}{12}$ from $\frac{1}{11}$, $\frac{1}{4}$ from $\frac{1}{3}$.
7. Take $\frac{1}{5}$ from $\frac{1}{3}$, $\frac{1}{5}$ from $\frac{1}{4}$, $\frac{1}{9}$ from $\frac{1}{6}$, $\frac{1}{12}$ from $\frac{1}{8}$.

QUESTIONS. — Art. 151. How do you subtract one fraction from another when both fractions have a unit for a numerator? What is the reason for this process?

MULTIPLICATION OF COMMON FRACTIONS.

ART. 152. MULTIPLICATION of Fractions is the process of taking one number as many times as there are units in another, when one or both of the numbers are fractions.

ART. 153. To multiply a fraction by a whole number.

Ex. 1. Multiply $\frac{7}{8}$ by 4. Ans. $3\frac{1}{2}$.

FIRST OPERATION.

$\frac{7}{8} \times 4 = \frac{28}{8} = 3\frac{1}{2}$

In the first operation we multiply the numerator of the fraction by the whole number, and obtain $3\frac{1}{2}$ for the answer. It is evident that the fraction $\frac{7}{8}$ is *multiplied* by *multiplying* its *numerator* by 4, since the parts taken are 4 times as many as before, while the parts into which the number or thing is divided remain the same. Therefore,

Multiplying the numerator of a fraction by any number multiplies the fraction by that number.

SECOND OPERATION.

$\frac{7}{8} \times 4 = \frac{7}{2} = 3\frac{1}{2}$

In the second operation we divide the denominator of the fraction by the whole number, and obtain $3\frac{1}{2}$ for the answer, as before. It is evident, also, that the fraction $\frac{7}{8}$ is *multiplied* by *dividing* its *denominator* by 4, since the parts into which the number or thing is divided are only $\frac{1}{4}$ as many, and consequently 4 times as *large*, as before, while the parts taken remain the same. Therefore,

Dividing the denominator of a fraction by any number multiplies the fraction by that number.

RULE. — *Multiply the numerator of the fraction by the whole number.* Or,

Divide the denominator of the fraction by the whole number, when it can be done without a remainder.

EXAMPLES FOR PRACTICE.

2. Multiply $\frac{5}{7}$ by 9. Ans. $6\frac{3}{7}$.
3. Multiply $\frac{8}{15}$ by 5. Ans. $2\frac{2}{3}$.
4. Multiply $\frac{16}{27}$ by 3. Ans. $1\frac{7}{9}$.
5. Multiply $\frac{49}{85}$ by 85. Ans. 49.

QUESTIONS. — Art. 152. What is multiplication of fractions? — Art. 153. How is a fraction multiplied, by the first operation? Give the reason of the operation. What inference is drawn from it? How is a fraction multiplied, by the second operation? What is the reason of the operation? What inference is drawn from it? What is the rule for multiplying a fraction by a whole number?

6. Multiply $\frac{11}{12}$ by 83. Ans. $76\frac{1}{12}$.
7. Multiply $\frac{15}{17}$ by 189. Ans. $166\frac{13}{17}$.
8. Multiply $\frac{113}{117}$ by 365. Ans. $352\frac{61}{117}$.
9. Multiply $\frac{87}{96}$ by 48. Ans. $43\frac{1}{2}$.
10. If a man receive $\frac{3}{8}$ of a dollar for one day's labor, what will he receive for 21 days' labor? Ans. $\$7\frac{7}{8}$.
11. What cost 56lb. of chalk at $\frac{3}{4}$ of a cent per lb.? Ans. $0.42.
12. What cost 396lb. of copperas at $\frac{9}{11}$ of a cent per lb.? Ans. $3.24.
13. What cost 79 bushels of salt at $\frac{7}{8}$ of a dollar per bushel? Ans. $\$69\frac{1}{8}$.

ART. 154. To multiply a whole number by a fraction.

Ex. 1. Multiply 15 by $\frac{3}{5}$. Ans. 9.

FIRST OPERATION.

$$5\,)\,15$$
$$3 \times 3 = 9$$

In the first operation we divide the *whole number* by the denominator of the fraction, and obtain $\frac{1}{5}$ of it. We then multiply this quotient by 3, the numerator of the fraction, and thus obtain $\frac{3}{5}$ of it, which is 9.

SECOND OPERATION.

$$15$$
$$3$$
$$45 \div 5 = 9$$

In the second operation we multiply the *whole number* by the numerator of the fraction, and divide the product by the denominator, and obtain 9 for the answer, as before. Therefore,

Multiplying by a fraction is taking the part of the multiplicand denoted by the multiplier.

RULE. — *Divide the whole number by the denominator of the fraction, when it can be done without a remainder, and multiply the quotient by the numerator.* Or,

Multiply the whole number by the numerator of the fraction, and divide the product by the denominator.

EXAMPLES FOR PRACTICE.

2. Multiply 36 by $\frac{7}{9}$. Ans. 28.
3. Multiply 144 by $\frac{11}{18}$. Ans. 88.
4. Multiply 375 by $\frac{13}{15}$. Ans. 325.
5. Multiply 2277 by $\frac{70}{99}$. Ans. 1610.
6. Multiply 376 by $\frac{11}{17}$. Ans. $243\frac{5}{17}$.

QUESTIONS. — Art. 154. How do you multiply a whole number by a fraction, according to the first operation? How by the second? What inference is drawn from the operation? What is the rule for multiplying a whole number by a fraction?

7. Multiply 471 by $\frac{2}{117}$. Ans. $8\frac{2}{39}$.
8. Multiply 871 by $\frac{1}{37}$. Ans. $23\frac{20}{37}$.
9. Multiply 867 by $\frac{1}{136}$. Ans. $6\frac{51}{136}$.

ART. **155.** To multiply a whole and mixed number together.

Ex. 1. Multiply 17 by $6\frac{3}{4}$. Ans. $114\frac{3}{4}$.

OPERATION.

$$\begin{array}{rr} & 17 \\ & 6\frac{3}{4} \\ \hline & 102 \\ \frac{3}{4} \text{ of } 17 = & 12\frac{3}{4} \\ \hline & 114\frac{3}{4} \end{array}$$

We first multiply 17 by 6, the whole number of the multiplier, and then by the fractional part, $\frac{3}{4}$, which is simply taking $\frac{3}{4}$ of it, and add the two products.

Ex. 2. Multiply $7\frac{3}{5}$ by 4. Ans. $30\frac{2}{5}$.

OPERATION.

$$\begin{array}{rr} & 7\frac{3}{5} \\ & 4 \\ \hline \frac{3}{5} \text{ of } 4 = & 2\frac{2}{5} \\ & 28 \\ \hline & 30\frac{2}{5} \end{array}$$

We first multiply $\frac{3}{5}$ in the multiplicand by 4, the multiplier; thus, 4 times $\frac{3}{5}$ are $\frac{12}{5}$, equal to $2\frac{2}{5}$, which is in effect taking $\frac{3}{5}$ of the multiplier, 4. We then multiply the whole number by 4, and add the two products. Hence, in performing like examples,

Multiply the fractional part and the whole number separately, and add the products.

EXAMPLES FOR PRACTICE.

3. Multiply $9\frac{3}{8}$ by 5. Ans. $46\frac{7}{8}$.
4. Multiply $12\frac{3}{5}$ by 7. Ans. $88\frac{1}{5}$.
5. Multiply 9 by $8\frac{11}{12}$. Ans. $80\frac{1}{4}$.
6. Multiply 10 by $7\frac{1}{9}$. Ans. $71\frac{1}{9}$.
7. Multiply $11\frac{6}{7}$ by 8. Ans. $94\frac{6}{7}$.
8. What cost $7\frac{6}{11}$lb. of beef at 5 cents per pound? Ans. $\$0.37\frac{8}{11}$.
9. What cost $23\frac{7}{12}$bbl. of flour at \$6 per barrel? Ans. $\$141\frac{1}{2}$.
10. What cost $8\frac{3}{8}$yd. of cloth at \$5 per yard? Ans. $\$41\frac{7}{8}$.
11. What cost 9 barrels of vinegar at $\$6\frac{3}{8}$ per barrel? Ans. $\$57\frac{3}{8}$.

QUESTIONS. — Art. 155. What is the rule for multiplying a whole and mixed number together? Does it make any difference which is taken for the multiplier?

12. What cost 12 cords of wood at $6.37\frac{1}{2}$ per cord?
Ans. $76.50.
13. What cost 11cwt. of sugar at $9\frac{3}{8}$ per cwt.?
Ans. $103\frac{1}{8}$.
14. What cost $4\frac{3}{8}$ bushels of rye at $1.75 per bushel?
Ans. $7.65\frac{5}{8}$.
15. What cost 7 tons of hay at $11\frac{7}{8}$ per ton?
Ans. $83\frac{1}{8}$.
16. What cost 9 doz. of adzes at $10\frac{5}{8}$ per doz.?
Ans. $95\frac{5}{8}$.
17. What cost 5 tons of timber at $3\frac{1}{8}$ per ton?
Ans. $15\frac{5}{8}$.
18. What cost 15cwt. of rice at $7.62\frac{1}{2}$ per cwt.?
Ans. $114.37\frac{1}{2}$.
19. What cost 40 tons of coal at $8.37\frac{1}{2}$ per ton?
Ans. $335.

ART. **156.** To multiply a fraction by a fraction.

Ex. 1. Multiply $\frac{7}{9}$ by $\frac{3}{4}$. Ans. $\frac{7}{12}$.

OPERATION.

$$\frac{7}{9} \times \frac{3}{4} = \frac{21}{36} = \frac{7}{12}$$

OPERATION BY CANCELLATION

$$\frac{7}{\not{9}_{3}} \times \frac{\not{3}}{4} = \frac{7}{12}$$

To multiply $\frac{7}{9}$ by $\frac{3}{4}$ is to take $\frac{3}{4}$ of the multiplicand, $\frac{7}{9}$ (Art 154). Now, to obtain $\frac{3}{4}$ of $\frac{7}{9}$, we simply multiply the numerators together for a new numerator, and the denominators together for a new denominator (Art. 138). Therefore,

Multiplying one fraction by another is the same as reducing compound fractions to simple ones.

RULE. — *Multiply the numerators together for a new numerator, and the denominators together for a new denominator.*

NOTE. — We can cancel all the common factors in the numerators, and denominators, and then multiply the remaining factors together as before.

EXAMPLES FOR PRACTICE.

2. Multiply $\frac{7}{8}$ by $\frac{8}{11}$. Ans. $\frac{7}{11}$.
3. Multiply $\frac{5}{11}$ by $\frac{11}{20}$. Ans. $\frac{1}{4}$.

QUESTIONS. — Art. 156. What is the first rule for multiplying one fraction by another? How does it appear that this operation multiplies the fraction of the multiplicand? What is the inference drawn from it? What is the note?

4. Multiply $\frac{8}{13}$ by $\frac{13}{24}$. Ans. $\frac{1}{3}$.
5. Multiply $\frac{18}{19}$ by $\frac{19}{90}$. Ans. $\frac{1}{5}$.
6. Multiply $\frac{15}{17}$ by $\frac{17}{60}$. Ans. $\frac{1}{4}$.
7. Multiply $\frac{1}{9}$ by $\frac{8}{17}$. Ans. $\frac{8}{153}$.
8. Multiply $\frac{6}{23}$ by $\frac{23}{36}$. Ans. $\frac{1}{6}$.
9. What cost $\frac{7}{8}$ of a bushel of corn at $\frac{8}{9}$ of a dollar per bushel? Ans. $\frac{7}{9}$ of a dollar.

10. If a man travels $\frac{8}{11}$ of a mile in an hour, how far would he travel in $\frac{11}{32}$ of an hour? Ans. $\frac{1}{4}$ of a mile.

11. If a bushel of corn will buy $\frac{7}{10}$ of a bushel of salt, how much salt might be bought for $\frac{3}{4}$ of a bushel of corn? Ans. $\frac{21}{40}$ of a bushel.

12. If $\frac{2}{3}$ of $\frac{3}{8}$ of a dollar buy one bushel of corn, what will $\frac{7}{9}$ of $\frac{9}{11}$ of a bushel cost? Ans. $\frac{7}{44}$ of a dollar.

13. If $\frac{3}{9}$ of $\frac{4}{7}$ of $\frac{9}{11}$ of an acre of land cost one dollar, how much may be bought with $\frac{2}{3}$ of \$18? Ans. $1\frac{67}{77}$ acres.

ART. **157.** To multiply a mixed number by a mixed number, it is only necessary to reduce them to improper fractions, and then proceed as in the foregoing rule.

Ex. 1. Multiply $4\frac{3}{5}$ by $6\frac{2}{3}$. Ans. $30\frac{2}{3}$.

OPERATION.

$$4\tfrac{3}{5} = \tfrac{23}{5}; \quad 6\tfrac{2}{3} = \tfrac{20}{3}.$$

$$\frac{23}{\not{5}} \times \frac{\overset{4}{\not{2}\not{0}}}{3} = \frac{92}{3} = 30\tfrac{2}{3}$$

EXAMPLES FOR PRACTICE.

2. Multiply $7\frac{1}{8}$ by $8\frac{3}{7}$. Ans. $60\frac{3}{56}$.
3. Multiply $4\frac{7}{8}$ by $9\frac{1}{4}$. Ans. $45\frac{3}{32}$.
4. Multiply $11\frac{2}{7}$ by $8\frac{4}{5}$. Ans. $99\frac{11}{35}$.
5. Multiply $12\frac{3}{4}$ by $11\frac{5}{9}$. Ans. $147\frac{1}{3}$.
6. What cost $7\frac{3}{4}$ cords of wood at $\$5\frac{3}{8}$ per cord? Ans. $\$41\frac{21}{32}$.
7. What cost $7\frac{3}{8}$yd. of cloth at $\$3\frac{1}{2}$ per yard? Ans. $25\frac{13}{16}$.
8. What cost $6\frac{3}{7}$ gallons of molasses at $23\frac{3}{4}$ cents per gallon? Ans. $\$1.52\frac{19}{28}$.

9. If a man travel $3\frac{4}{9}$ miles in one hour, how far will he travel in $9\frac{7}{8}$ hours? Ans. $34\frac{1}{72}$.

QUESTION. — Art. 157. How do you multiply a mixed number by a mixed number?

10. What cost $361\frac{1}{40}$ acres of land at $\$25\frac{3}{8}$ per acre?
Ans. $\$9167\frac{113}{320}$.

11. How many square rods of land in a garden, which is $97\frac{5}{16}$ rods long, and $49\frac{3}{7}$ rods wide? Ans. $4810\frac{1}{56}$ rods.

DIVISION OF COMMON FRACTIONS.

ART. **158.** DIVISION of Fractions is the process of dividing when the divisor or dividend, or both, are fractions.

ART. **159.** To divide a fraction by a whole number.

Ex. 1. Divide $\frac{8}{9}$ by 4. Ans. $\frac{2}{9}$.

FIRST OPERATION.

$$\frac{8}{9} \div 4 = \frac{2}{9}$$

In this operation we divide the numerator of the fraction by 4, and write the quotient, 2, over the denominator.

It is evident this process divides the fraction by 4, since the number and size of the parts into which the whole number is divided remain the same, while only $\frac{1}{4}$ of the number of parts is expressed by the fraction. Therefore,

Dividing the numerator of a fraction by any number divides the fraction by that number.

Ex. 2. Divide $\frac{5}{7}$ by 9. Ans. $\frac{5}{63}$.

SECOND OPERATION.

$$\frac{5}{7} \div 9 = \frac{5}{63}$$

We multiply the denominator of the fraction by the divisor, 9, and write the product under the numerator.

It is evident this process divides the fraction, since multiplying the denominator by 9 makes the number of parts into which the whole number is divided 9 times as many as before, and consequently each part can have but $\frac{1}{9}$ of its former value. Now, if each part has but $\frac{1}{9}$ of its former value, while only the same number of parts is expressed by the fraction, it is plain the fraction has been divided by 9. Therefore,

Multiplying the denominator of a fraction by any number divides the fraction by that number.

RULE. — *Divide the numerator of the fraction by the whole number, when it can be done without a remainder, and write the quotient over the denominator.* Or,

Multiply the denominator of the fraction by the whole number, and write the product under the numerator.

QUESTIONS. — Art. 158. What is division of common fractions? — Art. 159. How is the fraction divided by the first operation? What inference may be drawn from this operation? How is a fraction divided by the second operation? What inference is drawn from this operation? What is the rule?

EXAMPLES FOR PRACTICE.

3. Divide $\frac{6}{13}$ by 3. Ans. $\frac{2}{13}$.
4. Divide $\frac{18}{19}$ by 6. Ans. $\frac{3}{19}$.
5. Divide $\frac{7}{11}$ by 12. Ans. $\frac{7}{132}$.
6. Divide $\frac{11}{12}$ by 8. Ans. $\frac{11}{96}$.
7. Divide $\frac{27}{43}$ by 9. Ans. $\frac{3}{43}$.
8. Divide $\frac{75}{98}$ by 15. Ans. $\frac{5}{98}$.
9. Divide $\frac{450}{533}$ by 75. Ans. $\frac{6}{533}$.
10. Divide $\frac{7}{9}$ by 12. Ans. $\frac{7}{108}$.
11. John Jones owns $\frac{5}{7}$ of a share in a railroad valued at $117; this he bequeaths to his five children. What part of a share will each receive? Ans. $\frac{1}{7}$.
12. Divide $\frac{9}{23}$ by 15. Ans. $\frac{3}{115}$.
13. Divide $\frac{6}{17}$ by 28. Ans. $\frac{3}{238}$.
14. James Page's estate is valued at $10,000, and he has given $\frac{3}{7}$ of it to the Seamen's Society; $\frac{1}{3}$ of the remainder he gave to his good minister; and the remainder he divided equally among his 4 sons and 3 daughters. What sum will each of his children receive? Ans. 680\frac{40}{147}$.

ART. 160. To divide a whole number by a fraction.

Ex. 1. How many times will 13 contain $\frac{3}{7}$? Ans. $30\frac{1}{3}$.

OPERATION.

$$13 \div \frac{3}{7} = \frac{13 \times 7}{3} = \frac{91}{3} = 30\tfrac{1}{3}.$$

For convenience, we invert the terms of the divisor, and then multiply the whole number by the original denominator, and divide the product by the numerator.

The reason of this operation is evident, since 13 will contain $\frac{1}{7}$ as many times as there are *sevenths* in 13, equal 91 sevenths. Now, if 13 contain 1 seventh 91 times, it will contain $\frac{3}{7}$ as many times as 91 will contain 3, equal to $30\frac{1}{3}$.

RULE. — *Multiply the whole number by the denominator of the fraction, and divide the product by the numerator.*

EXAMPLES FOR PRACTICE.

2. Divide 18 by $\frac{7}{8}$. Ans. $20\frac{4}{7}$.
3. Divide 27 by $\frac{11}{12}$. Ans. $29\frac{5}{11}$.
4. Divide 23 by $\frac{1}{4}$. Ans. 92.

QUESTIONS. — Art. 160. What is the rule for dividing a whole number by a fraction? Give the reason for the rule.

5. Divide 5 by $\frac{1}{5}$. Ans. 25.
6. Divide 12 by $\frac{3}{4}$. Ans. 16.
7. Divide 16 by $\frac{1}{2}$. Ans. 32.
8. Divide 100 by $\frac{17}{19}$. Ans. $111\frac{13}{17}$.
9. I have 50 square yards of cloth; how many yards, $\frac{3}{5}$ of a yard wide, will be sufficient to line it? Ans. $83\frac{1}{3}$ yards.
10. A. Poor can walk $3\frac{7}{11}$ miles in 60 minutes; Benjamin can walk $\frac{9}{11}$ as fast as Poor. How long will it take Benjamin to walk the same distance? Ans. $73\frac{1}{3}$ minutes.

ART. **161.** To divide a mixed number by a whole number.

Ex. 1. Divide $17\frac{3}{8}$ by 6. Ans. $2\frac{43}{48}$.

OPERATION.

$$6)\underline{17\frac{3}{8}}$$
$$2, \quad 5\frac{3}{8}=\frac{43}{8}; \quad \frac{43}{8\times6}=\frac{43}{48};$$
$$2+\frac{43}{48}=2\frac{43}{48}.$$

Having divided the whole number as in simple division, we have a remainder of $5\frac{3}{8}$, which we reduce to an improper fraction, and divide it by the divisor, as in Art. 159. Annexing this fraction to the quotient 2, we obtain $2\frac{43}{48}$ for the answer. Hence, to divide a mixed number by a whole number,

Divide the integral part of the mixed number; and the remainder, reduced if necessary to a simple fraction, divide as in Art. 159.

EXAMPLES FOR PRACTICE.

2. Divide $17\frac{3}{5}$ by 7. Ans. $2\frac{18}{35}$.
3. Divide $18\frac{3}{7}$ by 8. Ans. $2\frac{17}{56}$.
4. Divide $27\frac{11}{12}$ by 9. Ans. $3\frac{11}{108}$.
5. Divide $31\frac{1}{10}$ by 11. Ans. $2\frac{91}{110}$
6. Divide $78\frac{4}{5}$ by 12. Ans. $6\frac{17}{30}$.
7. Divide $189\frac{11}{15}$ by 4. Ans. $47\frac{13}{30}$.
8. Divide $107\frac{1}{12}$ by 3. Ans. $35\frac{25}{36}$.
9. Divide $\$14\frac{3}{7}$ among 7 men. Ans. $\$2\frac{3}{49}$.
10. Divide $\$106\frac{7}{9}$ among 8 boys. Ans. $\$13\frac{25}{72}$.
11. What is the value of $\frac{25}{72}$ of a dollar? Ans. $\$0.34\frac{13}{18}$.
12. Divide $\$107\frac{7}{11}$ among 4 boys and 3 girls, and give the girls twice as much as the boys.
Ans. Boy's share, $\$10\frac{42}{55}$; Girl's share, $21\frac{29}{55}$.
13. If \$14 will purchase $\frac{17}{20}$ of a ton of copperas, what quantity will \$1 purchase? Ans. $1\frac{3}{14}$cwt.

QUESTION.—Art. 161. How do you divide a mixed number by a whole number?

ART. 162. To divide a whole number by a mixed number.

Ex. 1. Divide 25 by $4\frac{3}{5}$. Ans. $5\frac{10}{23}$.

OPERATION.

$4\frac{3}{5}$ | 25
5 | 5

23)125($5\frac{10}{23}$
115
10

We first reduce the divisor and dividend to fifths, and then divide as in whole numbers.

The reason why the answer is a mixed number, and not in fifths, is because the divisor and dividend were both multiplied by the same number, 5, and therefore their relation to each other is the same as before, and the quotient will not be altered. Hence,

Reduce the divisor and dividend to the same parts as are denoted by the denominator of the fraction in the divisor, and then divide as in whole numbers.

EXAMPLES FOR PRACTICE.

2. Divide 36 by $9\frac{7}{8}$. Ans. $3\frac{51}{79}$.
3. Divide 97 by $13\frac{11}{12}$. Ans. $6\frac{163}{167}$.
4. Divide 113 by $21\frac{1}{7}$. Ans. $5\frac{51}{148}$.
5. Divide 342 by $14\frac{47}{131}$. Ans. $23\frac{9}{11}$.
6. There is a board 19 feet in length, which I wish to saw into pieces $2\frac{3}{7}$ feet long; what will be the number of pieces, and how many feet will remain? Ans. $7\frac{14}{17}$ pieces.

ART. 163. To divide a fraction by a fraction.

Ex. 1. Divide $\frac{7}{8}$ by $\frac{4}{9}$. Ans. $1\frac{31}{32}$.

OPERATION.

$\frac{7}{8} \div \frac{4}{9} = \frac{7}{8} \times \frac{9}{4} = \frac{63}{32} = 1\frac{31}{32}$.

In this operation, we invert the terms of the divisor, and then proceed as in Art. 156.

The reason of this process will be seen, when we consider that the divisor, $\frac{4}{9}$, is an expression denoting that 4 is to be divided by 9. Now, regarding 4 as a whole number, we divide the fraction $\frac{7}{8}$ by it, by multiplying the denominator; thus, $\frac{7}{8 \times 4} = \frac{7}{32}$. But the divisor, 4, is 9 times too great, since it was to be divided by 9, as seen in the original fraction; therefore the quotient, $\frac{7}{32}$, is 9 times too small, and must be multiplied by 9; thus, $\frac{7 \times 9}{32} = \frac{63}{32} = 1\frac{31}{32}$. By this operation, we have multiplied the denominator of the dividend by the numerator of the divisor, and the numerator of the dividend by the denominator of the divisor. Hence the

---

QUESTIONS. — Art. 162. How do you divide a whole by a mixed number? How does it appear that this process does not alter the quotient? — Art. 163. How do you divide a fraction by a fraction? Give the reason why this process divides the fraction of the dividend.

RULE. — *Invert the divisor, and then proceed as in multiplication of fractions.*

NOTE 1. — Factors common to numerator and denominator should be cancelled.

NOTE 2. — When the divisor and dividend have a common denominator their denominators cancel each other, and the division may be performed by simply dividing the numerator of the dividend by that of the divisor.

EXAMPLES FOR PRACTICE.

2. Divide $\frac{7}{9}$ by $\frac{4}{7}$. Ans. $1\frac{13}{36}$.
3. Divide $\frac{7}{8}$ by $\frac{1}{4}$. Ans. $3\frac{1}{2}$.
4. Divide $\frac{13}{15}$ by $\frac{11}{12}$. Ans. $\frac{52}{55}$.
5. Divide $\frac{2}{3}$ by $\frac{3}{10}$. Ans. $2\frac{2}{9}$.
6. Divide $\frac{9}{10}$ by $\frac{1}{7}$. Ans. $6\frac{3}{10}$.
7. Divide $\frac{4}{5}$ by $\frac{2}{11}$. Ans. $4\frac{2}{5}$.
8. Divide $\frac{9}{13}$ by $\frac{3}{26}$. Ans. 6.
9. Divide $\frac{19}{20}$ by $\frac{7}{20}$. Ans. $2\frac{5}{7}$.
10. Divide $\frac{2}{3}$ of $\frac{7}{8}$ by $\frac{1}{7}$ of $\frac{2}{9}$. Ans. $18\frac{3}{8}$.
11. Divide $\frac{4}{9}$ of $\frac{6}{11}$ of $\frac{7}{16}$ by $\frac{2}{3}$ of $\frac{7}{4}$ of $\frac{1}{9}$. Ans. $\frac{9}{11}$.
12. Divide $\frac{3}{4}$ of $\frac{5}{7}$ of $\frac{4}{9}$ by $\frac{2}{3}$ of $\frac{6}{7}$ of $\frac{2}{18}$. Ans. $3\frac{3}{4}$.

ART. 164. To divide a mixed number by a mixed number, it is only necessary to reduce them to improper fractions, and proceed as in the foregoing rule. (Art. 163.)

Ex. 1. Divide $7\frac{3}{4}$ by $3\frac{3}{7}$. Ans. $2\frac{25}{96}$.

OPERATION.

$$7\frac{3}{4} = \frac{31}{4};\ 3\frac{3}{7} = \frac{24}{7}.$$
$$\frac{31}{4} \times \frac{7}{24} = \frac{217}{96} = 2\frac{25}{96}.$$

EXAMPLES FOR PRACTICE.

2. Divide $7\frac{3}{8}$ by $4\frac{1}{2}$. Ans. $1\frac{23}{36}$.
3. Divide $3\frac{1}{2}$ by $7\frac{1}{2}$. Ans. $\frac{7}{15}$.
4. Divide $11\frac{1}{4}$ by $5\frac{3}{7}$. Ans. $2\frac{11}{152}$.
5. Divide $4\frac{3}{7}$ by $1\frac{7}{9}$. Ans. $2\frac{55}{112}$.
6. Divide $116\frac{3}{7}$ by $14\frac{1}{7}$. Ans. $8\frac{23}{99}$.
7. Divide $81\frac{1}{7}$ by $9\frac{1}{5}$. Ans. $8\frac{132}{161}$.
8. Divide $\frac{3}{5}$ of $5\frac{1}{2}$ of 7 by $\frac{5}{8}$ of $3\frac{3}{10}$. Ans. $11\frac{1}{5}$.

QUESTIONS. — What is the rule for dividing one fraction by another? How may fractions be divided when they have a common denominator? Does this process differ in principle from the other? — Art. 164. How do you divide a mixed number by a mixed number?

COMPLEX FRACTIONS.

ART. 165. To reduce complex to simple fractions.

Ex. 1. Reduce $\frac{\frac{1}{9}}{\frac{3}{8}}$ to a simple fraction. Ans. $\frac{8}{27}$.

OPERATION.

$\frac{\frac{1}{9}}{\frac{3}{8}} = \frac{1}{9} \times \frac{8}{3} = \frac{8}{27}$

Since the numerator of a fraction is the dividend, and the denominator the divisor (Art. 132), it will be seen by this operation that we simply divide the numerator, $\frac{1}{9}$, by the denominator, $\frac{3}{8}$, as in *division of fractions*. (Art. 163.)

Ex. 2. Reduce $\frac{8}{4\frac{1}{2}}$ to a simple fraction. Ans. $1\frac{7}{9}$.

OPERATION.

$\frac{8}{4\frac{1}{2}} = \frac{\frac{8}{1}}{\frac{9}{2}} = \frac{8}{1} \times \frac{2}{9} = \frac{16}{9} = 1\frac{7}{9}$

We reduce the numerator, 8, and the denominator, $4\frac{1}{2}$, to improper fractions, and then proceed as in Ex. 1.

Ex. 3. Reduce $\frac{\frac{3}{7}}{\frac{1}{2} \text{ of } \frac{2}{3}}$ to a simple fraction. Ans. $1\frac{2}{7}$.

OPERATION.

$\frac{\frac{3}{7}}{\frac{1}{2} \text{ of } \frac{2}{3}} = \frac{\frac{3}{7}}{\frac{2}{6}} = \frac{3}{7} \times \frac{6}{2} = \frac{18}{14} = 1\frac{2}{7}$

We here reduce the denominator, $\frac{1}{2}$ of $\frac{2}{3}$, to a simple fraction, and then proceed as before.

From the preceding illustrations we deduce the following

RULE. — *Reduce the terms of the complex fraction, if necessary, to the form of a simple fraction. Then divide the numerator of the complex fraction by its denominator.*

EXAMPLES FOR PRACTICE.

4. Reduce $\frac{12}{\frac{3}{7}}$ to a whole number. Ans. 28.

5. Reduce $\frac{\frac{3}{7}}{14}$ to a simple fraction. Ans. $\frac{3}{98}$.

6. Reduce $\frac{4\frac{7}{8}}{9}$ to a simple fraction. Ans. $\frac{13}{24}$.

QUESTIONS. — Art. 165. What is the rule for reducing complex to simple fractions? How does this process differ from division of fractions?

7. Reduce $\frac{\frac{3}{4}}{\frac{11}{12}}$ to a simple fraction. Ans. $\frac{9}{11}$.

8. Change $\frac{\frac{5}{6}}{7\frac{3}{4}}$ to a simple fraction. Ans. $\frac{10}{93}$.

9. Change $\frac{8\frac{3}{4}}{\frac{2}{5}}$ to a mixed number. Ans. $21\frac{7}{8}$.

10. Reduce $\frac{9\frac{3}{5}}{12\frac{1}{2}}$ to a simple fraction. Ans. $\frac{96}{125}$.

11. If 7 is the denominator of the following fraction $\frac{9\frac{1}{4}}{12\frac{7}{8}}$, what is its value when reduced to a simple fraction?
Ans. $\frac{74}{721}$.

12. If $\frac{3}{4}$ is the numerator of the following fraction, $\frac{\frac{8}{9}}{\frac{1}{2}}$, what is its value when reduced to a simple fraction? Ans. $\frac{27}{64}$.

ART. 166. Complex fractions, after being reduced to simple ones, may be added, subtracted, multiplied, and divided, according to the respective rules for simple fractions.

EXAMPLES FOR PRACTICE.

1. Add $\frac{\frac{1}{3}}{\frac{3}{7}}$ and $\frac{4\frac{1}{7}}{12\frac{1}{2}}$ together. Ans. $1\frac{172}{1575}$.

2. Add $\frac{7\frac{3}{4}}{\frac{4}{7}}$ and $\frac{7}{\frac{7}{12}}$ together. Ans. $25\frac{9}{16}$.

3. From $\frac{\frac{3}{7}}{8\frac{1}{2}}$ take $\frac{\frac{1}{9}}{\frac{9}{2}}$. Ans. $\frac{248}{9639}$.

4. From $\frac{6\frac{3}{4}}{\frac{3}{4}}$ take $\frac{\frac{1}{9}}{\frac{3}{8}}$. Ans. $8\frac{19}{27}$.

5. Multiply $\frac{3}{4}$ of $\frac{8\frac{4}{5}}{6\frac{2}{5}}$ by $\frac{4}{9}$ of $\frac{\frac{2}{7}}{16}$. Ans. $\frac{11}{1344}$.

6. Multiply $\frac{3\frac{1}{2}}{5\frac{3}{4}}$ by $\frac{6\frac{1}{4}}{2\frac{4}{9}}$. Ans. $1\frac{563}{1012}$.

7. Divide $\frac{\frac{7}{8}}{\frac{3}{11}}$ of $12\frac{1}{2}$ by $\frac{\frac{1}{3}}{7\frac{1}{2}}$ of $8\frac{3}{4}$. Ans. $103\frac{1}{8}$

QUESTION. — Art. 166. How do you add, subtract, multiply, and divide complex fractions?

GREATEST COMMON DIVISOR OF FRACTIONS.

ART. 167. To find the greatest common divisor of two or more fractions.

Ex. 1. What is the greatest common divisor of $\frac{4}{9}$, $\frac{2}{3}$, and $\frac{12}{15}$?

OPERATION.

$$\tfrac{4}{9}, \tfrac{2}{3}, \tfrac{12}{15} = \tfrac{20}{45}, \tfrac{30}{45}, \tfrac{36}{45}.$$

Greatest common divisor of the numerators $= 2$ } greatest common divisor required.
Least common denominator of the fractions $= 45$ }

Having reduced the fractions to equivalent fractions with the least common denominator (Art. 141), we find the greatest common divisor of the numerators 20, 30, and 36, to be 2. (Art. 124.) Now, since the 20, 30, and 36, are *forty-fifths*, their greatest common divisor is not 2, a whole number, but so many *forty-fifths*. Therefore we write the 2 over the common denominator 45, and have $\frac{2}{45}$ as the answer.

RULE. — *Reduce the fractions, if necessary, to the least common denominator. Then find the greatest common divisor of the numerators, which, written over the least common denominator, will give the greatest common divisor required.*

EXAMPLES FOR PRACTICE.

2. What is the greatest common divisor of $\frac{3}{4}$, $\frac{5}{6}$, and $1\frac{1}{8}$? Ans. $\frac{1}{24}$.

3. What is the greatest common divisor of $\frac{12}{13}$, $\frac{4}{7}$, $\frac{8}{21}$, and $\frac{16}{39}$? Ans. $\frac{4}{273}$.

4. What is the greatest common divisor of $\frac{15}{16}$, $2\frac{1}{4}$, 4, and $5\frac{1}{3}$? Ans. $\frac{1}{48}$.

5. There is a three-sided lot, of which one side is $166\frac{2}{3}$ft., another side $156\frac{1}{4}$ft., and the third side $208\frac{1}{3}$ft. What must be the length of the longest rails that can be used in fencing it, allowing the end of each rail to lap by the other $\frac{1}{2}$ft., and all the panels to be of equal length? Ans. $10\frac{1}{12}$ft.

LEAST COMMON MULTIPLE OF FRACTIONS.

ART. 168. To find the least common multiple of fractions.

Ex. 1. What is the least common multiple of $\frac{7}{12}$, $1\frac{1}{2}$, and $5\frac{1}{4}$? Ans. $\frac{21}{2} = 10\frac{1}{2}$.

QUESTIONS. — Art. 167. What is the rule for finding the greatest common divisor of fractions? Why, in the operation, was the divisor 2 written over the denominator 45?

OPERATION.

$\frac{2}{12}$, $1\frac{1}{2}$, $5\frac{1}{4} = \frac{1}{6}$, $\frac{3}{2}$, $\frac{21}{4}$.

Least common multiple of the numerators $= 21$ } least common multiple required.
Greatest common divisor of the denominator $= 2$

Having reduced the fractions to their lowest terms, we find the least common multiple of the numerators, 1, 3, and 21, to be 21. (Art. 128.) Now, since the 1, 3, and 21, are, from the nature of a fraction, dividends of which their respective denominators, 6, 2, and 4, are the divisors (Art. 132), the least common multiple of the fractions is not 21, a whole number, but so many fractional parts of the greatest common divisor of the denominators. This common divisor we find to be 2, which, written as the denominator of the 21, gives $\frac{21}{2} = 10\frac{1}{2}$ as the least number that can be exactly divided by the given fractions.

RULE. — *Reduce the fractions, if necessary, to their lowest terms. Then find the least common multiple of the numerators, which, written over the greatest common divisor of the denominators, will give the least common multiple required.*

EXAMPLES FOR PRACTICE.

2. What is the least common multiple of $\frac{10}{28}$, $\frac{6}{7}$, and $\frac{15}{35}$? Ans. $4\frac{2}{7}$.

3. What is the least number that can be exactly divided by $\frac{1}{15}$, $2\frac{1}{2}$, 5, $6\frac{1}{3}$, and $\frac{1}{11}$? Ans. 95.

4. What is the smallest sum of money for which I could purchase a number of bushels of oats, at $\$\frac{5}{16}$ a bushel; a number of bushels of corn, at $\$\frac{5}{8}$ a bushel; a number of bushels of rye, at $\$1\frac{1}{2}$ a bushel; or a number of bushels of wheat, at $\$2\frac{1}{4}$ a bushel; and how many bushels of each could I purchase for that sum?

Ans. $\$22\frac{1}{2}$; 72 bushels of oats; 36 bushels of corn; 15 bushels of rye; 10 bushels of wheat.

5. There is an island 10 miles in circuit, around which A can travel in $\frac{3}{4}$ of a day, and B in $\frac{7}{8}$ of a day. Supposing them each to start together from the same point to travel around it in the same direction, how long must they travel before coming together again at the place of departure, and how many miles will each have travelled?

Ans. $5\frac{1}{4}$ days; A 70 miles; B 60 miles.

QUESTIONS. — Art. 168. What is the rule for finding the least common multiple of fractions? Why is not the least common multiple of the numerators the least common multiple of the fractions?

MISCELLANEOUS EXERCISES IN FRACTIONS.

1. What are the contents of a field $76\frac{7}{25}$ rods in length, and $18\frac{3}{4}$ rods in breadth? Ans. 8A. 3R. $30\frac{1}{4}$p.

2. What are the contents of 10 boxes which are $7\frac{3}{4}$ feet long, $1\frac{3}{4}$ feet wide, and $1\frac{1}{4}$ feet in height?
Ans. $169\frac{17}{32}$ cubic feet.

3. From $\frac{7}{11}$ of an acre of land there were sold 20 poles and 200 square feet. What quantity remained?
Ans. 22075ft.

4. What cost $\frac{11}{13}$ of an acre at $1.75 per square rod?
Ans. $$236.92\frac{4}{13}$.

5. What cost $\frac{9}{19}$ of a ton at $$15\frac{3}{4}$ per cwt.?
Ans. $$49.73\frac{13}{19}$.

6. What is the continued product of the following numbers: $14\frac{2}{5}$, $11\frac{3}{7}$, $5\frac{4}{9}$, and $10\frac{1}{4}$? Ans. 9184.

7. From $\frac{7}{12}$ of a cwt. of sugar there was sold $\frac{4}{7}$ of it; what is the value of the remainder at $$0.12\frac{3}{4}$ per pound?
Ans. $$3.18\frac{3}{4}$.

8. What cost $19\frac{3}{7}$ barrels of flour at $$7\frac{3}{8}$ per barrel?
Ans. $$143\frac{2}{7}$.

9. Bought a piece of land that was $47\frac{5}{11}$ rods in length, and $29\frac{7}{16}$ in breadth; and from this land there were sold to Abijah Atwood 5 square rods, and to Hazen Webster a piece that was 5 rods square; how much remains unsold?
Ans. $1366\frac{83}{88}$ square rods.

10. From a quarter of beef weighing $175\frac{3}{5}$lb. I gave John Snow $\frac{3}{5}$ of it; $\frac{2}{3}$ of the remainder I sold to John Cloon. What is the value of the remainder at $8\frac{3}{4}$ cents per pound?
Ans. $$2.04\frac{13}{15}$.

11. Alexander Green bought of John Fortune a box of sugar containing 475lb. for $30. He sold $\frac{1}{3}$ of it at 8 cents per pound, and $\frac{2}{3}$ of the remainder at 10 cents per pound. What is the value of what still remains at $12\frac{1}{2}$ cents per pound, and what does Green make on his bargain?

Ans. { Value of what remains, $$13.19\frac{4}{9}$.
Green's bargain, $$16.97\frac{2}{9}$.

12. What cost $\frac{14}{101}$ of an acre at $$14\frac{3}{7}$ per acre?
Ans. $2.

13. Multiply $\frac{7}{8}$ of $\frac{8}{11}$ of $\frac{11}{14}$ by $\frac{5}{17}$ of $\frac{17}{19}$ of $\frac{19}{25}$. Ans. $\frac{1}{10}$.

14. What are the contents of a board $11\frac{3}{4}$ inches long, and $4\frac{1}{4}$ inches wide? Ans. $49\frac{15}{16}$ square inches.

15. Mary Brown had \$17.87$\frac{1}{2}$; half of tnis sum was given to the missionary society, and $\frac{3}{5}$ of the remainder she gave to the Bible society; what sum has she left? Ans. \$3.57$\frac{1}{2}$.

16. What number shall be taken from $12\frac{3}{4}$, and the remainder multiplied by $10\frac{4}{5}$, that the product shall be 50?

Ans. $8\frac{13}{108}$

17. What number must be multiplied by $7\frac{3}{8}$, that the product may be 20? Ans. $2\frac{42}{59}$.

18. What are the contents of a box $8\frac{5}{12}$ feet long, $3\frac{11}{12}$ feet wide, and $2\frac{1}{12}$ feet high? Ans. $68\frac{1171}{1728}$ feet.

19. On $\frac{2}{3}$ of my field I plant corn; on $\frac{2}{3}$ of the remainder I sow wheat; potatoes are planted on $\frac{3}{7}$ of what still remains; and I have left two small pieces, one of which is 3 rods square, and the other contains 3 square rods. How large is my field?

Ans. 1A. 0R. 29p.

REDUCTION OF FRACTIONS OF COMPOUND NUMBERS.

ART. **169.** To reduce a fraction of a higher denomination to a fraction of a lower.

Ex. 1. Reduce $\frac{1}{2160}$ of a pound to the fraction of a farthing.

Ans. $\frac{4}{9}$far.

OPERATION.

$$\frac{1}{2160} \times 20 = \frac{20}{2160}; \quad \frac{20}{2160} \times 12 = \frac{240}{2160}; \quad \frac{240}{2160} \times 4 = \frac{960}{2160}\text{far.} = \frac{4}{9}\text{far.}$$

OPERATION BY CANCELLATION.

$$\frac{1 \times \not{20} \times \not{12} \times 4}{\not{2160}_{\,9}} = \frac{4}{9}\text{far.}$$

Since 20s. make a pound, there must be 20 times as many parts of a shilling as parts of a pound; we therefore multiply $\frac{1}{2160}$ by 20, and obtain $\frac{20}{2160}$s.; and since 12d. make a shilling, there will be 12 times as many parts of a penny as parts of a shilling; hence we multiply $\frac{20}{2160}$ by 12, and obtain $\frac{240}{2160}$d. Again; since 4far. make a penny, there will be 4 times as many parts of a farthing as parts of a penny; we therefore multiply $\frac{240}{2160}$ by 4, and obtain $\frac{960}{2160}$far. = $\frac{4}{9}$far., Ans.

RULE. — *Multiply the given fraction by the same numbers that would be employed in reduction of whole numbers to the lower denomination required.*

QUESTIONS. — Art. 167. What is the rule for reducing a fraction of a higher denomination to the fraction of a lower? Will you explain the operations? Does this process differ in principle from reduction of whole compound numbers?

EXAMPLES FOR PRACTICE.

2. Reduce $\frac{1}{1400}$ of a pound to the fraction of a farthing. Ans. $\frac{24}{35}$.
3. What part of a penny is $\frac{4}{75}$ of a shilling? Ans. $\frac{16}{25}$.
4. What part of a grain is $\frac{1}{8640}$ of a pound Troy? Ans. $\frac{2}{3}$.
5. What part of an ounce is $\frac{1}{1728}$ of a cwt.? Ans. $\frac{25}{27}$.
6. Reduce $\frac{1}{1320}$ of a furlong to the fraction of a foot. Ans. $\frac{1}{2}$.
7. What part of a square foot is $\frac{1}{58080}$ of an acre? Ans. $\frac{3}{4}$.
8. What part of a second is $\frac{1}{89600}$ of a day? Ans. $\frac{27}{28}$.
9. What part of a peck is $\frac{3}{14}$ of a bushel? Ans. $\frac{6}{7}$.
10. What part of a pound is $\frac{1}{200}$ of a cwt.? Ans. $\frac{1}{2}$.

ART. 170. To reduce a fraction of a lower denomination to a fraction of a higher.

Ex. 1. Reduce $\frac{4}{9}$ of a farthing to the fraction of a pound. Ans. $\frac{1}{2160}$.

OPERATION.

$$\frac{4}{9} \times 4 = \frac{4}{36};\ \frac{4}{36} \times 12 = \frac{4}{432};\ \frac{4}{432} \times 20 = \frac{4}{8640}£. = \frac{1}{2160}£.$$

OPERATION BY CANCELLATION.

$$\frac{\not{4}}{9 \times \not{4} \times 12 \times 20} = \frac{1}{2160}£.$$

Since 4far. make a penny, there will be $\frac{1}{4}$ as many pence as farthings; therefore we divide the $\frac{4}{9}$ by 4, and obtain $\frac{4}{36}$d. And, since 12d. make a shilling, there will be $\frac{1}{12}$ as many shillings as pence; hence we divide $\frac{4}{36}$ by 12, and obtain $\frac{4}{432}$s. Again, since 20s. make a pound, there will be $\frac{1}{20}$ as many pounds as shillings; therefore we divide $\frac{4}{432}$ by 20, and obtain $\frac{4}{8640}£. = \frac{1}{2160}£.$ for the answer.

RULE. — *Divide the given fraction by the same numbers that would be employed in reduction of whole numbers to the higher denomination required.*

EXAMPLES FOR PRACTICE.

2. Reduce $\frac{4}{7}$ of a grain Troy to the fraction of a pound. Ans. $\frac{1}{10080}$.

QUESTIONS. — Art. 170. Do you multiply or divide to reduce a fraction of a lower denomination to the fraction of a higher? What is the rule?

3. What part of an ounce is $\frac{3}{10}$ of a scruple? Ans. $\frac{1}{80}$.
4. What part of a ton is $\frac{4}{5}$ of an ounce? Ans. $\frac{1}{40000}$.
5. What part of a mile is $\frac{8}{9}$ of a rod? Ans. $\frac{1}{360}$.
6. What part of an acre is $\frac{2}{3}$ of a square foot? Ans. $\frac{1}{65340}$.
7. What part of a day is $\frac{24}{25}$ of a second? Ans. $\frac{1}{90000}$.
8. What part of 3 acres is $\frac{4}{9}$ of a square foot? Ans. $\frac{1}{294030}$.
9. What part of 3hhd. is $\frac{4}{7}$ of a quart? Ans. $\frac{1}{1323}$.
10. What part of $\frac{1}{6}$ of a solid foot is a cube whose sides are each $\frac{1}{6}$ of a yard square? Ans. $\frac{3}{4}$.

ART. **171.** To find the value of a fraction in whole numbers of a lower denomination.

Ex. 1. What is the value of $\frac{7}{18}$ of 1£.?

Ans. 7s. 9d. $1\frac{1}{3}$far.

OPERATION.

```
      7
     20
18 ) 140 ( 7s.
     126
      14
      12
18 ) 168 ( 9d.
     162
       6
       4
18 ) 24 ( 1⅓far.
     18
```

$\frac{6}{18} = \frac{1}{3}$

The reason of this operation will be seen, if we analyze the question according to Art 169. Thus, $\frac{7}{18} \times 20 = \frac{140}{18}$s. $= 7\frac{14}{18}$s., and $\frac{14}{18} \times 12 = \frac{168}{18}$d. $= 9\frac{6}{18}$d.; and $\frac{6}{18} \times 4 = \frac{24}{18}$far. $= 1\frac{1}{3}$ far.

Ans. 7s. 9d. $1\frac{1}{3}$far.

RULE. — *Multiply the numerator of the given fraction by the number required to reduce it to the next lower denomination, and divide the product by the denominator.*

Then, if there is a remainder, proceed as before, until it is reduced to the denomination required.

QUESTION. — Art. 171. What is the rule for finding the value of a fraction in whole numbers of a lower denomination?

EXAMPLES FOR PRACTICE.

2. What is the value of $\frac{7}{9}$ of a cwt.? Ans. 3qr. 2lb. 12oz. $7\frac{1}{9}$dr.
3. What is the value of $\frac{7}{9}$ of a yard? Ans. 3qr. $0\frac{4}{9}$na
4. What is the value of $\frac{3}{7}$ of an acre? Ans. 1R. 28p. 155ft. $82\frac{2}{7}$in.
5. What is the value of $\frac{2}{9}$ of a mile? Ans. 1 fur. 31rd. 1ft. 10in.
6. What is the value of $\frac{3}{11}$ of an ell English? Ans. 1qr. $1\frac{5}{11}$na.
7. What is the value of $\frac{2}{7}$ of a hogshead of wine? Ans. 18gal.
8. What is the value of $\frac{7}{11}$ of a year? Ans. 232da. 10h. 21m. $49\frac{1}{11}$sec.

ART. 172. To reduce a simple or compound number to the fractional part of any other simple or compound number of the same kind.

Ex. 1. What part of 1£. is 3s. 6d. $2\frac{2}{3}$far. Ans. $\frac{8}{45}$£.

OPERATION.

$$\begin{array}{ll} \text{3s. 6d. } 2\frac{2}{3}\text{far.} = 512 \\ \text{1£.} \qquad = 2880 \end{array} \quad \frac{512}{2880} = \frac{8}{45}\text{£.}$$

In performing this operation, we reduce the 3s. 6d. $2\frac{2}{3}$far. to thirds of far., the lowest denomination in the question, for the numerator of the required fraction, and 1£. to the same denomination for the denominator. We then reduce this fraction to its lowest terms, and obtain $\frac{8}{45}$£. for the answer.

RULE. — *Reduce the given numbers to the lowest denomination mentioned in either of them. Then, write the number which is to become the fractional part for the numerator, and the other number for the denominator, of the required fraction.*

EXAMPLES FOR PRACTICE.

2. Reduce 4s. 8d. to the fraction of 1£. Ans. $\frac{7}{30}$.
3. What part of a ton is 4cwt. 3qr. 12lb.? Ans. $\frac{487}{2000}$.
4. What part of 2m. 3fur. 20rd. is 2fur. 30rd.? Ans. $\frac{11}{78}$.
5. What part of 2A. 2R. 32p. is 3R. 24p.? Ans. $\frac{1}{3}$.
6. What part of a hogshead of wine is 18gal. 2qt.? Ans. $\frac{37}{126}$.

QUESTION. — Art. 172. What is the rule for reducing a simple or compound number to the fractional part of any other simple or compound number of the same kind?

7. What part of 30 days are 8 days 17h. 20m.?

Ans. $\frac{157}{540}$.

8. From a piece of cloth containing 13yd. 0qr. 2na. there were taken 5yd. 2qr. 2na. What part of the whole piece was taken? Ans. $\frac{3}{7}$.

9. What part of 3 yards square are 3 square yards?

Ans. $\frac{1}{3}$.

ADDITION OF FRACTIONS OF COMPOUND NUMBERS.

ART. 173. To add fractions of compound numbers.

Ex. 1. Add $\frac{6}{7}$ of a pound to $\frac{7}{9}$ of a shilling.

Ans. 17s. 11d. $0\frac{4}{21}$far.

FIRST OPERATION.

	s.	d.	far.
Value of $\frac{6}{7}$£. =	17	1	$2\frac{6}{7}$
Value of $\frac{7}{9}$s. =		9	$1\frac{1}{3}$
	17	11	$0\frac{4}{21}$

We find the value of each fraction separately, and add the two values together, according to the rule for adding compound numbers. (Art. 101.)

SECOND OPERATION.

$\frac{7}{9 \times 20} = \frac{7}{180}$£.

$\frac{6}{7}$£. + $\frac{7}{180}$£. = $\frac{1129}{1260}$£. = 17s. 11d. $0\frac{4}{21}$far.

We first reduce the fraction of a shilling to the fraction of a pound, then add the two fractions together and find the value of their sum. (Art 171.)

EXAMPLES FOR PRACTICE.

2. Add $\frac{4}{11}$ of a pound to $\frac{5}{7}$ of a shilling.

Ans. 7s. 11d. $3\frac{29}{77}$far.

3. Add together $\frac{10}{11}$ of a ton, $\frac{7}{9}$ of a ton, and $\frac{4}{7}$ of a cwt.

Ans. 1T. 14cwt. 1qr. $5\frac{610}{693}$.

4. Add together $\frac{2}{3}$ of a yard, $\frac{8}{9}$ of a yard, $\frac{4}{11}$ of a quarter.

Ans. 1yd. 2qr. 2na. $0\frac{17}{22}$in.

5. Add together $\frac{4}{11}$ of a mile, $\frac{4}{9}$ of a mile, $\frac{3}{11}$ of a furlong, and $\frac{7}{11}$ of a yard. Ans. 6fur. 29rd. 3yd. 1ft. $0\frac{10}{11}$in.

6. Add together $\frac{9}{11}$ of an acre, $\frac{4}{5}$ of a rood, and $\frac{5}{7}$ of a square rod. Ans. 1A. 0R. 3p. 169ft. $102\frac{6}{7}$in.

7. Sold 4 house-lots; the first $\frac{4}{17}$ of an acre, the second $\frac{1}{7}$ of an acre, the third $\frac{2}{11}$ of an acre, and the fourth $\frac{3}{7}$ of an acre; what was the quantity of land in the four lots?

Ans. 3R. 38p. $45\frac{81}{238}$ft.

QUESTIONS. — Art. 173. What is the first method of adding fractions of compound numbers? What is the second?

SUBTRACTION OF FRACTIONS OF COMPOUND NUMBERS.

ART. **174.** To subtract fractional parts of compound numbers.

Ex. 1. From $\frac{6}{7}$ of a pound take $\frac{4}{11}$ of a pound.

Ans. 9s. 10d. $1\frac{59}{77}$far.

FIRST OPERATION.

	s.	d.	far.
Value of $\frac{6}{7}$£. =	17	1	$2\frac{6}{7}$
Value of $\frac{4}{11}$£. =	7	3	$1\frac{1}{11}$
	9	10	$1\frac{59}{77}$

We find the value of each fraction separately, and subtract one from the other, according to the rule for subtracting compound numbers. (Art 102.)

SECOND OPERATION.

$\frac{6}{7}$£. — $\frac{4}{11}$£. = $\frac{38}{77}$£. = 9s. 10d. $1\frac{59}{77}$far.

We first subtract the less fraction from the greater, and then find the value of their difference. (Art. 171.)

EXAMPLES FOR PRACTICE.

2. From $\frac{4}{7}$ of a ton take $\frac{6}{17}$ of a cwt.

Ans. 11cwt. 0qr. $7\frac{67}{119}$lb.

3. From $\frac{7}{9}$ of a mile take $\frac{7}{18}$ of a furlong.

Ans. 5fur. 33rd. 5ft. 6in.

4. From $\frac{10}{11}$ of an acre take $\frac{2}{9}$ of a rood.

Ans. 3R. 16p. 154ft.

5. From a hogshead of molasses containing 100 gallons, $\frac{3}{11}$ of it leaked out; $\frac{2}{3}$ of the remainder I kept for my family; what quantity remained for sale? Ans. 24gal. 0qt. $1\frac{31}{33}$pt.

6. The distance from Boston to Worcester is about 41 miles. A sets out from Worcester, and travels $\frac{3}{11}$ of this distance towards Boston; B then starts from Boston to meet A, and, having travelled $\frac{4}{7}$ of the remaining distance, it is required to find the distance between A and B.

Ans. 12m. 6fur. 9rd. 5ft. $9\frac{3}{7}$in.

7. A agrees to labor for B 365 days; but he was absent on account of sickness $\frac{1}{7}$ part of the time; he was also obliged to be employed in his own business $\frac{3}{11}$ of the remaining time; required the time lost. Ans. 137da. 11h. 13m. $14\frac{62}{77}$sec.

8. From 11 acres, 33 poles, $101\frac{1}{16}$ feet of land, I sold $\frac{3}{5}$ to A, $\frac{1}{3}$ of the remainder to B, and four house-lots, each 144 feet square, to C; what is the value of the remainder, at $8\frac{1}{3}$ cents per square foot? Ans. $3937.89\frac{7}{12}$.

QUESTIONS. — Art. 174. What is the first method of subtracting fractions of compound numbers? The second?

QUESTIONS TO BE PERFORMED BY ANALYSIS.

1. If one yard of cloth cost \$4.40, what will $\frac{4}{5}$ of a yard cost?

ILLUSTRATION. — If 1 yard cost \$4.40, $\frac{1}{5}$ of a yard will cost $\frac{1}{5}$ of \$4.40, equal to \$0.88; and $\frac{4}{5}$ will cost 4 times \$0.88, equal to \$3.52, Ans.

2. If a barrel of flour cost \$7.80, what will $\frac{3}{10}$ of a barrel cost? Ans. \$2.34.

3. If a load of hay cost \$17.84, what will $\frac{7}{8}$ of a load cost? Ans. \$15.61.

4. If \$786.63 are paid for a cargo of wheat, what is the cost of $\frac{11}{13}$ of the cargo? Ans. \$665.61.

5. What is $\frac{11}{12}$ of \$87.50? Ans. \$80.20$\frac{5}{6}$.

6. What is $\frac{3}{4}$ of 17£. 18s. 9d.? Ans. 13£. 9s. 0$\frac{3}{4}$d.

7. What is $\frac{4}{7}$ of 3T. 16cwt. 3qr. 23lb.? Ans. 2T. 3cwt. 3qr. 23$\frac{6}{7}$lb.

8. What is $\frac{4}{9}$ of 27A. 3R. 33p.? Ans. 12A. 1R. 28p.

9. If \$3.52 are paid for $\frac{4}{5}$ of a yard of cloth, what is the price of 1 yard? Ans. \$4.40.

ILLUSTRATION. — If $\frac{4}{5}$ of a yard cost \$3.52, $\frac{1}{5}$ will cost $\frac{1}{4}$ of \$3.52, equal to \$0.88; and $\frac{5}{5}$, or a whole yard, will cost 5 times \$0.88, equal to \$4.40, Ans.

10. If $\frac{3}{10}$ of a barrel of flour cost \$2.34, what will be the cost of a whole barrel? Ans. \$7.80.

11. When \$15.57$\frac{1}{2}$ are paid for $\frac{7}{8}$ of a ton of hay, what will 1 ton cost? Ans. \$17.80.

12. When $\frac{11}{13}$ of a cargo of flour cost \$665.50, what sum will pay for the whole cargo? Ans. \$786.50.

13. If \$73.60$\frac{5}{6}$ are paid for $\frac{11}{12}$ of a ton of potash, what sum must be paid for a ton? Ans. \$80.30.

14. Bought $\frac{3}{4}$ of a bale of broadcloth for 13£. 9s. 0$\frac{3}{4}$d.; what would have been the cost of the whole bale? Ans. 17£. 18s. 9d.

15. If $\frac{4}{17}$ of an acre produce 18cwt. 0qr. 12lb. of hay, what quantity will a whole acre produce? Ans. 77cwt. 0qr. 1lb.

16. Bought $\frac{4}{9}$ of a lot of land containing 12A. 1R. 30$\frac{7}{9}$p.; what were the contents of the whole lot? Ans. 27A. 3R. 39$\frac{1}{4}$p.

17. If $\frac{11}{12}$ of a ton of potash cost \$80.20$\frac{5}{6}$, what is the value of a ton? Ans. \$87.50.

18. If $\frac{3}{4}$ of a cwt. of sugar cost \$5.40, what is the value of $\frac{7}{9}$ of a cwt.?

ILLUSTRATION. — If $\frac{3}{4}$ of a cwt. cost \$5.40, $\frac{1}{4}$ will cost $\frac{1}{3}$ of \$5.40, equal to \$1.80; and $\frac{4}{4}$, or a cwt., will cost 4 times \$1.80, equal to \$7.20. Now, if 1cwt. cost \$7.20, $\frac{1}{9}$ of a cwt. will cost $\frac{1}{9}$ of \$7.20, equal to \$0.80; and $\frac{7}{9}$ will cost 7 times \$0.80, equal to \$5.60, Ans.

19. If $\frac{7}{11}$ of a pound of ipecacuanha cost \$2.52, what is the value of $\frac{4}{9}$ of a pound? Ans. \$1.76.

20. When \$80 are paid for $\frac{3}{4}$ of an acre of land, what cost $\frac{7}{8}$ of an acre? Ans. \$93.33$\frac{1}{3}$.

21. If $\frac{9}{16}$ of a carding-mill are worth \$631.89, what are $\frac{5}{14}$ of it worth? Ans. \$401.20.

22. If $\frac{4}{5}$ of a ship and cargo are valued at \$141.52, what are $\frac{5}{29}$ of them worth? Ans. \$30.50.

23. If the value of $\frac{3}{8}$ of a farm containing 178$\frac{4}{37}$ acres is \$1728, what is the price of $\frac{4}{5}$ of the remainder?

Ans. \$2304.

24. E. Carter's garden is 17$\frac{4}{11}$ rods long, and 11$\frac{9}{13}$ rods wide. He disposes of $\frac{4}{7}$ of it for \$82.80; what is the value of $\frac{2}{3}$ of the remainder? Ans. \$41.40.

25. When 26£. 12s. 6d. are paid for $\frac{5}{9}$ of a bale of cloth, what sum should be paid for $\frac{7}{8}$ of the remainder?

Ans. 18£. 12s. 9d.

26. If 7cwt. of sugar cost \$28.14, what will 9$\frac{5}{6}$cwt. cost?

ILLUSTRATION. — If 7cwt. cost \$28.14, 1cwt. will cost $\frac{1}{7}$ of \$28.14, equal to \$4.02. In 9$\frac{5}{6}$cwt. there are $\frac{59}{6}$cwt.; and if 1cwt. cost \$4.02, $\frac{1}{6}$cwt. will cost $\frac{1}{6}$ of \$4.02, equal to \$0.67, and $\frac{59}{6}$ will cost 59 times \$0.67, equal to \$39.53, Ans.

27. If three tons of hay cost \$49, what will 7$\frac{4}{11}$ tons cost?

Ans. \$120.27$\frac{3}{11}$.

28. Gave \$78.80 for 11 tons of coal; what should I give for 3$\frac{4}{9}$ tons? Ans. \$24.67$\frac{47}{99}$.

29. Paid 37£. 18s. 10d. for 3 bales of velvet; what was the cost of 5$\frac{3}{8}$ bales? Ans. 67£. 19s. 6$\frac{11}{12}$d.

30. Gave \$40 for 5 yards of broadcloth; what was the price of 19$\frac{4}{7}$ yards? Ans. \$156.57$\frac{1}{7}$.

31. Paid \$360 for 20 barrels of beer; what must be given for 43$\frac{5}{6}$ barrels? Ans. \$789.

32. If 7 bushels of rye cost \$8.75, what cost 18$\frac{7}{11}$ bushels?

Ans. \$23.29$\frac{6}{11}$.

33. Paid \$19.80 for 3 yards of broadcloth; what sum must be given for $11\frac{4}{7}$ yards? Ans. \$76.37$\frac{1}{7}$.

34. If $9\frac{5}{6}$cwt. of sugar cost \$39.53, what must be paid for 7cwt.?

ILLUSTRATION. — In $9\frac{5}{6}$cwt. there are $\frac{59}{6}$cwt. If $\frac{59}{6}$cwt. cost \$39.53, $\frac{1}{6}$cwt. will cost $\frac{1}{59}$ of \$39.53, equal to \$0.67; and $\frac{6}{6}$, or 1cwt., will cost 6 times \$0.67, equal to \$4.02; and 7cwt. will cost 7 times \$4.02, equal to \$28.14, Ans.

35. When \$$18\frac{7}{8}$ are paid for 3cwt. of sugar, how much may be purchased for \$1? How much for \$78? Ans. $12\frac{60}{151}$cwt.

36. If $3\frac{3}{7}$ tons of potash cost \$276.18, what will be the value of 1 ton? Of 75 tons? Ans. \$6041.43$\frac{3}{4}$.

37. If $7\frac{4}{11}$ acres of land cost \$875, what will one acre cost? What will 75 acres cost? Ans. \$8912.03$\frac{19}{27}$.

38. If $4\frac{3}{8}$ tons of coal cost \$70, what will 1 ton cost? What will 86 tons cost? Ans. \$1376.

39. For $27\frac{3}{4}$ acres of land there were paid \$375; what cost 1 acre? What cost 69 acres? Ans \$932.43$\frac{9}{37}$.

40. If $4\frac{3}{5}$ tons of hay cost \$80.50, what costs 1 ton? What cost 15 tons? Ans. \$262.50.

41. If $7\frac{4}{11}$cwt. of sugar cost \$62.37, what will 1cwt. cost? What cost 19cwt.? Ans. \$160.93.

42. If $7\frac{3}{4}$ yards of cloth cost \$13.95, what will be the value of $11\frac{4}{9}$ yards?

ILLUSTRATION. — In $7\frac{3}{4}$ yards there are $\frac{31}{4}$ of a yard. If $\frac{31}{4}$ of a yard cost \$13.95, $\frac{1}{4}$ will cost $\frac{1}{31}$ of \$13.95, equal to \$0.45; and $\frac{4}{4}$, or 1 yard will cost 4 times \$0.45, equal to \$1.80. In $11\frac{4}{9}$ yards there are $\frac{103}{9}$ of a yard. If 1 yard cost \$1.80, $\frac{1}{9}$ of a yard will cost $\frac{1}{9}$ of \$1.80 equal to \$0.20, and $\frac{103}{9}$ will cost 103 times \$0.20, equal to \$2.060, Ans.

43. When \$668.50 are paid for $17\frac{4}{11}$ acres, what would be the value of $89\frac{4}{5}$ acres? Ans. \$3457.30.

44. If \$1738 are given for $19\frac{3}{4}$ tons of iron, what will be the cost of $37\frac{4}{11}$ tons? Ans. \$3288.

45. Paid \$$11\frac{3}{4}$ for 1128 feet of boards; how many could I have purchased for \$$119\frac{7}{12}$? Ans. 11480 feet.

46. For $3\frac{5}{8}$ tons of potash I received 116cwt. of sugar; required the quantity of sugar that may be received for $11\frac{3}{4}$ tons of potash. Ans. 376cwt.

47. For $11\frac{3}{4}$ tons of potash I received 376cwt. of sugar

required the quantity of sugar that should be received for $3\frac{5}{8}$ tons. Ans. 116cwt.

48. When $8 are paid for $1\frac{3}{7}$ yards of broadcloth, how much must be given for $8\frac{3}{4}$ yards? Ans. $49.

49. Gave $414 for $20\frac{7}{10}$ acres of land; what shall be given for $11\frac{4}{5}$ acres? Ans. $236.

MISCELLANEOUS QUESTIONS BY ANALYSIS.

1. Sold a small farm for $896.50; what was received for $\frac{1}{11}$ of it? For $\frac{7}{11}$ of it? For $\frac{10}{11}$ of it? Ans. $815.

2. Gave $\$17\frac{3}{11}$ for 3 barrels of flour; what cost 1 barrel? What 37 barrels? Ans. $\$213.03\frac{1}{33}$.

3. Sold a house for $3687; what sum was received for $\frac{7}{8}$ of it? Ans. $\$3226.12\frac{1}{2}$.

4. Bought $17\frac{7}{12}$ tons of hay for $\$187\frac{3}{8}$; what is the cost of $\frac{5}{7}$ of a ton? Ans. $\$7.61\frac{253}{1477}$.

5. Bought a hogshead of molasses for $\$13\frac{7}{8}$; what cost $\frac{1}{5}$ of it? What cost $\frac{7}{5}$? What cost $\frac{11}{5}$? Ans. $\$30.52\frac{1}{2}$.

6. When $\$37\frac{3}{11}$ are paid for 100 gallons of molasses, what cost $\frac{4}{7}$ of a gallon? Ans. $\$0.21\frac{23}{77}$.

7. When 12 cents are paid for $\frac{4}{11}$ of a gallon of molasses, what will $48\frac{7}{13}$ gallons cost? Ans. $\$16.01\frac{10}{13}$.

8. If $\frac{4}{9}$ of a barrel of flour cost $\$3\frac{2}{7}$, what will $6\frac{3}{5}$ barrels cost? Ans. $\$48\frac{111}{140}$.

9. When $236 are paid for $11\frac{4}{5}$ acres, what will be paid for $20\frac{7}{10}$ acres? Ans. $414.

10. Paid in Liverpool $97\frac{4}{7}$£. for 3 bales of cloth; how many bales should be received for $1073\frac{2}{7}$£.? Ans. 33 bales.

11. If $6\frac{3}{5}$ barrels of flour cost $\$48\frac{111}{140}$, what will $\frac{4}{9}$ of a barrel cost? Ans. $\$3.28\frac{4}{7}$.

12. If $3\frac{2}{3}$ pounds of coffee cost 34 cents, what sum must be paid for $74\frac{1}{2}$ pounds? Ans. $\$6.90\frac{9}{11}$.

13. If $2\frac{5}{7}$ tons of hay cost $63, what will be the cost of $16\frac{4}{9}$ tons? Ans. $\$381\frac{13}{19}$.

14. If a piece of land 3 rods square cost $\$17\frac{4}{11}$, what will be the cost of 4 square rods? Ans. $\$7\frac{71}{99}$.

15. Paid $\$31\frac{4}{7}$ for $2\frac{5}{6}$cwt. of iron; required the sum to be paid for $689\frac{4}{13}$cwt. Ans. $\$7680\frac{6}{7}$.

16. For $6\frac{2}{3}$ cords of wood J. Holt paid $63; what sum must be paid for 18 cords? Ans. $170.10.

17. Gave $\$243\frac{1}{11}$ for 96 barrels of tar; what quantity could be purchased for $1000? Ans $394\frac{1222}{1337}$ barrels.

18. Paid \$7888.30 for $83\frac{9}{16}$ acres of wild land; what sum did I pay for each acre, and what would be the cost of 7 acres? Ans. \$660.80.

19. Gave 132£. 12s. for $7\frac{5}{9}$ tons of starch; what cost $12\frac{7}{9}$ tons? Ans. 224£. 5s.

20. For $17\frac{2}{3}$ days' work I paid \$25.44; what should be paid for $89\frac{1}{3}$ days' labor? Ans. \$128.64.

21. Sold $7\frac{7}{12}$ bushels of apples for \$7.28; what should I receive for $19\frac{11}{12}$ bushels? Ans. \$19.12.

22. Paid \$4355.52 for $49\frac{6}{7}$ pieces of carpeting; what did $37\frac{5}{7}$ pieces cost? Ans. \$3294.72.

23. If $\frac{1}{4}$ of $\frac{3}{5}$ of the cost of the Capitol at Washington was \$300,000, what was the whole cost? Ans. \$2,000,000.

24. Purchased $7\frac{6}{13}$ thousand of boards for \$135.80; what must be paid for $19\frac{3}{4}$ thousand? Ans. \$359.45.

25. My wood-pile contains 6 cords and 76 cubic feet. If I dispose of $\frac{3}{7}$ of it, what is the value of the remainder at $4\frac{4}{5}$ cents per cubic foot? Ans. $\$23.14\frac{34}{35}$.

26. I have a field 30 rods square, and having sold 18 square rods to S. Brown, and 82 square rods to J. Smith, what part of the field remained unsold? Ans. $\frac{8}{9}$.

27. Bought 7T. 12cwt. 3qr. 18lb. of iron, and having sold 3T. 18cwt. 1qr. 20lb., what is the value of $\frac{3}{5}$ of the remainder at $5\frac{3}{7}$ cents per lb.? Ans. $\$242.59\frac{1}{5}$.

28. Bought 37 tons of iron at \$68.50 per ton, for $\frac{3}{4}$ of which I paid in coffee at \$8.50 per cwt., and for the remainder I paid cash. Required the amount of cash paid, and also the value of the coffee.

Ans. Cash, $\$633.62\frac{1}{2}$; Value of the coffee, $\$1900.87\frac{1}{2}$.

29. A man, having received a legacy of \$7896, spent $\frac{3}{4}$ of it in speculations, and the remainder he put in the savings bank, where it continued 15 years. It was then found that the sum deposited had doubled. Required the sum in the bank. Ans. \$3948.

30. Bought a piece of broadcloth for \$88, and sold $\frac{4}{13}$ of it to J. Smith, and $\frac{5}{13}$ of the remainder to O. Lake; what is the value of the part unsold? Ans. $\$37.49\frac{19}{169}$.

31. A gentleman gave $\frac{3}{4}$ of his estate to his wife, $\frac{2}{3}$ of the remainder to his oldest son, and $\frac{3}{4}$ of what then remained to his daughter, who received \$750; required the whole estate. Ans. \$12,000.

32. From an acre of land I sold two house-lots, each 100 feet square; what is the value of the remainder, at 8 cents per square foot? Ans. \$1884.80.

§ XIX. DECIMAL FRACTIONS.

ART. **175.** A DECIMAL FRACTION is a fraction whose denominator is ten, or the product of a number of tens.

Decimal fractions are commonly expressed by writing the numerator only, with a point (.) before it, called the *decimal point* or *separatrix;* thus,

$\frac{9}{10}$ is written .9.
$\frac{99}{100}$ " .99.
$\frac{999}{1000}$ " .999.

By examining the foregoing fractions, it will be seen that $\frac{9}{10}$ = .9 can occupy only *one* place while it remains a proper fraction; $\frac{99}{100} = .99$, only *two* places; and $\frac{999}{1000} = .999$, only *three* places; for, if their numerators are increased by $\frac{1}{10} = .1$, $\frac{1}{100} = .01$, $\frac{1}{1000} = .001$, respectively, each fraction becomes a unit or whole number. Hence,

The first figure or place of any decimal on the right of the point is tenths, the second hundredths, the third thousandths, &c.

NOTE. — When a decimal place has no significant figure it must be filled with a cipher.

ART. **176.** The denominator of $\frac{9}{10} = .9$ is 1 with *one* cipher annexed; the denominator of $\frac{99}{100} = .99$ is 1 with *two* ciphers annexed; the denominator of $\frac{999}{1000} = .999$ is 1 with *three* ciphers annexed. Hence,

The denominator of a decimal fraction is 1 *with as many ciphers annexed as the numerator has places.*

ART. **177.** Decimal fractions originate from dividing the *unit*, first, into 10 equal parts, and then each of these parts into 10 other equal parts, and so on indefinitely. Thus, $1 \div 10 = \frac{1}{10} = .1$; $\frac{1}{10} \div 10 = \frac{1}{100} = .01$; $\frac{1}{100} \div 10 = \frac{1}{1000} = .001$. Hence,

The unit in decimal fractions is divided into 10, 100, 1000, *&c., equal parts.*

QUESTIONS. — Art. 175. What is a decimal fraction? How are decimal fractions commonly expressed? What is the first figure or place of any decimal? The second? The third? &c. Why? What must be done when a decimal place has no significant figure to fill it? — Art. 176. What is the denominator of a decimal fraction? — Art. 177. How do decimal fractions originate?

ART. **178.** If ciphers are placed on the left hand of decimal figures, between them and the decimal point, those figures change their places, each cipher removing them one place to the right; thus, $.3 = \frac{3}{10}$, but $.03 = \frac{3}{100}$, and $.003 = \frac{3}{1000}$. Hence,

Every cipher placed on the left of decimal figures, between them and the decimal point, decreases the value represented by them the same as dividing by ten.

ART. **179.** If ciphers are placed on the right hand of decimal figures, their places are not changed; thus, $.3 = \frac{3}{10}$, and $.30 = \frac{30}{100} = \frac{3}{10} = .3$. Hence,

Ciphers placed on the right hand of decimals do not alter the value represented by them.

NUMERATION OF DECIMAL FRACTIONS.

ART. **180.** The relation of decimals to whole numbers and to each other, and also the names of their different orders and places, may be learned from the following

TABLE.

Figure	Name	Order	
7	Millions.	7th order or place,	Whole Numbers.
6	Hunds. of Thousands.	6th order or place,	
5	Tens of Thousands.	5th order or place,	
4	Thousands.	4th order or place,	
3	Hundreds.	3d order or place,	
2	Tens.	2d order or place,	
1	Units.	1st order or place,	
.	Separatrix.		
2	Tenths.	1st order or place,	Decimals.
3	Hundredths.	2d order or place,	
4	Thousandths.	3d order or place,	
5	Ten Thousandths.	4th order or place,	
6	Hundred Thousandths.	5th order or place,	
7	Millionths.	6th order or place,	
8	Ten Millionths.	7th order or place,	
9	Hundred Millionths.	8th order or place,	
3	Billionths.	9th order or place,	

QUESTIONS. — Art. 178. What effect have ciphers placed at the left hand of decimals? Why? — Art. 179. What effect if placed at the right hand? Why? — Art. 180. What may be learned from the table?

The preceding table consists of a *whole number* and *decimal*, which, taken together, are called a *mixed number*. The part on the left of the separatrix is the whole number, and that on the right the decimal. The decimal part is numerated from the left to the right, and its value is expressed in words thus: Two hundred thirty-four millions five hundred sixty-seven thousand eight hundred ninety-three billionths. And the mixed number thus: Seven millions six hundred fifty-four thousand three hundred twenty-one, and two hundred thirty-four millions five hundred sixty-seven thousand eight hundred ninety-three billionths.

From the table and explanations we have the following rule for reading decimals:

RULE. — *Read the decimal as though it were a whole number, adding the name of the right-hand order.*

The pupil may write in words, or read orally, the following numbers:

1.	.5	5.	.3001	9.	.72859
2.	.42	6.	.0984	10.	12.02003
3.	.01	7.	.00013	11.	121.000386
4.	.908	8.	.82007	12.	2.3058217

NOTATION OF DECIMAL FRACTIONS.

ART. **181.** By examining the decimal table we perceive that tenths occupy the first place, hundredths the second, &c., and that each figure takes its value by its distance from the place of units; therefore, to write decimals, we have the following

RULE. — *Write the decimal as though it were a whole number, supplying with ciphers such places as have no significant figures.*

The pupil may write in figures the following numbers:

1. Three hundred seven, and twenty-five hundredths.
2. Forty-seven, and seven tenths.

QUESTIONS. — Of what does it consist? What is the number called, when taken together? What is the part on the left hand of the separatrix? The part on the right? What is the value of the decimal? What is the value of the mixed number? What is the rule for reading decimals? — Art. 181. Upon what does the value of a decimal figure depend? What is the rule for writing decimals?

3. Eighteen, and five hundredths.
4. Twenty-nine, and three thousandths
5. Forty-nine ten thousandths.
6. Eight, and eight millionths.
7. Seventy-five, and nine tenths.
8. Two thousand, and two thousandths.
9. Eighteen, and eighteen thousandths.
10. Five hundred five, and one thousand six millionths.
11. Three hundred, and forty-two ten millionths.
12. Twenty-five hundred, and thirty-seven billionths.

ART. **182.** It will be seen that decimals increase from right to left, and decrease from left to right, by the same law as do simple whole numbers; hence they may be added, subtracted, multiplied, and divided, in the same manner.

ADDITION OF DECIMALS.

ART. **183.** Ex. 1. Add together 5.018, 171.16, 88.133, 1113.6, .00456, and 14.178. Ans. 1392.09356.

OPERATION.
```
    5.018
  171.16
   88.133
 1113.6
     .00456
   14.178
 ----------
 1392.09356
```

We write the numbers so that figures of the same decimal place shall stand in the same column, and then, beginning at the right hand, add them as whole numbers, and place the decimal point in the result directly under those above.

RULE. — *Write the numbers so that figures of the same decimal place shall stand in the same column.*

Add as in whole numbers, and point off, in the sum, from the right hand as many places for decimals as equal the greatest number of decimal places in any of the numbers added.

Proof. — The proof is the same as in addition of simple numbers.

EXAMPLES FOR PRACTICE.

2. Add together 171.61111, 16.7101, .00007, 71.0006, and 1.167895. Ans. 260.489775.

3. Add together .16711, 1.766, 76111.1, 167.1, .000007, and 1476.1. Ans. 77756.233117.

QUESTIONS. — Art. 182. How do decimals increase and decrease? How may they be added, subtracted, multiplied, and divided? — Art. 183. How are decimals arranged for addition? What is the rule for addition of decimals? What is the proof?

4. Add together 151.01, 611111.01, 16.5, 6.7, 46.1, and .67896. Ans. 611331.99896.

5. Add fifty-six thousand, and fourteen thousandths; nineteen, and nineteen hundredths; fifty-seven, and forty-eight ten thousandths; twenty-three thousand five, and four tenths; and fourteen millionths. Ans. 79081.608814.

6. What is the sum of forty-nine, and one hundred and five ten thousandths; eighty-nine, and one hundred seven thousandths; one hundred twenty-seven millionths; forty-eight ten thousandths? Ans. 138.122427.

7. What is the sum of three, and eighteen ten thousandths; one thousand five, and twenty-three thousand forty-three millionths; eighty-seven, and one hundred seven thousandths; forty-nine ten thousandths; forty-seven thousand, and three hundred nine hundred thousandths? Ans. 48095.139833.

SUBTRACTION OF DECIMALS.

ART. 184. Ex. 1. From 74.806 take 49.054.

Ans. 25.752.

OPERATION.
```
74.806
49.054
------
25.752
```

Having written the less number under the greater, so that figures of the same decimal place stand in the same column, we subtract as in whole numbers, and place the decimal point in the result, as in addition of decimals.

RULE. — *Write the less number under the greater, so that figures of the same decimal place shall stand in the same column.*

Subtract as in whole numbers, and point off the remainder as in addition of decimals.

Proof. — The proof is the same as in subtraction of simple numbers.

EXAMPLES FOR PRACTICE.

2.	3.	4.	5.
11.078	47.117	46.13	87.107
9.81	8.78195	7.8915	1.11986
1.268	38.33505	38.2385	85.98714

6. From 81.35 take 11.678956. Ans. 69.671044.
7. From 1 take .876543. Ans. .123457.
8. From 100 take 99.111176. Ans. .888824.
9. From 87.1 take 5.6789. Ans. 81.4211.

QUESTIONS. — Art. 184. What is the rule for subtraction of decimals? What is the proof?

10. From 100 take .001. Ans. 99.999.

11. From seventy-three, take seventy-three thousandths. Ans. 72.927.

12. From three hundred sixty-five take forty-seven ten thousandths. Ans. 364.9953.

13. From three hundred fifty-seven thousand take twenty-eight, and four thousand nine ten millionths. Ans. 356971.9995991.

14. From .875 take .4. Ans. .475

15. From .3125 take .125. Ans. .1875

16. From .95 take .44. Ans. .51.

17. From 3.7 take 1.8. Ans. 1.9.

18. From 8.125 take 2.6875. Ans. 5.4375.

19. From 9.375 take 1.5. Ans. 7.875.

20. From .666 take .041. Ans. .625.

MULTIPLICATION OF DECIMALS.

ART. 185. Ex. 1. Multiply 18.72 by 7.1. Ans. 132.912.

OPERATION.

```
    18.72
      7.1
   ------
     1872
   13104
   ------
  132.912
```

We multiply as in whole numbers, and point off on the right of the product as many figures for decimals as there are decimal figures in the multiplicand and multiplier counted together.

The reason for pointing off decimals in the product as above will be seen, if we convert the multiplicand and multiplier into common fractions, and multiply them together. Thus, $18.72 = 18\frac{72}{100} = \frac{1872}{100}$; and $7.1 = 7\frac{1}{10} = \frac{71}{10}$. Then $\frac{1872}{100} \times \frac{71}{10} = \frac{132912}{1000} = 132\frac{912}{1000} = 132.912$, Ans., the same as in the operation.

Ex. 2. Multiply 5.12 by .012.

OPERATION.

```
    5.12
    .012
   -----
    1024
    512
   -----
  .06144 Ans.
```

Since the number of figures in the product is not equal to the number of decimals in the multiplicand and multiplier, we supply the deficiency by placing a cipher on the left hand.

The reason of this process will appear, if we perform the question thus: $5.12 = 5\frac{12}{100} = \frac{512}{100}$, and $.012 = \frac{12}{1000}$. Then $\frac{512}{100} \times \frac{12}{1000} = \frac{6144}{100000} = .06144$, Ans., the same as before. Hence we deduce the following

QUESTIONS. — Art. 185. In multiplication of decimals how do you point off the product? Will you give the reason for it? When the number of figures in the product is not equal to the number of decimals in the multiplicand and multiplier, what must be done?

RULE. — *Multiply as in whole numbers, and point off as many figures for decimals, in the product, as there are decimals in the multiplicand and multiplier.*

If there be not so many figures in the product as there are decimal places in the multiplicand and multiplier, supply the deficiency by prefixing ciphers.

NOTE. — When a decimal number is to be multiplied by 10, 100, 1000, &c., remove the decimal point as many places to the right as there are ciphers in the multiplier; and if there be not figures enough in the number, annex ciphers. Thus, $1.25 \times 10 = 12.5$; and $1.7 \times 100 = 170$.

Proof. — The proof is the same as in multiplication of simple numbers.

EXAMPLES FOR PRACTICE.

3. Multiply 18.07 by .007. Ans. .12649.
4. Multiply 18.46 by 1.007. Ans. 18.58922.
5. Multiply .00076 by .0015. Ans. .00000114.
6. Multiply 11.37 by 100. Ans. 1137.
7. Multiply 47.01 by .047. Ans. 2.20947.
8. Multiply .0701 by .0067. Ans. .00046967.
9. Multiply 47 by .47. Ans. 22.09.
10. Multiply eighty-seven thousandths by fifteen millionths. Ans. .000001305.
11. Multiply one hundred seven thousand, and fifteen ten thousandths by one hundred seven ter thousandths. Ans. 1144.90001605.
12. Multiply ninety-seven ten thousandths by four hundred, and sixty-seven hundredths. Ans. 3.886499.
13. Multiply ninety-six thousandths by ninety-six hundred thousandths. Ans. .00009216.
14. Multiply one million by one millionth. Ans. 1.
15. Multiply one hundred by fourteen ten thousandths. Ans. .14.
16. Multiply one hundred one thousandths by ten thousand one hundred one hundred thousandths. Ans. .01020201.
17. Multiply one thousand fifty, and seven ten thousandths by three hundred five hundred thousandths. Ans. 3.202502135.
18. Multiply two million by seven tenths. Ans. 1400000.

QUESTIONS. — What is the rule for multiplication of decimals? What is the proof? How do you multiply a decimal by 10, 100, 1000, &c.?

19. Multiply four hundred, and four thousandths by thirty, and three hundredths. Ans. 12012.12012.

20. What cost 46lb. of tea at \$1.125 per pound? Ans. \$51.75.

21. What cost 17.125 tons of hay at \$18.875 per ton? Ans. \$323.234375.

22. What cost 18lb. of sugar at \$0.125 per pound? Ans. \$2.25.

23. What cost 375.25bu. of salt at \$0.62 per bushel? Ans. \$232.655.

DIVISION OF DECIMALS.

ART. **186.** Ex. 1. Divide 45.625 by 12.5. Ans. 3.65.

```
       OPERATION.
12.5 ) 4 5.6 2 5 ( 3.6 5
       3 7 5
       -----
         8 1 2
         7 5 0
         -----
           6 2 5
           6 2 5
           -----
```

We divide as in whole numbers, and since the divisor and quotient are the two factors, which, being multiplied together, produce the dividend, we point off two decimal figures in the quotient, to make the number in the two factors equal to the product or dividend.

The reason for pointing off will also be seen by performing the question with the decimals in the form of common fractions. Thus, 45.625 = $45\frac{625}{1000} = \frac{45625}{1000}$, and $12.5 = 12\frac{5}{10} = \frac{125}{10}$. Then $\frac{45625}{1000} \div \frac{125}{10} = \frac{45625}{1000} \times \frac{10}{125} = \frac{456250}{125000} = \frac{365}{100} = 3\frac{65}{100} = 3.65$, Ans., as before.

Ex. 2. Divide .175 by 2.5. Ans. .07.

```
      OPERATION.
2.5 ) .1 7 5 ( .0 7
        1 7 5
        -----
```

We divide as in whole numbers, and since we have but one figure in the quotient, we place a cipher before it, which removes it to the place of hundredths, and thus makes the decimal places in the divisor and quotient equal to those of the dividend.

The reason for *prefixing* the cipher will appear more obvious by solving the question with the decimals in the form of common fractions. Thus, $.175 = \frac{175}{1000}$, and $2.5 = 2\frac{5}{10} = \frac{25}{10}$. Then $\frac{175}{1000} \div \frac{25}{10} = \frac{175}{1000} \times \frac{10}{25} = \frac{1750}{25000} = \frac{7}{100} = .07$, Ans., as before. Hence the following

---

QUESTIONS. — Art. 186. In division of decimals how do you point off the quotient? What is the reason for it? If the decimal places of the divisor and quotient are not equal to the dividend, what must be done?

RULE. — *Divide as in whole numbers, and point off as many decimals in the quotient as the decimals in the dividend exceed those of the divisor; but if there are not as many, supply the deficiency by prefixing ciphers.*

NOTE 1. — When the decimal places in the divisor exceed those in the dividend, make them equal by annexing ciphers to the dividend, and the quotient will be a whole number.

NOTE 2. — When there is a remainder after dividing the dividend, ciphers may be annexed, and the division continued, the ciphers thus annexed being regarded as decimals of the dividend; to indicate in any case that the division does not terminate, the sign plus (+) can be used.

NOTE 3. — When a decimal number is to be divided by 10, 100, 1000, &c., remove the decimal point as many places to the left as there are ciphers in the divisor, and if there be not figures enough in the number, prefix ciphers. Thus 1.25 ÷ 10 = .125; and 1.7 ÷ 100 = .017.

Proof. — The proof is the same as in division of simple numbers.

EXAMPLES FOR PRACTICE.

3. Divide 183.375 by 489. Ans. .375.
4. Divide 67.8632 by 32.8. Ans. 2.069.
5. Divide 67.56785 by .035. Ans. 1930.51.
6. Divide .567891 by 8.2. Ans. .069255.
7. Divide .1728 by 10. Ans. .01728.
8. Divide 13.50192 by 1.38. Ans. 9.784.
9. Divide 783.5 by 6.25. Ans. 125.36.
10. Divide 983 by 6.6. Ans. 148.939 +.
11. Divide 172.8 by 1.2. Ans.
12. Divide 1728 by .12. Ans.
13. Divide .1728 by .12. Ans.
14. Divide 1.728 by 12. Ans.
15. Divide 17.28 by 1.2. Ans.
16. Divide 1728 by .0012. Ans.
17. Divide .001728 by 12. Ans.
18. Divide 116.31 by 1000. Ans. .11631.
19. Divide one hundred forty-seven, and eight hundred twenty-eight thousandths by nine, and seven tenths. Ans. 15.24.
20. Divide seventy-five, and sixteen hundredths by five, and forty-two thousand eight hundred one hundred thousandths. Ans. 13.846+.

QUESTIONS. — What is the rule for division of decimals? What is note 1? Note 2? Note 3? What is the proof?

21. Divide six hundred seventy-eight thousand seven hundred sixty-seven millionths, by three hundred twenty-eight thousandths. Ans. 2.069+.

REDUCTION OF DECIMALS.

ART. 187. To reduce a common fraction to a decimal.

Ex. 1. Reduce $\frac{5}{8}$ to a decimal. Ans. .625.

OPERATION.

```
8 ) 5.0 ( 6 tenths.
    4 8
    ---
  8 ) 2 0 ( 2 hundredths.
      1 6
      ---
    8 ) 4 0 ( 5 thousandths.
        4 0
        ---  Ans. .625.

Or thus: 8 ) 5.0 0 0
             -------
              .6 2 5
```

Since we cannot divide the numerator, 5, by 8, we reduce it to tenths by annexing a cipher, and then dividing, we obtain 6 tenths and a remainder of 2 tenths. Reducing this remainder to hundredths by annexing a cipher, and dividing, we obtain 2 hundredths and a remainder of 4 hundredths, which being reduced to thousandths by annexing a cipher, and then dividing again, gives a quotient of 5 thousandths. The sum of the several quotients, .625, is the answer.

To prove that 625 is equal to $\frac{5}{8}$, we change it to the form of a common fraction, by writing its denominator (Art. 176), and reduce it to its lowest terms. Thus, $\frac{625}{1000} = \frac{5}{8}$, Ans. Hence the following

RULE. — *Annex ciphers to the numerator, and divide by the denominator. Point off in the quotient as many decimal places, as there have been ciphers annexed.*

EXAMPLES FOR PRACTICE.

2. Reduce $\frac{3}{4}$ to a decimal. Ans. .75.
3. Reduce $\frac{7}{8}$ to a decimal. Ans. .875.
4. What decimal fraction is equal to $\frac{7}{16}$? Ans. .4375.
5. Reduce $\frac{4}{17}$ to a decimal. Ans. .235294+.
6. Reduce $\frac{4}{11}$ to a decimal. Ans. .363636+.
7. Reduce $\frac{5}{12}$ to a decimal. Ans. .416666+.

NOTE. — In reducing a common fraction to a decimal, when the denominator contains other prime factors than 2 and 5, there cannot be an exact division of the numerator; but, on continuing the division, some figure or figures of the quotient will be continually repeated.

A decimal, of which there is a continual repetition of the same figure or figures, is called an *infinite or circulating decimal.*

The figures that repeat are called *repetends.* When the repetend is pre-

---

QUESTIONS. — Art 187. How do you reduce a common fraction to a decimal? How can you prove the answer correct? What is the rule for reducing a common fraction to a decimal?

ceded by another decimal, the whole is called a *mixed* repetend, and the part not repeating is called the *finite* part. To mark a repetend a dot (.) is placed over the first and last of the repeating figures. Thus, the answer to example sixth, $.\dot{3}\dot{6}$, is a repetend ; and the answer to example seventh, $.41\dot{6}$, is a mixed repetend, of which the figure 6 is the repetend, and the figures 41 the finite part.

To change an infinite decimal to an equivalent common fraction, we *write the repetend for the numerator, and as many nines as the repetend has figures for the denominator.* Thus, $.\dot{3}\dot{6} = \frac{36}{99} = \frac{4}{11}$; and the mixed repetend, $.41\dot{6} = \frac{41\frac{6}{9}}{100} = \frac{375}{900} = \frac{5}{12}$.

A decimal other than a repetend is changed to the form of a common fraction, simply *by writing the denominator under the given numerator.* (Art. 176.) Thus, $.75 = \frac{75}{100} = \frac{3}{4}$; $.005 = \frac{5}{1000} = \frac{1}{200}$.

8. Reduce .875 to a common fraction.

9. Change .4375 to the form of a common fraction.

10. Change $.\dot{7}\dot{2}$ to a common fraction. Ans. $\frac{8}{11}$.

11. Change $.\dot{1}3\dot{5}$ to a common fraction. Ans. $\frac{5}{37}$.

12. What common fraction is equivalent to $.235\dot{6}\dot{2}$? Ans. $\frac{23327}{99000}$.

13. Change $.09\dot{3}$ to an equivalent common fraction. Ans. $\frac{7}{75}$.

ART. 188. To reduce a compound number to a decimal of a higher denomination.

Ex. 1. Reduce 8s. 6d. 3far. to the decimal of a pound. Ans. .428125.

OPERATION.

```
 4 | 3.0 0
   -------
12 | 6.7 5 0 0
   -----------
20 | 8.5 6 2 5 0 0
   ---------------
      .4 2 8 1 2 5
```

We commence with the 3far., and first reduce them to hundredths, by annexing two ciphers; and then, to reduce these to the decimal of a penny, we divide by 4, since there will be $\frac{1}{4}$ as many hundredths of a penny as of a farthing, and obtain .75d. Annexing this decimal to the 6d., we divide by 12, since there will be $\frac{1}{12}$ as many shillings as pence ; and then the 8s. and this quotient by 20, since there will be $\frac{1}{20}$ as many pounds as shillings, and obtain .428125£. for the answer Hence the following

RULE. — *Divide the lowest denomination, annexing ciphers if necessary, by that number which will reduce it to one of the next higher denomination. Then divide as before, and so continue dividing till the decimal is of the denomination required.*

QUESTIONS. — What is an infinite decimal? What is a repetend? What is a mixed repetend? How is an infinite decimal changed to the form of a common fraction? — Art. 188. What is the rule for reducing a compound number to a decimal of a higher denomination?

NOTE. — A compound number may also be reduced to a decimal by first reducing it to a common fraction (Art. 170), and then this fraction to a decimal. (Art. 187.) Thus, 2s. 6d. $= \frac{30}{240} = \frac{1}{8} =$ .125£.

EXAMPLES FOR PRACTICE.

2. Reduce 15s. 6d. to the fraction of a pound. Ans. .775.
3. Reduce 5cwt. 2qr. 14lb. to the decimal of a ton. Ans. .282.
4. Reduce 3qr. 21lb. to the decimal of a cwt. Ans. .96.
5. Reduce 6fur. 8rd. to the decimal of a mile. Ans. .775.
6. Reduce 3R. 19p. 167ft. 72in. to the decimal of an acre. Ans. .872595+.

ART. **189.** To find the value of a decimal in whole numbers of a lower denomination.

Ex. 1. What is the value of .9875 of a pound? Ans. 19s. 9d.

OPERATION.

```
  .9 8 7 5
       2 0
 ---------
 1 9.7 5 0 0
       1 2
 ---------
  9.0 0 0 0
```

There will be 20 times as many ten thousandths of a shilling as of a pound; therefore, we multiply the decimal, .9875, by 20, and reduce the improper fraction to a mixed number by pointing off four figures on the right, which is dividing by its denominator, 10000. The figures on the left of the point are shillings, and those on the right decimals of a shilling. This decimal of a shilling we multiply by 12, and, pointing off as before, obtain 9d., which, added to the 19s., gives 19s. 9d. for the answer.

RULE. — *Multiply the decimal by that number which will reduce it to the lower denomination, and point off as in multiplication of decimals.*

Then, multiply the decimal part of the product, and point off as before. So continue till the decimal is reduced to the denominations required.

The several whole numbers of the successive products will be the answer.

EXAMPLES FOR PRACTICE.

2. What is the value of .628125 of a pound? Ans. 12s. $6\frac{3}{4}$d.
3. What is the value of .778125 of a ton? Ans. 15cwt. 2qr. 6lb. 4oz.
4. What is the value of .75 of an ell English? Ans. 3qr. 3na.
5. What is the value of .965625 of a mile? Ans. 7fur. 29rd.

QUESTION. — Art. 189. What is the rule for finding the value of a decimal in whole numbers of a lower denomination?

6. What is the value of .94375 of an acre?
Ans. 3R. 31p.

7. What is the value of .815625 of a pound Troy?
Ans. 9oz. 15pwt. 18gr.

8. What is the value of .5555 of a pound apothecaries' weight? Ans. 6℥ 5ʒ 0℈ 19$\frac{17}{25}$gr

MISCELLANEOUS EXERCISES IN DECIMALS.

1. What is the value of 15cwt. 3qr. 14lb. of coffee at $9.50 per cwt.? Ans. $150.95½.

2. What cost 17T. 18cwt. 1qr. 7lb. of potash at $53.80 per ton? Ans. $963.88+.

3. What cost 37A. 3R. 16p. of land at $75.16 per acre?
Ans. $2844.80+.

4. What cost 15yd. 3qr. 2na. of cloth at $3.75 per yard?
Ans. $59.53+.

5. What cost 15⅜ cords of wood at $4.62½ per cord?
Ans. $71.10+.

6. What cost the construction of 17m. 6fur. 36rd. of railroad at $3765.60 per mile? Ans. $67263.03+.

7. What cost 27hhd. 21gal. of temperance wine at $15.37½ per hogshead? Ans. $420.24+.

8. What are the contents of a pile of wood, 18ft. 9in. long, 4ft. 6in. wide, and 7ft. 3in. high? Ans. 611ft. 1242in.

9. What are the contents of a board 12ft. 6in. long, and 2ft. 9in. wide? Ans. 34ft. 54in.

10. Bought a cask of vinegar containing 25gal. 3qt. 1pt. at $0.37½ per gallon; what was the amount? Ans. $9.70+.

11. Bought a farm containing 144A. 3R. 30p. at $97.62½ per acre; what was the cost of the farm?
Ans. $14149.52+.

12. Sold Joseph Pearson 3T. 18cwt. 21lb. of salt hay at $9.37½ per ton. He having paid me $20.25, what remains due? Ans. $16.41+.

13. If ⅞ of a cord of wood cost $5.50, what cost one cord? What cost 7¾ cords? Ans. $48.71+.

14. If 4¾ yards of cloth cost $12⅝, what cost 17⅜ yards?
Ans. $46.18+.

15. The ship Constantine cost $35000; ¼ of it was sold to Captain Sampson for $9000; ⅓ of the remainder to T. Lamb for $9200, and the balance to another person at a profit of $500; what was gained in the sale of the whole ship? Ans. $1200.

§ XX. PERCENTAGE.

ART. **190.** PERCENTAGE and *per cent.* are terms derived from Latin words, *per centum*, which signify *by the hundred.* Percentage, therefore, is any rate or sum on a hundred, or it is any number of hundredths. Thus, if an article is bought for $100, and sold for $105, the gain is 5 per cent., because $5 are $\frac{5}{100}$ of $100, or of the original cost. Again, if an article is bought for $25, and sold for $30, the gain is 20 per cent., because $5 are $\frac{5}{25} = \frac{20}{100}$ of $25, or of the original cost.

Since per cent. is any number of hundredths, it is a decimal written in the same manner as hundredths in decimal fractions. Thus, 5 per cent., 25 per cent., &c., are written .05, .25, respectively. (Art. 175.)

When the per cent. is more than 100 it is an improper fraction, and if expressed decimally it becomes a mixed number; thus, 103 per cent., equal to $\frac{103}{100}$, is written 1.03.

If the per cent. is less than 1, or a part of *one* hundredth, to be expressed decimally, it must be written at the right of hundredths in the place of thousandths, &c. Thus, $\frac{1}{2}$ of 1 per cent., $\frac{3}{4}$ of 1 per cent., $12\frac{1}{5}$ per cent., are written .005, .0075, .122, respectively.

EXAMPLES.

Write decimally 2 per cent.; 3 per cent.; 5 per cent.; 6 per cent.; 7 per cent.; 8 per cent.; 10 per cent.; 12 per cent., 15 per cent.; 25 per cent.; 30 per cent.; 40 per cent.; 50 per cent.; 60 per cent.; 75 per cent.; 100 per cent.; 105 per cent., 115 per cent.; $6\frac{1}{2}$ per cent.; $8\frac{3}{4}$ per cent.; $20\frac{1}{5}$ per cent.; $\frac{1}{4}$ of 1 per cent.; $\frac{1}{2}$ of 1 per cent.; $\frac{4}{5}$ of 1 per cent.; $\frac{9}{10}$ of 1 per cent.; $\frac{7}{8}$ of 1 per cent.

ART. **191.** To find the percentage on any sum or quantity.

Ex. 1. Bought a house for $625, and sold it at 6 per cent. advance; what did I gain by the sale? Ans. $37.50.

QUESTIONS. — Art. 190. From what are the terms percentage and per cent. derived, and what do they signify? How, then, will you define percentage? How will you illustrate it? How is per cent. written when less than 100? How, when more than 100? If the per cent. is a fraction, or contains a fraction, what is the fraction, and, if expressed decimally, what place must it occupy?

OPERATION.	
Sum,	\$6 2 5
Rate per cent.,	.0 6
Per cent., or gain,	\$3 7.5 0

Since 6 per cent. is $\frac{6}{100} = .06$ of the original cost, we multiply \$625 by the decimal expression .06, and point off as in multiplication of decimal fractions.

RULE. — *Multiply the given quantity or number by the rate per cent., considered as a decimal, and point off the product as in multiplication of decimal fractions.* (Art. 185.)

NOTE. — If the per cent. contains a fraction that cannot be expressed decimally, or, if thus expressed, would require several figures, it is more convenient to multiply by it as a mixed number. (Art. 155.)

EXAMPLES FOR PRACTICE.

2. What is 2 per cent. of \$325? Ans. \$6.50.
3. What is 5 per cent. of \$789? Ans. \$39.45.
4. What is 6 per cent. of \$856.49? Ans. \$51.38,9.
5. What is $7\frac{1}{2}$ per cent. of 765 tons? Ans. 57.375 tons.
6. What is $9\frac{4}{5}$ per cent. of \$5000? Ans. \$490.
7. What is $\frac{7}{8}$ per cent. of \$1728? Ans. \$15.12.
8. What is $4\frac{1}{2}$ per cent. of 587 yards of cloth? Ans. 26.415 yards.
9. I lost 10 per cent. of \$975; how much have I remaining? Ans. \$877.50.
10. Sent to Liverpool 5000 bushels of wheat, which cost me \$1.25 per bushel; but 25 per cent. of the wheat was thrown overboard in a storm, and the remainder was sold at \$2 per bushel; what was gained on the wheat? Ans. \$1250.
11. T. Page received a legacy of \$8000; he gave 19 per cent. of it to his wife, 37 per cent. of the remainder to his sons, and \$2000 to his daughters; what sum had he remaining? Ans. \$2082.40.
12. My tailor informs me it will take 10 square yards of cloth to make me a full suit of clothes. The cloth I am about to purchase is $1\frac{3}{4}$ yards wide, and on sponging it will shrink 5 per cent. in width and 5 per cent. in length. How many yards of the above cloth must I purchase for my "new suit"? Ans. 6yd. 1qr. $1\frac{773}{2527}$na.
13. A man having \$10000, lost 15 per cent. of it in speculation; what sum had he remaining? Ans. \$8500.

QUESTIONS. — Art. 191. Will you explain the operation for finding the percentage on any sum or quantity? Give the reason for the process. What is the rule?

§ XXI. SIMPLE INTEREST.

ART. 192. INTEREST is the compensation which the borrower of money makes to the lender.

The *rate per cent.* is the sum paid for the use of $100, 100 cents, 100£., &c., for one year.

The *principal* is the sum lent, on which interest is computed.

The *amount* is the interest and principal added together.

Legal interest is the rate per cent. established by law.

Usury is a higher rate per cent. than is allowed by law.

The legal rate per cent. varies in the different States and in different countries.

In Maine, New Hampshire, Vermont, Massachusetts, Rhode Island, Connecticut, New Jersey, Pennsylvania, Delaware, Maryland, Virginia, North Carolina, Tennessee, Kentucky, Ohio, Indiana, Illinois, Iowa, Missouri, Arkansas, Mississippi, Florida, District of Columbia, and on debts or judgments in favor of the United States, it is 6 per cent.

In New York, Michigan, Wisconsin, Minnesota, Georgia, and South Carolina, it is 7 per cent.

In Alabama and Texas, it is 8 per cent.

In California, it is 10 per cent.

In Louisiana, it is 5 per cent.

In Canada, Nova Scotia, and Ireland, it is 6 per cent.

In England and France, it is 5 per cent.

NOTE. — The legal rate, as above, in some of the States, is only that which the law allows, when no particular rate is mentioned. By special agreement between parties, in Ohio, Indiana, Michigan, Illinois, Iowa, and Arkansas, interest can be taken as high as 10 per cent. ; in Florida and Louisiana, as high as 8 per cent. ; in Texas and Wisconsin, as high as 12 per cent. ; and in California, any per cent. In New Jersey, by a special law, 7 per cent. may be taken in the city of Paterson, and in the counties of Essex and Hudson.

ART. 193. To find the interest of $1 at 6 per cent. for any given time.

Since the interest of $1 is 6 cents, or $\frac{6}{100}$ of the principal, for 1 year, or 12 months, for 1 month it will be $\frac{1}{12}$ of 6 cents, or $\frac{1}{2}$ a cent, equal to 5 mills, or $\frac{1}{200}$ of the principal; and for 2

QUESTIONS. — Art. 192. What is interest? What is rate per cent.? What is the principal? What is the amount? What is legal interest? What is usury? What is the legal rate per cent. in the different States? In Canada, Nova Scotia, and Ireland? In England and France?

months, twice 5 mills, or 1 cent, or $\frac{1}{100}$ of the principal. Now, since the interest for 1 month, or 30 days, is 5 mills, the interest for 6 days, or $\frac{1}{5}$ of 30 days, will be one mill, or $\frac{1}{1000}$ of the principal. And as 1 day, 2 days, &c., are $\frac{1}{6}$, $\frac{2}{6}$, &c., of 6 days, the interest for any number of days less than 6 will be as many sixths of a mill, or six thousandths of the principal, as there are days. Also, since the interest for 2 months is 1 cent, or $\frac{1}{100}$ of the principal, for 100 times 2 months, or 200 months, or 16 years 8 mo., it will be 100 cents, or equal the whole principal; and in the same proportion for any other length of time. Hence, the

INTEREST OF $1, AT 6 PER CENT., FOR

1 yr., or 12 mo., is 6 cents, or $0.06, equal $\frac{6}{100}$ of the prin.
$\frac{1}{6}$ of a yr., or 2 mo., is 1 cent, or $0.01, equal $\frac{1}{100}$ of the prin.
$\frac{1}{12}$ of a yr., or 1 mo., is 5 mills, or $0.005. equal $\frac{1}{200}$ of the prin.
$\frac{1}{5}$ of a mo., or 6 da., is 1 mill, or $0.001, equal $\frac{1}{1000}$ of the prin.
$\frac{1}{30}$ of a mo., or 1 da., is $\frac{1}{6}$ of a mill, or 0.000\frac{1}{6}$, equal $\frac{1}{6000}$ prin.

ALSO,

16 yr. 8 mo., or 200 mo., is 100 cts., or $1.00, equal the whole prin.
8 yr. 4 mo., or 100 mo., is 50 cts., or $0.50, equal $\frac{1}{2}$ of the prin.
5 yr. 6$\frac{2}{3}$ mo., or 66$\frac{2}{3}$ mo., is 33$\frac{1}{3}$ cts., or 0.333\frac{1}{3}$, equal $\frac{1}{3}$ of prin.
4 yr. 2 mo., or 50 mo., is 25 cts., or $0.25, equal $\frac{1}{4}$ of prin.
3 yr. 4 mo., or 40 mo., is 20 cts., or $0.20, equal $\frac{1}{5}$ of prin.
2 yr. 9$\frac{1}{3}$ mo., or 33$\frac{1}{3}$ mo., is 16$\frac{2}{3}$ cts., or 0.166\frac{2}{3}$, equal $\frac{1}{6}$ of prin.
2 yr. 1 mo., or 25 mo., is 12$\frac{1}{2}$ cts., or $0.125, equal $\frac{1}{8}$ of prin.
1 yr. 8 mo., or 20 mo., is 10 cts., or $0.10, equal $\frac{1}{10}$ of prin.
1 yr. 4$\frac{2}{3}$ mo., or 16$\frac{2}{3}$ mo., is 8$\frac{1}{3}$ cts., or 0.083\frac{1}{3}$, equal $\frac{1}{12}$ of prin.
$\frac{5}{6}$ of a yr., or 10 mo., is 5 cts., or $0.05, equal $\frac{1}{20}$ of prin.
$\frac{5}{9}$ of a yr., or 6$\frac{2}{3}$ mo., is 3$\frac{1}{3}$ cts., or 0.033\frac{1}{3}$, equal $\frac{1}{30}$ of prin.
$\frac{5}{12}$ of a yr., or 5 mo., is 2$\frac{1}{2}$ cts., or $0.025, equal $\frac{1}{40}$ of prin.
$\frac{1}{3}$ of a yr., or 4 mo., is 2 cts., or $0.02, equal $\frac{1}{50}$ of prin.

Ex. 1. What is the interest of $1 for 2yr. 7mo. 20da.?
Ans. 0.158\frac{1}{3}$.

FIRST OPERATION.

Interest for 2y. = .1 2
" " 7mo. = .0 3 5
" " 20da. = .0 0 3$\frac{1}{3}$
Ans. $ 0.1 5 8$\frac{1}{3}$

The interest for 2 years will be twice as much as for 1 year, equal 12 cents; and since the interest for 2 months is 1 cent, for 7 months it will be 3$\frac{1}{2}$ cents. And as the interest for 6 days is 1 mill, for 20 days it will be 3$\frac{1}{3}$ mills. Adding the several sums together, we have 0.158\frac{1}{3}$ for the answer.

QUESTION. — Art. 193. How do you explain the operation?

SECOND OPERATION.

Principal,	$ 1.0 0	
$\frac{1}{8}$ of the prin.,	.1 2 5	Int. for 2yr. 1mo.
$\frac{1}{30}$ of the prin.,	.0 3 3$\frac{1}{3}$	Int. for 6mo. 20da.
	$ 0.1 5 8$\frac{1}{3}$	Int. for 2y. 7mo. 20da.

The time, 2y 7mo. 20da., is equal to 2y. 1mo. + 6mo. 20da. Now, since the interest on any sum, at 6 per cent., in 200 months equals the principal, for 2y. 1mo., or $\frac{1}{8}$ of 200 months, it will equal $\frac{1}{8}$ of the principal. We, therefore, take $\frac{1}{8}$ of the principal, $1.00, equal 12 cents and 5 mills, the interest for 2y. 1mo. The balance of time, 6mo. 20da., or 6$\frac{2}{3}$mo., being $\frac{1}{30}$ of 200 months, we take $\frac{1}{30}$ of the principal, equal 3 cents and 3$\frac{1}{3}$ mills, as the interest for the 6mo. 20da. We add together the interest for the parts of the whole, and obtain, as by first operation, 0.158\frac{1}{3}$ as the whole interest.

RULE 1. — *Reckon* 6 *cents for every* YEAR, 1 *cent for every* TWO MONTHS, 5 *mills for the odd month*, 1 *mill for every* 6 *days; and for any number of days less than six, as many sixths of a mill as there are days.* Or,

Reduce the years and months to months, and call half the number of months cents, and one sixth the number of days mills. Or,

RULE 2. — *Take such fractional part or parts of the principal as the number expressing the time is of* 200 *months.*

EXAMPLES FOR PRACTICE.

2. What is the interest of $1 for 1y. 4mo. 6da.? Ans. $0.081.
3. What is the interest of $1 for 1y. 9mo. 12da.? Ans. $0.107.
4. What is the interest of $1 for 3y. 8mo. 19da.? Ans. 0.223\frac{1}{6}$.
5. What is the interest of $1 for 2y. 1mo. 20da.? Ans. 0.128\frac{1}{3}$.
6. What is the interest of $1 for 7y. 15da.? Ans. 0.422\frac{1}{2}$.
7. What is the interest of $1 for 3mo. 28d.? Ans. 0.019\frac{2}{3}$.
8. What is the interest of $1 for 4y. 2mo. 5da.? Ans. 0.250\frac{5}{6}$.
9. What is the interest of $1 for 4mo. 3da.? Ans. 0.020\frac{1}{2}$.

QUESTIONS. — How do you explain the second operation? What is the first rule? What is the second rule?

ART. **194.** To find the interest on any sum of money at 6 per cent. for any given time.

Ex. 1. What is the interest of $926 for 3y. 11mo. 15da.? What is the amount?

Ans. Interest, $219.925; Amount, $1145.925.

OPERATION.

Principal,	$9 2 6
Interest of $1,	.2 3 7½
	6 4 8 2
	2 7 7 8
	1 8 5 2
	4 6 3
Int.	$2 1 9.9 2 5
Prin.	9 2 6
Amt.	$1 1 4 5.9 2 5

We find the interest of $1 for the given time to be $0.237½. (Art. 193.) Now, since the interest of $1 is $0.237½, the interest of $926 will be 926 times as much; therefore we multiply them together. To find the amount, we add the principal to the interest. Hence the following

RULE. — *Find the interest of* $1 *for the given time; then multiply the principal by the number denoting this interest, and point off as in multiplication of decimal fractions.* (Art. 185.)

To find the amount, add the principal to the interest.

NOTE. — If the interest of $1 contains a common fraction, the fraction may be reduced to a decimal, if preferred. The interest may also be multiplied by the number denoting the principal, when it is more convenient.

EXAMPLES FOR PRACTICE.

2. What is the interest of $197 for 1 year? Ans. $11.82.
3. What is the interest of $1728 for 3 years? Ans. $311.04.
4. What is the interest of $69 for 2 years? Ans. $8.28.
5. What is the interest of $1728 for 1 year, 6 months? Ans. $155.52.
6. What is the interest of $16.87 for 1 year, 8 months? Ans. $1.687.
7. Required the interest of $118.15 for 2 years, 6 months. Ans. $17.722.
8. Required the interest of $97.16 for 1 year, 5 months. Ans. $8.258.
9. Required the interest of $789.87 for 1 year, 11 months. Ans. $90.835.

QUESTIONS. — Art. 194. Explain the operation for finding the interest on any sum of money at 6 per cent. for any given time. What is the rule? How do you find the amount?

10. Required the amount of $978.18 for 2 years, 3 months. Ans. $1110.234.

11. Required the amount of $87.96 for 1 month. Ans. $88.399.

12. Required the amount of $81.81 for 8 years, 4 months. Ans. $122.715.

13. Required the amount of $0.87 for 7 years, 3 months. Ans. $1.248.

14. What is the interest of $1.71 for 2 years, 2 days? Ans. $0.205.

15. Required the interest of $100 for 8 years, 4 months, 1 day. Ans. $50.016.

16. Required the interest of $3.05 for 2 months, and 2 days. Ans. $0.031.

17. What is the interest of $761.75 for 1 year, 2 months, 18 days? Ans. $55.607.

18. What is the interest of $1728.19 for 1 year, 5 months 10 days? Ans. $149.776.

19. What is the interest of $88.96 for 1 year, 4 months, 6 days? Ans. $7.205.

20. What is the interest of $107.50 for 1 month, 29 days? Ans. $1.057.

ART. **195.** To find the interest of any sum of money at any rate per cent. for any given time.

Ex. 1. What is the interest of $26.25 for 2 years, 4 months, at 7 per cent.? Ans. $4.2875.

OPERATION.

Principal,	$ 2 6.2 5
Interest of $1 at 6 per cent.,	.1 4
	1 0 5 0 0
	2 6 2 5
Interest at 6 per cent.,	$ 3.6 7 5 0
$\frac{1}{6}$ of interest at 6 per cent.,	.6 1 2 5
Interest at 7 per cent.,	$ 4.2 8 7 5

We first find the interest on the given sum at 6 per cent., and then add to this interest the fractional part of itself, denoted by the excess of the rate above 6 per cent. The excess is 1 per cent.; therefore we add $\frac{1}{6}$ of the interest at 6 per cent. to itself, for the answer. If the rate per cent. had been less than 6, we should have subtracted the fractional part.

QUESTION. — Art. 195. Explain the operation for finding the interest on any sum of money at any rate per cent.

RULE. — *Find the interest of the given sum at 6 per cent., and then add to this interest, or subtract from it, such a fractional part of itself as the given rate is greater or less than 6 per cent.*

NOTE 1. — If the rate per cent. is 12 per cent., the interest at 6 per cent. must be doubled.

NOTE 2. — If the interest is for years only, it may be found by multiplying the principal by the interest of $1 for the given time and rate.

EXAMPLES FOR PRACTICE.

1. What is the interest of $144 for one year at 7 per cent.? Ans. $10.08.
2. What is the interest of $850 for 1 year, 7 months, 18 days, at 7 per cent.? Ans. $97.18.
3. What is the interest of $865.75 for 3 years, 9 months, 24 days, at 7 per cent.? Ans. $231.299.
4. What is the interest of $960.18 for 1 year, 2 months, at 7 per cent.? Ans. $78.414.
5. What is the interest of $1728.19 for 3 years, 8 months, 10 days, at 7 per cent.? Ans. $446.929.
6. What is the interest of $17.90 for 8 months, 4 days, at 7 per cent.? Ans. $0.849.
7. What is the interest of $1165.50 for 5 years, 3 months, 9 days, at 7 per cent.? Ans. $430.36.
8. What is the interest of $1237.90 for 1 year, 7 months, 3 days, at 7 per cent.? Ans. $137.922.
9. What is the interest of $156.80 for 3 years and 3 days, at 3 per cent.? Ans. $14.151.
10. What is the interest of $579.75 for 1 year, 2 months, 2 days, at 5 per cent.? Ans. $33.979.
11. What is the interest of $7671.09 for 2 years, 8 months, 5 days, at 8 per cent.? Ans. $1645.02.
12. What is the interest of $943.11 for 1 month, 29 days, at 9 per cent.? Ans. $13.91.
13. What is the interest of $975.06 for 2 years, 7 months, 9 days, at $8\frac{1}{4}$ per cent.? Ans. $209.82.
14. What is the amount of $1000 for 3 years, 3 months, 29 days, at $5\frac{1}{2}$ per cent.? Ans. $1183.18.
15. What is the interest of $765 for 2 years, 9 months, at 1 per cent.? Ans. $21.037.
16. What is the interest of $979.15 for 3 years, 2 months, 4 days, at $12\frac{1}{2}$ per cent.? Ans. $388.94.

QUESTIONS. — What is the rule? What is note first? Note second?

ART. **196.** Second method of finding the interest of any sum of money, at any rate per cent., for any time.

Ex. 1. What is the interest of $26.25 for 3 years, 5 months and 15 days, at 8 per cent.? Ans. $7.262.

OPERATION.

Principal,	$2 6.2 5
Rate per cent.,	.0 8
Interest for 1 year,	2.1 0 0 0
	3
Interest for 3 years,	6.3 0 0 0
Interest for 4mo., $\frac{1}{3}$ of 1y.,	.7 0 0 0
Interest for 1mo., $\frac{1}{4}$ of 4mo.,	.1 7 5 0
Interest for 15da., $\frac{1}{2}$ of 1mo.,	.0 8 7 5
Int. for 3y. 5mo. 15da.,	$7.2 6 2 5

Having found the interest for 1 year and then for 3 years, the interest for 5 months is obtained by first taking $\frac{1}{3}$ of 1 year's interest, for 4 months, and then $\frac{1}{4}$ of this last interest, for 1 month.

And since 15 days are $\frac{1}{2}$ of 1 month, we take $\frac{1}{2}$ of 1 month's interest for the interest of 15 days, and add the several sums together for the answer.

RULE. — *First find the interest for one year by multiplying the principal by the rate per cent.; and for two or more years multiply this product by the number of years.*

Find the interest for months by taking the most convenient fractional part or parts of ONE *year's interest.*

Find the interest for days by taking the most convenient fractional part or parts of ONE *month's interest.*

NOTE. — Many practical men prefer this method of casting interest to any other, but in most questions it is not so expeditious as the preceding. The pupil may be required to solve the questions in interest by both methods.

EXAMPLES FOR PRACTICE.

2. What is the interest of $1775 for 7 years? Ans. $745.50.

3. What is the interest of $987 for 3 years, 6 months? Ans. $207.27.

4. Required the interest of $69.17 for 4 years, 9 months. Ans. $19.713.

5. Required the interest of $96.87 for 10 years, 7 months, 15 days. Ans. $61.754.

QUESTIONS. — Art. 196. Explain the operation for finding the interest of any sum of money, at any rate per cent., for any time. What is the rule?

6. Required the interest of $1.95 for fifteen years, 11 months, 20 days. Ans. $1.868.

7. Required the interest of $1789 for 20 years, 1 month, 25 days. Ans. $2163.199.

8. Required the interest of $666.66 for 6 years, 10 months, 13 days. Ans. $274.775.

9. What is the amount of $98.50 for 5 years, 8 months? Ans. $131.99.

10. What is the amount of $168.13 for 8 years, 5 months, 3 days? Ans. $253.119.

11. What is the amount of $75.75 for 4 years, 2 months, 27 days? Ans. $95.028.

12. Required the amount of $675.50 for 30 years, 3 months, 23 days. Ans. $1904.121.

ART. **197.** To find the interest on pounds, shillings, pence, and farthings, at any rate per cent., for any time.

Ex. 1. What is the interest of 25£. 2s. 6d. for 2 years, 6 months, at 6 per cent.? Ans. 3£. 15s. 5d. 2far.

OPERATION.

```
25£. 2s. 6d. = 2 5.1 2 5 £.
Interest of 1£.      .1 5
                 ----------
                 1 2 5 6 2 5
                 2 5 1 2 5
                 ----------
                 3.7 6 8 7 5 £.=
                 3£. 15s. 4d. 2far.
```

We reduce the 2s. 6d. to the decimal of a pound (Art. 188) and, annexing it to the pounds multiply this principal by the interest of 1£. for the given time. The product is the answer in pounds and the decimal of a pound, which must be reduced to shillings, pence, and farthings. (Art. 189.)

RULE. — *Reduce the shillings, pence, and farthings, to the decimal of a pound, and annex it to the pounds; then proceed as in United States money, and reduce the decimal in the result to a compound number.*

EXAMPLES FOR PRACTICE.

2. What is the interest of 26£. 10s. for 2 years, 4 months, at 5 per cent.? Ans. 3£. 1s. 10d.

3. What is the interest of 42£. 18s. for 1 year, 9 months, 25 days, at 6 per cent.? Ans. 4£. 13s. 7$\frac{3}{4}$d.

4. What is the interest of 94£. 12s. 6d. for 4 years, 6 months 7 days, at 8 per cent.? Ans. 34£. 4s. 2$\frac{3}{4}$d.

QUESTIONS. — Art. 197. How do you find the interest on pounds, shillings, pence, and farthings? Repeat the rule.

MISCELLANEOUS EXERCISES IN INTEREST.

1. What is the interest of $172.50 from Sept. 25, 1850, to July 9, 1852? Ans. $18.515.

2. What is the interest of $169.75 from Dec. 10, 1848, to May 5, 1851? Ans. $24,472.

3. What is the interest of $17.18 from July 29, 1847, to Sept. 1, 1851? Ans. $4.214.

4. What is the interest of $67.07 from April 7, 1849, to Dec. 11, 1851? Ans. $10.775.

5. Required the interest of $117.75 from Jan. 7, 1849, to Dec. 19, 1851. Ans. $20.841.

6. Required the interest of $847.15 from Oct. 9, 1849, to Jan. 11, 1853. Ans. $165.476.

7. Required the interest of $7.18 from March 1, 1851, to Feb. 11, 1852. Ans. $0.406.

8. What is the interest of $976.18 from May 29, 1852, to Nov. 25, 1855? Ans. $204.347.

9. I have John Smith's note for $144, dated July 25, 1849; what is due March 9, 1852? Ans. $166.656.

10. George Cogswell has two notes against J. Doe; the first is for $375.83, and is dated Jan. 19, 1850; the other is for $76.19, dated April 23, 1851; what is the amount of both notes Jan. 1, 1852? Ans. $499.141.

11. What is the interest of $68.19, at 7 per cent., from June 5, 1850, to June 11, 1851? Ans. $4.852.

12. Required the amount of $79.15 from Feb. 17, 1849, to Dec. 30, 1852, at $7\frac{1}{2}$ per cent. Ans. $102.119.

13. What is the amount of $89.96 from June 19, 1850, to Dec. 9, 1851, at $8\frac{1}{4}$ per cent.? Ans. $100.886.

14. A. Atwood has J. Smith's note for $325, dated June 5, 1849; what is due, at $7\frac{1}{4}$ per cent., July 4, 1851? Ans. $374.022.

15. J. Ayer has D. How's note for $1728, dated Dec. 29, 1849; what is the amount Oct. 9, 1852, at 9 per cent.? Ans. $2160.

16. What is the interest of $976.18 from Jan. 29, 1851, to July 4, 1852, at 12 per cent.? Ans. $167.577.

17. What is the amount of $175.08 from May 7, 1851, to Sept. 25, 1853, at 7 per cent.? Ans. $204.289.

18. What is the amount of $160 from Dec. 11, 1853, to Sept. 9, 1854, at 7 per cent.? Ans. $168.337.

PARTIAL PAYMENTS.

ART. **198.** A NOTE, or, as it is often called, a promissory note, or note of hand, is an engagement, in writing, to pay a specified sum, either to a person named in the note, or to his order, or to the bearer.

A *joint* note is one signed by two or more persons, who together are holden for its payment; and a *joint* and *several* note is one signed by two or more persons, who separately and together are holden for its payment.

A *negotiable* note is one so made that it can be sold or transferred from one person to another.

The *maker* or *drawer* of a note is the person who signs it; the *payee*, *promisee*, or *holder*, the person to whom it is to be paid; an *endorser*, the person who writes his name upon its back to transfer it, or as guarantee of its payment; and the *face* of a note, the sum for which it is given.

Partial payments are part payments of a note or other obligation; and being receipted for by an entry on the back of the note or obligation, are called *endorsements*.

ART. **199.** When notes are paid within one year from the time they become due, it has been the usual custom to compute the interest by the following

RULE. — *Find the amount of the principal from the time it became due until the time of payment. Then find the amount of each endorsement from the time it was paid until settlement, and subtract their sum from the amount of the principal.*

Ex. 1. $1234. *Boston, Jan.* 1, 1853.

For value received, I promise to pay John Smith, or order, on demand, one thousand two hundred thirty-four dollars, with interest. *John Y. Jones.*

Attest, Samuel Emerson.

On this note are the following endorsements:

March 1, 1853, received ninety-eight dollars. June 7, 1853, received five hundred dollars. Sept. 25, 1853, received two hundred ninety dollars. Dec. 8, 1853, received one hundred dollars.

What remains due at the time of payment, Jan. 1, 1854?

Ans. $293.12.

QUESTIONS. — Art. 198. What is a note? What a negociable note? A joint note? Who is the maker of a note? Who the payee? Who the endorser? What are partial payments? — Art. 199. What is the rule for computing the interest when there are partial payments, and all are made within one year?

OPERATION.

Principal, - - - - - - -		$1234.00
Int. from Jan. 1, 1853, to Jan. 1, 1854 (1y.), -		74.04
Amount, - - - - - -		1308.04
First payment, March 1, 1853, - -	$98.00	
Int. from March 1, 1853, to Jan. 1, 1854 (10mo.),	4.90	
Second payment, June 7, 1853, - - -	500.00	
Int. from June 7, 1853, to Jan. 1, 1854 (6mo. 24da.),	17.00	
Third payment, Sept. 25, 1853, - -	290.00	
Int. from Sept. 25, 1853, to Jan. 1, 1854 (3mo. 6da.),	4.64	
Fourth payment, Dec. 8, 1853, - -	100.00	
Int. from Dec. 8, 1853, to Jan. 1, 1854 (23da.),	.38	
Amount of payments to be deducted, - - -		$1014.92
Balance remains due Jan. 1, 1854, - - -		$293.12

2. $987.75. *Trenton, Jan.* 11, 1852.

For value received, we jointly and severally promise to pay James Dayton, or order, on demand, two months from date, nine hundred eighty-seven dollars seventy-five cents, with interest after two months. *John T. Johnson.*
Attest, Isaiah Webster. *Samuel Jones.*

On this note are the following endorsements:

May 1, 1852, received three hundred dollars. June 5, 1852, received four hundred dollars. Sept. 25, 1852, received one hundred and fifty dollars.

What is due Dec. 13, 1852? Ans. $156.94.

3. $800. *Indianapolis, July* 4, 1852.

For value received, I promise to pay Leonard Johnson, or order, on demand, eight hundred dollars, with interest.
Attest, Charles True. *Samuel Neverpay.*

On this note are the following endorsements:

Aug. 10, 1852, received one hundred forty-four dollars. Nov. 1, 1852, received ninety dollars. Jan. 1, 1853, received four hundred dollars. March 4, 1853, received one hundred dollars.

What remains due June 1, 1853? Ans. $88.02.

ART. **200.** In the United States court, and in most of the courts of the several States, the following rule is adopted for computing the interest on notes and bonds, when partial payments have been made.

QUESTION. — How do you explain the operation?

Compute the interest on the principal to the time when the first payment was made, which equals, or exceeds, either alone or with preceding payments, the interest then due.

Add that interest to the principal, and from the amount subtract the payment or payments thus far made.

The remainder will form a new principal; on which compute the interest, proceeding as before.

NOTE. — This rule is on the principle, that neither interest nor payment should draw interest.

This rule is illustrated in the following question:

Ex. 1. $365.50. *Wilmington, Jan. 1, 1852.*

For value received, I promise to pay to John Dow, or order, on demand, three hundred sixty-five dollars fifty cents, with interest. *John Smith.*

Attest, Samuel Webster.

On this note are the following endorsements:

June 10, 1852, received fifty dollars. Dec. 8, 1852, received thirty dollars. Sept. 25, 1853, received sixty dollars. July 4, 1854, received ninety dollars. Aug. 1, 1855, received ten dollars. Dec. 2, 1855, received one hundred dollars.

What remains due Jan. 7, 1857? Ans. $92.53.

OPERATION.

Principal carrying interest from Jan. 1, 1852, to June 10, 1852, - - - - - - -	$365.50
Interest from Jan. 1, 1852, to June 10, 1852 (5 months, 9 days). - - - - - -	9.68
Amount, - - - - - - -	375.18
First payment, June 10, 1852, - - - - -	50.00
Balance for new principal, - - - - -	325.18
Interest from June 10, 1852, to Dec. 8, 1852 (5 months, 28 days), - - - - - -	9.64
Amount, - - - - - - -	334.82
Second payment, Dec. 8, 1852, - - - - -	30.00
Balance for new principal, - - - - -	304.82
Int. for Dec. 8, 1852, to Sept. 25, 1853 (9mo. 17 days),	14.58
Amount. - - - - - -	319.40
Third payment, Sept. 25, 1853, - - - - -	60.00
Balance for new principal, - - - - -	259.40
Interest from Sept. 25, 1853, to July 4, 1854 (9 months, 9 days), - - - - - - -	12.06
Amount. - - - - - -	271.46

QUESTION. — Art. 200. What is the rule generally adopted by the several States for computing the interest on notes and bonds, when partial payments have been made?

Amount brought up,		271.46
Fourth payment, July 4, 1854, - - -		90.00
Balance for new principal, - - - - -		181.46
Interest from July 4, 1854, to Aug. 1, 1855 (12 mo. 27 days),		11.70
Interest from Aug. 1, 1855, to Dec. 2, 1855 (4 mo. 1 day),		3.66
Amount,		196.82
Fifth payment, Aug. 1, 1855 (a sum less than the interest), - - - - - -	$10.00	
Sixth payment, Dec. 2, 1855 (a sum greater than the interest),	100.00	
		110.00
Balance for new principal,		86.82
Interest from Dec. 2, 1855, to Jan. 7, 1857 (13 months, 5 days), - - - - - - -		5.71
Remains due Jan. 7, 1857,		$92.53

2. $1666. *Philadelphia, June* 5, 1848.

For value received, I promise to pay J. B. Lippincott & Co., or order, on demand, without defalcation, one thousand six hundred sixty-six dollars, with interest. *John J. Shellenberger.*

Attest, T. Webster.

On this note are the following indorsements:

July 4, 1849, received one hundred dollars. Jan. 1, 1850, received ten dollars. July 4, 1850, received fifteen dollars. Jan. 1, 1851, received five hundred dollars. Feb. 7, 1852, received six hundred and fifty-six dollars.

What is due Jan. 1, 1853? Ans. $767.08.

3. $960. *Detroit, Oct.* 23, 1850.

On demand, I promise to pay S. S. St. John, or order, nine hundred sixty dollars, for value received, with interest at seven per cent. *John Q. Smith.*

Attest, H. F. Wilcox.

On this note are the following indorsements:

Sept. 25, 1851, received one hundred forty dollars. July 7, 1852, received eighty dollars. Dec. 9, 1852, received seventy dollars. Nov. 8, 1853, received one hundred dollars.

What is due Oct. 23, 1854? Ans. $807.76.

4. $1000. *New York, Jan.* 1, 1849.

Two months after date I promise to pay to S. Durand, or

QUESTION — Explain the operation.

order, one thousand dollars, for value received, with interest after, at seven per cent. *Paul Sampson, Jr*
Attest, William S. Hall.

On this note are the following indorsements:

March 1, 1850, received one hundred dollars. Sept. 25, 1851, received two hundred dollars. Oct. 9, 1852, received one hundred fifty dollars. July 4, 1853, received twenty dollars. Oct 9, 1853, received three hundred dollars.

What is due Dec. 1, 1854? Ans. $567.49.

ART. 201. The following is the rule established by the Supreme Court of the State of Connecticut.

1. "Compute the interest to the time of the first payment; if that be one year or more from the time the interest commenced, add it to the principal, and deduct the payment from the sum total. If there be after payments made, compute the interest on the balance due to the next payment, and then deduct the payment as above; and in like manner from one payment to another, till all the payments are absorbed; provided the time between one payment and another be one year or more."

2. "But if any payments be made before one year's interest hath accrued, then compute the interest on the principal sum due on the obligation for one year,* add it to the principal, and compute the interest on the sum paid from the time it was paid up to the end of the year; add it to the sum paid, and deduct that sum from the principal and interest added together."

3. "If any payments be made of a less sum than the interest arisen at the time of such payment, no interest is to be computed, but only on the principal sum for any period."

Ex. 1. $500. *Hartford, July* 1, 1854.

For value received, I promise to pay J. Dow, or order, on demand, five hundred dollars, with interest. *D. P. Page.*

On this are the following indorsements:

Sept. 1, 1855, received one hundred dollars. April 1, 1856, received one hundred forty-four dollars. Jan. 1, 1857, received ninety dollars, fifty cents. Dec. 1, 1858, received one hundred sixty-eight dollars, five cents.

What is due Oct. 1, 1859? Ans. $92.40.

* If a year extends beyond the time when the note becomes due, find the amount of the remaining principal to the time of *settlement*; find also the amount of the indorsement or indorsements, if any, from the time they were paid to the time of settlement, and subtract their sum from the amount of the principal.

QUESTION. — Art. 201. What is the Connecticut rule for computing interest on notes and bonds, when partial payments have been made?

PROBLEMS IN INTEREST.

ART. **202.** A PROBLEM in arithmetic is a question proposed which requires solution.

ART. **203.** In the preceding questions in interest, five terms or things have been mentioned; namely, the Interest, Amount, Rate per cent., Time, and Principal. The investigation of these involves five problems: I. To find the interest; II. To find the amount; III. To find the rate per cent.; IV. To find the time; V. To find the principal.

With one exception, any *three* of the preceding terms being given, a *fourth* may be found by the rules deduced from the solution of the problems. But if the rate per cent., time, and amount, are given, an additional rule is necessary to find the principal, which will form a sixth problem; but, from its connection with *Discount*, its solution will be deferred until that subject is considered.

The Problems I. and II. have already been examined (Art. 194), and we now proceed to an examination of those remaining.

ART. **204.** Problem III. To find the rate per cent., the principal, interest, and time, being given.

Ex. 1. The interest of $300 for 2 years is $48; what is the rate per cent.? Ans. 8 per cent.

OPERATION.

```
$3 0 0
   .0 2
$6.0 0 ) 4 8.0 0 ( 8 per cent.
         4 8.0 0
```

We find the interest on the principal for 2 years at 1 per cent., and divide the given interest by it.

Since the interest of $1 at 1 per cent. for 2 years is 2 cents, the interest of $300 will be 300 times as much, equal to $6. Now, if $6 is 1 per cent., $48 will be as many per cent. as $6 is contained times in $48, which gives 8 per cent. for the answer.

RULE. — *Divide the given interest by the interest of the given sum at* 1 *per cent. for the given time, and the quotient will be the rate per cent. required.*

QUESTIONS. — Art. 202. What is a problem in arithmetic? — Art. 203. How many terms or things have been given in the preceding questions in interest? Name them. What does an investigation of these terms involve? Name them. How many terms are given in each problem in order to find a fourth? What two problems have been examined? — Art. 204. What is Problem III.? Explain the operation. What is the rule for finding the rate per cent., the principal, interest, and time, being given?

EXAMPLES FOR PRACTICE.

2. The interest of $250 for 1 year, 3 months, is $28.125; what is the rate per cent.? Ans. 9 per cent.

3. If I pay $8.82 for the use of $72 for 1 year, 9 months, what is the rate per cent.? Ans. 7 per cent.

4. A note of $500, being on interest 2 years, 6 months amounted to $550; what was the rate per cent.? Ans. 4 per cent.

5. The interest for $700 for 1 year, 6 months, is $63; what is the rate per cent.? Ans. 6 per cent.

6. If I pay 53.78\frac{1}{3}$ for the use of $922 for 1 year, 2 months, what is the rate per cent.? Ans. 5 per cent.

ART. **205.** Problem IV. To find the time, the principal, interest, and rate per cent., being given.

Ex. 1. For how long a time must $300 be on interest at 6 per cent. to gain $36? Ans. 2 years.

OPERATION.

```
  $3 0 0
     .0 6
$1 8.0 0 ) 3 6.0 0 ( 2 years.
           3 6.0 0
```

We find the interest on the given principal for 1 year, by which we divide the given interest.

Since the interest of $1 for 1 year is 6 cents, the interest of $300 will be 300 times as much, equal to $18. Now, if it require 1 year for the given principal to gain $18, it will require as many years to gain $36 as $18 is contained times iu $36, which gives 2 years for the answer.

RULE. — *Divide the given interest by the interest of the given principal for* 1 *year, and the quotient will be the time.*

EXAMPLES FOR PRACTICE.

2. If the interest of $140 at 6 per cent. is $42, for how long a time was it on interest? Ans. 5 years.

3. How long a time must $165 be on interest at six per cent. to gain $14.85? Ans. 1 year, 6 months.

4. How long must $98 be on interest at 8 per cent. to gain $25.48? Ans. 3 years, 3 months.

5. A note of $680 being on interest at 4 per cent. amounted to $727.60; how long was it on interest? Ans. 1 year, 9 months.

QUESTIONS. — Art. 205. What is Problem IV.? Explain the operation. What is the rule for finding the time, the principal, interest, and rate per cent., being given?

ART. 206. Problem V. To find the principal, the interest, time, and rate per cent., being given.

Ex. 1. What principal at 6 per cent. will gain $36 in 2 years?
Ans. $300.

OPERATION.

.0 6 int. of $1 for 1y.
2
.1 2) 3 6.0 0 ($3 0 0 principal.

We find the interest of $1 for 2 years, by which we divide the given interest.

Since it requires 2 years for a principal of $1 to gain 12 cents, it will require a principal of as many dollars to gain $36 as .12 is contained times in $36. Thus, $36.00 ÷ .12 = $300 for the answer.

RULE. — *Divide the given interest or amount by the interest or amount of* $1 *for the given rate and time, and the quotient will be the principal.*

EXAMPLES FOR PRACTICE.

2. What principal will gain $24.225 in 4 years, 3 months, at six per cent.? Ans. $95.

3. What principal will gain $5.11 in 3 years, 6 months, at 8 per cent.? Ans. $18.25.

4. The interest on a certain note at 9 per cent. in 1 year and 8 months amounted to $42; what was the full amount of the note? Ans. $280.

---

§ XXII. COMPOUND INTEREST.

ART. 207. COMPOUND INTEREST is interest on the principal and interest together, when the interest is not paid at the end of the year, or when due.

The law specifies that the borrower of money shall pay the lender a certain sum for the use of $100 for a year. Now, if he does not pay this sum at the end of the year, it is no more than just that he should pay interest for the use of it as long as he shall keep it in his possession. The computation of compound interest is based upon this principle.

QUESTIONS. — Art. 206. What is Problem V.? Explain the operation. What is the rule for finding the principal, the interest, time, and rate per cent., being given? — Art. 207. What is compound interest? On what prin ciple is it based?

ART. 208. To find the compound interest of any sum of money at any rate per cent. for any given time.

Ex. 1. What is the compound interest of $500 for 3 years, 7 months, and 12 days, at 6 per cent.? Ans. $117.541.

OPERATION.

Principal,	$500
Interest of $1 for 1 year,	.06
Interest for 1st year,	30.00
	500
Amount for 1st year,	530.00
	.06
Interest for 2d year,	31.8000
	530.00
Amount for 2d year,	561.80
	.06
Interest for 3d year,	33.7080
	561.80
Amount for 3d year,	595.508
Interest of $1 for 7mo. 12da.,	.037
	4168556
	1786524
Interest for 7mo. 12da.,	22.033796
	595.508
Amount for 3y. 7mo. 12da.,	617.541796
Principal subtracted,	500
Compound interest,	$117.541796

We first multiply the principal by the interest of $1 for 1 year, and add the interest thus found to the principal for the amount, or new principal. We then find the interest on this amount for 1 year, and proceed as before; and so also with the third year. For the months and days we find the interest on the amount for the last year, and, adding it as before, we subtract the original principal from the last amount for the answer.

RULE. — *Find the interest of the given sum for one year, and add it to the principal; then find the amount of this amount for the next year; and so continue, until the time of settlement.*

If there are months and days in the given time, find the amount for them on the amount for the last year.

Subtract the principal from the last amount, and the remainder is the compound interest.

QUESTIONS. — Art. 208. Explain the operation in computing compound interest. What is the rule?

NOTE.—1. If the interest is to be paid semi-annually, quarterly, monthly, or daily, it must be computed for the half-year, quarter-year, month, or day, and added to the principal, and then the interest computed on this. and each succeeding amount thus obtained, up to the time of settlement.

2. When partial payments have been made on notes at compound interest, the rule is like that adopted in Art. 199.

EXAMPLES FOR PRACTICE.

2. What is the compound interest of $761.75 for 4 years? Ans. $199.941.

3. What is the amount of $67.25 for 3 years, at compound interest? Ans. $80.095.

4. What is the amount of $78.69 for 5 years, at 7 per cent.? Ans. $110.364.

5. What is the amount of $128 for 3 years, 5 months, and 18 days, at compound interest? Ans. $156.717.

6. What is the compound interest of $76.18 for 2 years, 8 months, 9 days? Ans. $12.967.

ART. **209.** There is a more expeditious method of computing compound interest than the preceding, by means of the following

TABLE.

Showing the amount of $1, or £1, for any number of years not exceeding 20, at 3, 4, 5, 6, and 7 per cent., compound interest.

Years.	3 per cent.	4 per cent.	5 per cent.	6 per cent.	7 per cent.	Years.
1	1.030000	1.040000	1.050000	1.060000	1.070000	1
2	1.060900	1.081600	1.102500	1.123600	1.144900	2
3	1.092727	1.124864	1.157625	1.191016	1.225043	3
4	1.125508	1.169858	1.215506	1.262476	1.310796	4
5	1.159274	1.216652	1.276281	1.338225	1.402552	5
6	1.194052	1.265319	1.340095	1.418519	1.500730	6
7	1.229873	1.315931	1.407100	1.503630	1.605781	7
8	1.266770	1.368569	1.477455	1.593848	1.718186	8
9	1.304773	1.423311	1.551328	1.689478	1.838459	9
10	1.343916	1.480244	1.628894	1.790847	1.967151	10
11	1.384233	1.539454	1.710339	1.898298	2.104852	11
12	1.425760	1.601032	1.795856	2.012196	2.252191	12
13	1.468533	1.665073	1.885649	2.132928	2.409845	13
14	1.512589	1.731676	1.979931	2.260903	2.578534	14
15	1.557967	1.800943	2.078928	2.396558	2.750032	15
16	1.604706	1.872981	2.182874	2.540351	2.952164	16
17	1.652847	1.947900	2.292018	2.692772	3.158815	17
18	1.702433	2.025816	2.406619	2.854339	3.379932	18
19	1.753506	2.106849	2.526950	3.025599	3.616527	19
20	1.806111	2.191123	2.653297	3.207135	3.869685	20

QUESTIONS.—If the interest is to be paid semi-annually, quarterly, &c., how is it computed? How, when partial payments have been made?

Ex. 1. What is the interest of $240 for 6 years, 4 months, and 6 days, at 6 per cent.? Ans. $107.593.

OPERATION.

Amount of $1 for 6 years,	1.418519
Principal,	240
	56740760
	2837038
Amount of principal for 6 years,	340.444560
Interest of $1 for 4mo. 6da.,	.021
	34044456
	68088912
Interest of amount for 4mo. 6da.,	7.14933576
Amount added,	340.444560
Amount for 6y. 4mo. 6da.,	347.59389576
Principal subtracted,	240
Interest for given time,	$107.59389576

We find the amount of $1 for 6 years in the table, and multiply it by the principal, and obtain the amount for 6 years. We then find the interest on this amount for the 4 months and 6 days, and add it to the first amount, and from this sum subtract the principal for the answer. Hence, to find the compound interest by use of the table,

Multiply the amount of $1 for the given rate and time, as found in he table, by the principal, and the product will be the amount. Subtract the principal from the amount, and the remainder will be the compound interest. If there are months and days, proceed as in the foregoing rule.

EXAMPLES FOR PRACTICE.

2. What is the interest of $884 for 7 years, at 4 per cent.? Ans. $279.283.

3. What is the interest of $721 for 9 years, at 5 per cent.? Ans. $397.507.

4. What is the amount of $960 for 12 years, 6 months, at 3 per cent.? Ans. $1389.26.

5. What is the amount of $25.50 for 20 years, 2 months, and 12 days, at 7 per cent.? Ans. $100.058.

6. What is the amount of $12 for 6 months, the interest to be added each month? Ans. $12.364+.

7. What is the amount of $100 for 6 days, the interest to be added daily? Ans. $100.10004.

§ XXIII. DISCOUNT.

ART. **210.** DISCOUNT is an allowance or deduction made for the payment of money before it is due.

The *present worth* of any sum is the principal, which, being put at interest, will amount to the given sum in the time for which the discount is made. Thus, $100 is the *present worth* of $106, due one year hence at 6 per cent.; for $100 at 6 per cent. will amount to $106 in this time; and $6 is the discount.

NOTE. — Business men, however, often deduct five per cent., or more, from the face of a bill due in six months, or a percentage greater than the legal rate of interest.

ART. **211.** The interest or percentage of any sum cannot *properly* be taken for the discount; for we see, from the preceding illustration, that the *interest* for one year is the fractional part of the sum at interest, denoted by the rate per cent. for the numerator, and 100 for the denominator; and the *discount* for one year is the fractional part of the sum on which discount is to be made, denoted by the rate per cent. for the numerator, and 100 plus the rate per cent. for the denominator. Thus, if the rate per cent. of interest is 6, the *interest* for one year is $\frac{6}{100}$ of the sum at interest; but if the rate per cent. of discount is 6, the *discount* for one year is $\frac{6}{106}$ of the sum on which discount is made.

ART. **212.** In discount, the rate per cent., time, and the *sum* on which the discount is made, are given to find the *present worth.* These terms correspond precisely to Problem VI. in interest, in which the time, rate per cent., and *amount*, are given to find the *principal.* (Art. 203.)

ART. **213.** To find the present worth and the discount on any sum, at any rate per cent., for any given time.

Ex. 1. What is the present worth of $25.44, due one year hence, discounting at 6 per cent.? What is the discount?

Ans. $24 present worth; $1.44 discount.

QUESTIONS. — Art. 210. What is discount? What is the present worth of any sum of money? How illustrated? — Art. 211. Are interest and discount the same? Explain the difference. Which is the greater, the interest or discount on any sum, for a given time? — Art. 212. What terms are given in discount, and what is required? To what do these correspond in interest?

OPERATION.

```
Amount of $1,   1.0 6 ) 2 5.4 4 ( $2 4, present worth.
                        2 1 2
                        -----
                          4 2 4   $2 5.4 4, sum or amount.
                          4 2 4    2 4.0 0, present worth.
                          -----   --------
                                   $1.4 4, discount.
```

We find the amount of $1 for the given time, by which we divide the given sum, and obtain the present worth. Then, subtracting the present worth from the given sum, we obtain the discount.

Since the present worth of $1.06, due one year hence, at 6 per cent., is $1, it is evident the present worth of $25.44 is as many dollars as $1.06 is contained times in $25.44. $25.44 ÷ $1.06 = $24. Hence the following

RULE. — *Find the amount of* $1 *for the given time and rate; by which divide the given sum, and the quotient will be the present worth.*

The present worth subtracted from the given sum will give the discount.

NOTE. — The discount may be found directly by making the *interest* of $1 for the given rate and time the *numerator* of a fraction, and the *amount* of $1 for the given rate and time the *denominator*, and then multiply the given sum by this fraction.

EXAMPLES FOR PRACTICE.

2. What is the present worth of $152.64, due 1 year hence? Ans. $144.

3. What is the present worth of $477.71, due 4 years hence? Ans. $385.25.

4. What is the discount of $172.86, due 3 years, 4 months hence? Ans. $28.81.

5. What is the discount of $800, due 3 years, 7 months, and 18 days hence? Ans. $143.186.

6. Samuel Heath has given his note for $375.75, dated Oct. 4, 1852, payable to John Smith, or order, Jan. 1, 1854; what is the real value of the note at the time given? Ans. $349.697.

7. Bought a chaise and harness of Isaac Morse for $125.75, for which I gave him my note, dated Oct. 5, 1852, to be paid in 6 months; what is the present value of the note, Jan. 1, 1853? Ans. $123.81.

QUESTIONS. — Art. 213. Explain the operation for finding the present worth and discount. Give the reason of the operation. What is the rule? What other method is given?

§ XXIV. COMMISSION, BROKERAGE, AND STOCKS.

ART. **214.** *Commission* is the percentage paid to an agent factor, or commission merchant, for buying or selling goods, or transacting other business.

Brokerage is the percentage paid to a dealer in money and stocks, called a broker, for making exchanges of money, negotiating different kinds of bills of credit, or transacting other business.

Stocks is a general name given to government funds, state bonds, and to the capital of moneyed incorporations, such as banks, insurance, railroad, manufacturing, and mining companies. Stocks are usually divided into equal *shares*, the market value of which is often variable.

When stocks sell for their original value they are said to be *at par*; when for more than their original value, *above par*, or in advance; when for less than their original value, *below par*, or at a discount.

The premium, or advance, and the discount on stocks, are generally computed at a certain per cent. on the original value of the shares.

The rate per cent. of commission or brokerage is not regulated by law, but varies in different places, and with the nature of the business transacted.

Commission and brokerage are computed in the same manner.

ART. **215.** To find the commission or brokerage on any sum of money.

Ex. 1. A commission merchant sells goods to the amount of $879; what is his commission at 3 per cent.?

Ans. $26.37.

Since commission is a percentage on the given sum, the commission on $879, at 3 per cent., will be $\$879 \times .03 = \$26.37$.

RULE. — *Find the percentage on the given sum at the given rate per cent., and the result is the commission or brokerage.* (Art. 191.)

QUESTIONS. — Art. 214. What is commission? What is brokerage? What is stock? Into what are stocks divided? When are stocks at par? When above par? When below par? How is the premium or discount on stocks computed? How is commission and brokerage computed? — Art. 215. What is the rule?

EXAMPLES FOR PRACTICE

2. What is the commission on the sale of a quantity of cotton goods valued at $5678, at 3 per cent.? Ans. $170.34.

3. A commission merchant sells goods to the amount of $7896, at 2 per cent.; what is his commission? Ans. $157.92.

4. My agent in Chicago has purchased wheat for me to the amount of $1728; what is his commission, at $1\frac{1}{2}$ per cent.? Ans. $25.92.

5. My factor advises me that he has purchased, on my account, 97 bales of cloth, at $15.50 per bale; what is his commission, at $2\frac{1}{2}$ per cent.? Ans. $37.587.

6. My agent at New Orleans informs me that he has disposed of 500 barrels of flour at $6.50 per barrel, 88 barrels of apples at $2.75 per barrel, and 56cwt. of cheese at $10.60 per cwt.; what is his commission, at $3\frac{3}{4}$ per cent.? Ans. $153.21.

7. A broker negotiates a bill of exchange of $2500 at $\frac{1}{2}$ per cent. commission; what is his commission? Ans. $12.50.

8. A broker in New York exchanged $46256 on the Canal Bank, Portland, at $\frac{1}{8}$ of 1 per cent.; what did he receive for his trouble? Ans. $57.82.

9. A broker in Baltimore exchanged $20500 on the State Bank of Indiana, at $\frac{1}{2}$ of 1 per cent.; what was the amount of his brokerage? Ans. $102.50.

ART. **216.** To find the commission or brokerage on any sum of money, when it is to be deducted from the given sum, and the balance invested.

Ex. 1. A merchant in Cincinnati sends $1500 to a commission merchant in Boston, with instructions to lay it out in goods, after deducting his commission of $2\frac{1}{2}$ per cent.; what is his commission? Ans. $36.586.

OPERATION.

$$\$1500 \div 1.025 = \$1463.414.$$
$$\$1500 - \$1463.414 = \$36.586.$$

Since the agent is entitled to $2\frac{1}{2}$ per cent. of the amount he lays out, it is evident he requires $1.02½ to purchase goods to the amount of $1. Hence, he can expend for goods as many dollars as 1.02½ is contained times in $1500. On dividing we find he can expend $1463.414, which, being subtracted from $1500, the amount sent him, leaves as his commission $36.586.

QUESTION. — How do you find the commission or brokerage when it is to be subtracted from the given sum?

RULE. — *Divide the given sum by* 1 *increased by the per cent. of commission, and the quotient will be the sum to be invested.*

Subtract the sum to be invested from the given sum, and the remainder will be the commission.

EXAMPLES FOR PRACTICE.

2. A town agent has \$2000 to invest in bank stock, after deducting his commission of $1\frac{1}{2}$ per cent.; what will be his commission, and what the sum invested?

Ans. \$29.557 commission; \$1970.443 sum invested.

3. A shoe-dealer sends \$5256 to his agent in Boston, which he wishes him to lay out for shoes, reserving his commission of 3 per cent.; what is his commission? Ans. \$153.088.

4. A broker expends \$3865.94 for merchandise, after deducting his commission of 4 per cent.; what was his commission, and what sum did he expend?

Ans. \$148.69 commission; \$3717.25 sum expended.

5. I have sent to my agent at Buffalo, N. Y., \$10000, which I wish him to expend for flour, after deducting his commission of $3\frac{1}{4}$ per cent.; what will be his commission, and also the value of the flour purchased?

Ans. \$314.76+ commission; \$9685.23+ value of flour.

ART. **217.** To find the value of stocks, when at an advance or at a discount.

Ex. 1. What is the value of \$2150 railroad stock, at 7 per cent. advance? Ans. \$2300.50.

OPERATION.

$$\$2150 \times .07 = \$150.50; \ \$2150 + \$150.50 = \$2300.50.$$

RULE. — *Find the percentage on the given sum, and add or subtract, according as the stock is at an advance or at a discount.* (Art. 191.)

EXAMPLES FOR PRACTICE.

2. What must be given for 10 shares in the Boston and Maine Railroad, at 15 per cent. advance, the shares being \$100 each? Ans. \$1150.

3. What must be given for 75 shares in the Lowell Railroad, at 25 per cent. advance, the original shares being \$100 each?

Ans. \$9375.

QUESTIONS. — What is the rule? — Art. 217. How do you find the value of stocks, when at an advance or at a discount? What is the rule?

4. What is the purchase of $8979 bank stock, at 12 per cent. advance? Ans. $10056.48.

5. What is the purchase of $1789 bank stock, at 9 per cent. below par? Ans. $1627.99.

6. A stockholder in the Illinois Central Railroad sells his right of purchase on 5 shares of $100 each at 12 per cent. advance; what is the premium? Ans. $60.

7. What is the value of 20 shares canal stock, at $12\frac{1}{2}$ per cent. discount, the original shares being $100 each?

Ans. $1750.

8. What is the value of 15 shares in the Livingston County Bank, at $8\frac{1}{4}$ per cent. advance, the original shares being $100 each? Ans. $1623.75.

9. Bought 87 shares in a certain corporation, at 12 per cent. below par, and sold the same at $19\frac{1}{2}$ per cent. above par; what sum did I gain, the original shares being $175 each?

Ans. 4795.87\frac{1}{2}$.

---

§ XXV. BANKING.

ART. **218.** A BANK is a joint stock company, established for the purpose of receiving deposits, loaning money, dealing in exchange, or issuing bank notes or bills, as a circulating medium, redeemable in specie at its place of business.

The *capital* of a bank is the money paid in by its stockholders, as the basis of business.

Banking is the general business commonly transacted at banks.

NOTE. — The persons chosen by the stockholders to manage the affairs of the bank are called its *board* of *directors*, who select one of their own number as *president*, and some person as *cashier*. The president and cashier sign the bills issued, which also are, in some instances, countersigned by some state officer. The cashier superintends the bank accounts; and another person, called the *teller*, usually receives and pays out money. A *check* is an order drawn on the cashier of the bank for money.

QUESTIONS. — Art. 218. What is a bank? What is the capital of a bank? What is banking? Who choose the directors? Who choose the president and cashier? Who sign the bills issued? Who superintends the accounts? Who receives and pays out the money? What is a check?

BANK DISCOUNT.

ART. **219.** BANK DISCOUNT is the simple interest of a note draft, or bill of exchange, deducted from it in advance, or before it becomes due.

The interest is computed, not only for the specified time, but also for *three* days additional, called *days of grace.* Thus, if a note is given at the bank for 60 days, the interest, which is called the discount, is computed for 63 days; and if the note is paid within this time, the debtor complies with the requirements of the law.

The legal rate of discount is usually the same as the legal rate of interest; and the difference between *bank discount* and *true discount* is the same as the difference between interest and *true discount.*

A note is said to be discounted at a bank, when it is received as security for the money that is paid for it, after deducting the interest for the time it was given. The sum paid is called the *avails* or *present worth* of the note.

ART. **220.** To find the present worth and the bank discount of any note or sum of money for any rate per cent. and time.

Ex. 1. What is the bank discount on $842 for 90 days, at 6 per cent.? What is the present worth?

Ans. $13.051 discount; $828.949 present worth.

OPERATION.

Sum discounted,	$842	$842.000	
Interest of $1,	.0155	13.051	
	4210	$828.949	present worth.
	4210		
	842		
Bank discount,	$13.0510		

We find the interest of $1 for 93 days, by which we multiply the sum discounted, and the product is the discount. We then subtract the discount from the given sum, and obtain the present worth.

QUESTIONS. — Art. 219. What is bank discount? When is it paid? Is interest computed fc- more than the specified time? What are these three additional days called? How will you illustrate this? What is the legal rate of discount? What is the difference between bank discount and true discount? When is a note said to be discounted at a bank? What is the sum paid for it called? — Art. 220. Explain the operation for finding the bank discount on any sum.

RULE. — *Find the interest on the note, or sum discounted, for the given rate and time, including* THREE *days of grace, and this interest is the discount.*

Subtract the discount from the face of the note or sum discounted, and the remainder is the present worth.

EXAMPLES FOR PRACTICE.

2. What is the bank discount on $478 for 60 days?
Ans. $5.019.

3. What is the bank discount on $780 for 30 days?
Ans. $4.29.

4. What is the bank discount on $1728 for 90 days?
Ans. $26.784.

5. How much money should be received on a note of $1000, payable in 4 months, discounting at a bank where the interest is 6 per cent.? Ans. $979.50.

6. What sum must a bank pay for a note of $875.35, payable in 7 months and 15 days, discounting at 7 per cent.?
Ans. $836.542.

7. What are the avails of a note of $596.24, payable in 8 months and 9 days, discounted at a bank at 8 per cent.?
Ans. $562.85.

8. What is the bank discount of a draft of $1350.50, payable in 1 year, 4 months, at 5 per cent.? Ans. $90.596.

ART. 221. To find the amount for which a note must be given at a bank, to obtain a specified sum for any given time.

Ex. 1. For what amount must a note be given, payable in 90 days, to obtain $500 from a bank, discounting at 6 per cent.?
Ans. $507.872.

OPERATION.

	$1.0 0 0 0
Int. of $1 for 93da.,	.0 1 5 5
Present worth of $1,	.9 8 4 5

$500 ÷ .9845 = $507.872

We subtract the interest of $1 for 93 days, at six per cent., from $1, and divide the given sum by the remainder, for the answer.

Since $0.9845 requires $1 principal for the given time, $500 will require as many dollars principal as $0.9845 is contained times in $500; and $500 ÷ $0.9845 = $507.872. Hence the

QUESTIONS. — What is the rule? — Art. 221. Explain the operation for finding the amount for which a note must be given at a bank to obtain a specified sum for a given time.

RULE. — *Divide the given sum by the present worth of $1 for th given time and rate per cent. of bank discount, including* THREE *day of grace, and the quotient will be the answer.*

EXAMPLES FOR PRACTICE.

2. For what sum must I give my note at a bank, payable in 4 months, at 6 per cent. discount, to obtain $300? Ans. $306.278.

3. A merchant sold a quantity of lumber, and received a note payable in 6 months; he had his note discounted at a bank, at 6 per cent., and received $4572.40. What was the amount of his note? Ans. $4716.245.

4. A gentleman wishes to take $1000 from the bank; for what sum must he give his note, payable in 5 months, at 6 per cent. discount? Ans. $1026.167.

5. The avails of a note, discounted at the bank for 8 months, at 7½ per cent., were $483.56; what was the face of the note? Ans. $509.345.

---

§ XXVI. INSURANCE.

ART. 222. INSURANCE is indemnity obtained, by paying a certain sum, against such losses of property or of life as are agreed upon.

The party taking the risk is called the *insurer* or *underwriter*, and the party protected, the *insured.*

The *policy* is the written obligation, or contract, entered into between the parties.

Premium is the amount of percentage paid on the property insured for one year, or any specified time.

As a security against fraud, property is not usually insured for its whole value, nor is the insurer or underwriter bound to indemnify the insured for a loss more than is specified in the policy.

QUESTIONS. — What is the rule? — Art. 222. What is insurance? What is the party called that takes the risk? What is the party called that is protected? What is the policy? What is the premium? Is property usually insured to its whole value?

ART. **223.** To find the premium on any amount of property insured.

Ex. 1. What is the premium on $485 at 2 per cent.?
Ans. $9.70

OPERATION.

$$\$485 \times .02 = \$9.70.$$

RULE. — *Find the percentage on the given sum, and the result is the premium.* (Art. 191.)

EXAMPLES FOR PRACTICE.

2. What is the premium on $868 at 12 per cent.? Ans. $104.16.

3. What is the premium on $1728 at 15 per cent.? Ans. $259.20.

4. A house, valued at $3500, is insured at $1\frac{3}{4}$ per cent.; what is the premium? Ans. $61.25.

5. A vessel and cargo, valued at $35000, are insured at $3\frac{3}{4}$ per cent.; now, if this vessel should be destroyed, what will be the actual loss to the insurance company? Ans. $33687.50.

6. A cotton factory and its machinery, valued at $75000, are insured at $2\frac{1}{2}$ per cent.; what is the yearly premium? and if it should be destroyed, what loss would the insurance company sustain? Ans. $1875 premium; $73125 loss.

---

§ XXVII. CUSTOM-HOUSE BUSINESS.

ART. **224.** DUTIES are sums of money required by government to be paid on imported goods.

All goods from foreign countries brought into the United States are required to be landed at particular places, called ports of entry, where are custom-houses, at which the duties or revenue is collected.

Duties are either *specific* or *ad valorem.*

A *specific duty* is a certain sum paid on a ton, hundred weight, yard, gallon, &c.

QUESTIONS. — Art. 223. What is the rule for finding the premium on any amount of property insured? — Art. 224. What are duties? Where are duties collected? What is a specific duty?

An *ad valorem duty* is a certain per cent. paid on the actua cost of the goods in the country from which they are imported

Draft is an allowance for *waste* made in the weight of goods.

Tare is an allowance made for the weight of the cask, box &c., containing the commodity.

Leakage is an allowance for waste made on liquors.

Gross weight is the weight of the commodity, together with the cask, box, bag, &c., containing it.

Net weight is what remains after all allowances have been made.

By the present tariff, all duties are levied on the *ad valorem* principle.

It has been decided that no allowances for tare, draft, breakage, &c., are applicable to imports subject to ad valorem duties, except actual tare, or weight of a cask, or package, and the actual drainage, leakage, or damage.

The collector may cause these to be ascertained, when he has any doubts as to what they are.

ART. 225. To find the ad valorem duty on goods or merchandise.

Ex. 1. At 25 per cent., what is the ad valorem duty on 165 yards of broadcloth, at $5 per yard? Ans. $206.25.

OPERATION.

165 × $5 = $825; $825 × .25 = $206.25, duty.

RULE. — *Find the percentage on the cost of the goods, and the result is the ad valorem duty.* (Art. 191.)

EXAMPLES FOR PRACTICE.

2. What is the duty on 1728lb. of copper sheathing, invoiced at $3200, at 20 per cent. ad valorem? Ans. $640.

3. What is the duty on 2231lb. of Russian iron, at 30 per cent. ad valorem; the cost of the iron being 4 cents per lb.? Ans. $26.772, duty.

4. What is the duty on 1691lb. of lead, at 20 per cent. ad valorem; the value of the lead being 5 cents per pound? Ans. $16.91, duty.

QUESTIONS. — What is an ad valorem duty? What is draft? Tare? Gross weight? Net weight? — Art. 225. What is the rule for finding the ad valorem duty?

5. What is the duty on 10 hogsheads of molasses, each hogshead gauging 150 gallons gross, the actual wants being 5 gallons to each hogshead, and the cost of the molasses 25 cents per gallon; duty 20 per cent. ad valorem? Ans. $72.50, duty.

6. What are the net weight and duty, at 30 per cent. ad valorem, on 13 boxes of sugar, weighing gross 450 pounds each; actual tare 15 per cent., and the cost of the sugar being 8 cents per pound? Ans. 4972½lbs., net weight; $119.34, duty.

7. What is the duty on an invoice of woollen goods, which cost in Liverpool 1376£. sterling, at 30 per cent. ad valorem; the pound sterling being $4.84? Ans. $1997.95+.

8. What is the duty on an invoice of goods, which cost in Paris $2340, at 80 per cent. ad valorem?

Ans. $1872.

---

§ XXVIII. ASSESSMENT OF TAXES.

ART. **226.** A TAX is a sum of money assessed by government for public purposes, on property, and in most states on persons.

Taxes may be either *direct* or *indirect.* A direct tax is one imposed on the income or property of an individual; an indirect tax is one imposed on the articles for which the income or property is expended.

A *poll* or *capitation* tax is one without regard to property, on the person of each male citizen, liable by law to assessment. A person so liable is termed a *poll.*

Immovable property, such as lands, houses, &c., is called *real estate.* All other property, such as money, notes, cattle, furniture, &c., is called *personal property.*

The method of assessing taxes is not precisely the same in all the states, yet the principle is virtually the same.

The following is the law regulating taxation in Massachusetts (Revised Statutes, p. 79):

QUESTIONS. — Art. 226. What is a tax? What is a direct tax? What an indirect tax? What is real estate? What is personal property? What is a poll or capitation tax? What is a poll? Is the method of assessing taxes the same in all the states?

"The assessors shall assess upon the *polls*, as nearly as the same can be conveniently done, one sixth part of the whole sum to be raised; provided the whole poll tax assessed in any one year upon any individual for town and county purposes, except highway taxes, shall not exceed one dollar and fifty cents; and the residue of said whole sum to be raised shall be apportioned upon property;" that is, on the real and personal estate of individuals which is taxable.

ART. 227. To assess a town or other tax.

Ex. 1. The tax to be assessed on a certain town is $2200. The real estate of the town is valued at $60000, and the personal property at $30000. There are 400 polls, each of which is taxed $1.00. What is the tax on $1.00? What is A's tax, whose real estate is valued at $2000, and his personal property at $1200, and who pays for 2 polls?

OPERATION.

$1.00 × 400 = $400, amount assessed on the polls.
$2200 — $400 = $1800, am't to be assessed on the property.
$60000 + $30000 = $90000, amount of taxable property.
$1800 ÷ $90000 = $0.02, tax on $1.00.
$2000 × .02 = $40, A's tax on real estate.
$1200 × .02 = $24, A's tax on personal property.
$1.00 × 2 = $2, A's tax on 2 polls.
$40 + $24 + $2 = $66, amount of A's tax.

Hence, in assessing taxes, it is necessary to have an inventory of the taxable property, and, if a levy on the polls is to be included, there should be also a complete list of taxable polls. Having these, we then

Multiply the tax on each poll by the number of taxable polls, and subtract the product from the whole sum to be raised, which gives the sum to be raised on the property.

The sum to be raised on property divided by the whole taxable property, will give the sum to be paid on each dollar of property taxed.

Each man's taxable property multiplied by the sum to be paid on $1, with his poll tax added to the product, will give the amount of his tax.

EXAMPLES FOR PRACTICE.

2. The town of L. is taxed $3600. The real estate of the town is valued at $560,000, and the personal property at $152,500. There are 600 polls, each of which is taxed $1.25. What is the per cent. or tax on $1.00? and what is B's tax,

---

QUESTIONS. — What is the law regulating taxation in Massachusetts? — Art. 227. What is the rule for assessing taxes?

whose real estate is valued at $4100, and his personal property at $1,800, he paying for four polls?

Ans. $.004, tax on $1; $28.60, B's tax.

3. What is the tax of a nonresident, having property in the same town, worth $15800? Ans. $63.20.

4. What is D's tax, who pays for 3 polls, and whose real estate is valued at $40000, and his personal property at $23600?

Ans. $258.15.

ART. 228. The operation of assessing taxes may be facilitated by the use of a table, which can be easily made after having found the tax on $1.

Ex. 1. A tax of $3900 is to be assessed on the town of P. The real estate is valued at $840000, and the personal property at $210000; and there are 500 polls, each of which is taxed $1.50. What is the assessment on $1? Ans. $.003.

Having found the tax on $1 to be $.003, before proceeding to make the assessment on the inhabitants of the town, we find the tax on $2, $3, &c., and arrange the numbers as in the following

TABLE.

$1	gives	$.003	$20	gives	$.06	$300	gives	$ .90
2	"	.006	30	"	.09	400	"	1.20
3	"	.009	40	"	.12	500	"	1.50
4	"	.012	50	"	.15	600	"	1.80
5	"	.015	60	"	.18	700	"	2.10
6	"	.018	70	"	.21	800	"	2.40
7	"	.021	80	"	.24	900	"	2.70
8	"	.024	90	"	.27	1000	"	3.00
9	"	.027	100	"	.30	2000	"	6.00
10	"	.030	200	"	.60	3000	"	9.00

2. What is E's tax, by the above table, whose property, real and personal, is valued at $1860, and who pays 3 polls?

Ans. $10.08.

OPERATION.

Tax	on	$1000	is	$3.00
"	"	800	"	2.40
"	"	60	"	.18
"	"	3 polls	"	4.50
Valuation,		$1860		$10.08, Tax,

We find the tax on $1000 in the table, and then on $800, and then on $60, and to these sums add the tax on the 3 polls for the answer.

QUESTIONS. — Art. 228. How may the operation of assessing taxes be facilitated? How is the above table formed?

3. What is F's tax, whose real estate is valued at $6535, and his personal property at $3175, and who pays for 6 polls? Ans. $38.13.

4. What is Mrs. G's tax, who has property to the amount of $7980? Ans. $23.94.

5. If H pays for 2 polls, and has property to the amount of $4790, what is his tax? Ans. $17.37.

6. M's real estate is valued at $9280, and his personal property at $3600; what is his tax, if he pays for 4 polls? Ans. $44.64.

---

§ XXIX. EQUATION OF PAYMENTS.

ART. **229.** EQUATION OF PAYMENTS is the process of finding the *average* or *mean* time when the payment of several sums, due at different times, may all be made at one time, without loss either to the debtor or creditor.

ART. **230.** To find the average or mean time of payment, when the several sums have the same date.

Ex. 1. John Jones owes Samuel Gray $100; $20 of which is to be paid in 2 months, $40 in 6 months, $30 in 8 months, and $10 in 12 months; what is the average time for the payment of the whole sum? Ans. 6mo. 12da.

OPERATION.

```
$20 ×  2 =  40
$40 ×  6 = 240
$30 ×  8 = 240
$10 × 12 = 120
-----------------
$100     100)640(6 mo.
             600
             ---
              40
              30
             ---
        100)1200(12 da.
            1200
            ----
```

It is evident that the interest of $20 for two months is the same as the interest of $1 for 40 months; and of $40 for 6 mo., the same as of $1 for 240 mo.; and of $30 for 8 mo., the same as of $1 for 240 mo.; and of $10 for 12 mo., the same as of $1 for 120 mo. Hence, the interest of all the sums to the times of their payment, is the same as the interest of $1 for 40 + 240 + 240 + 120 = 640 mo. Now, if $1 require 640 mo. to gain a certain sum, $20 +

QUESTIONS. — Art. 229. What is equation of payments? — Art. 230. Why in the example, do we multiply the $20 by 2?

\$40 + \$30 + \$10 = \$100 will require only $\frac{1}{100}$ of this time ; and 640 mo. ÷ 100 = 6 mo. 12 da., the average or mean time for the payment of the whole. Hence the following

RULE. — *Multiply each payment by its own time of credit, and divide the sum of the products by the sum of the payments.*

NOTE 1. — This is the rule usually adopted by merchants, but it is not *perfectly* correct ; for if I owe a man \$200, \$100 of which I was to pay *down*, and the other \$100 in two years, the equated time for the payment of both sums would be one year. It is evident that, for deferring the payment of the first \$100 for 1 year, I ought to pay the amount of \$100 for that time, which is \$106 ; but for the other \$100, which I pay a year *before* it is due, I ought to pay the *present worth* of \$100, which is \$94.33$\frac{51}{53}$; whereas, by equation of payments, I only pay \$200.

NOTE 2. — When a payment is to be made *down* it has no product, but it must be added with the other payments in finding the average time.

EXAMPLES FOR PRACTICE.

2. John Smith owes a merchant in Boston \$1000, \$250 of which is to be paid in 4 months, \$350 in 8 months, and the remainder in 12 months ; what is the average time for the payment of the whole sum? Ans. 8mo. 18da.

3. A gentleman purchased a house and lot for \$1560, $\frac{1}{4}$ of which is to be paid in 3 months, $\frac{1}{5}$ in 6 months, $\frac{1}{6}$ in 8 months, and the remainder in 10 months ; what is the average time of payment? Ans. 7$\frac{91}{780}$ months.

4. Samuel Church sold a farm for \$4000 ; \$1000 of which is to be paid down, \$1000 in one year, and the remainder in 2 years ; but he afterwards agreed to take a note for the whole amount ; for what time must the note be given?
Ans. 15 months.

5. A wholesale merchant in Boston sold a bill of merchandise to the amount of \$5000 to a retail merchant of Exeter, N. H. ; he is to pay $\frac{1}{4}$ of the money down, $\frac{1}{3}$ of the remainder in 6 months, $\frac{2}{5}$ of what then remains in 9 months, and the rest at the end of the year. If he wishes to pay the whole at once, what will be the average time of payment? Ans. 6mo. 27da.

ART. 231. To find the average or mean time of payment, when the several sums have different dates.

Ex. 1. Purchased of James Brown, at sundry times, and on

QUESTIONS. — What is the rule for equation of payments? Is the rule perfectly correct? Explain why it is not. When a payment is to be made down, what is to be done with it?

various terms of credit, as by the statement annexed. When is the *medium* time of payment?

Jan.	1,	a bill	amounting to	$360,	on 3 months' credit.
Jan.	15,	do.	do.	186,	on 4 months' credit.
March	1,	do.	do.	450,	on 4 months' credit.
May	15,	do.	do.	300,	on 3 months' credit.
June	20,	do.	do.	500,	on 5 months' credit.

Ans. July 25, or in 115 da.

OPERATION.

Due April 1, $360
May 15, $186 × 44 = 8184
July 1, $450 × 91 = 40950
Aug. 15, $300 × 136 = 40800
Nov. 20, $500 × 233 = 116500

1796) 206434 (114 $\frac{845}{898}$ days.
1796
2683
1796
8874
7184
1690

We first find the time when each of the bills will become due. Then, since it will shorten the operation and bring the same result, *we take the first time when any bill becomes due*, instead of its *date*, for the *period* from which to compute the average time. Now, since April 1 is the period from which the average time is computed, no time will be reckoned on the first bill, but the time for the payment of the second bill extends 44 days beyond April 1, and we multiply it by 44. (Art. 230.) Proceeding in the same manner with the remaining bills, we find the average time of payment to be 115 days nearly, from April 1, or on the 25th of July. Hence, in like cases,

Find the time when each of the sums becomes due. Multiply each sum by the number of days intervening between the date of its becoming due and the earliest date on which any sum becomes due. Then proceed as in the rule (Art. 230), *and the quotient will be the average time required, in days, from the earliest payment.*

NOTE. — In the work, if there be a fraction of a day less than ½, it may be rejected; but if more than ½, it may be reckoned as 1 day.

QUESTIONS. — Art. 231. What is the rule for finding the average time, when there are different dates? By what other method can you obtain nearly the same result?

EXAMPLES FOR PRACTICE.

2. I have purchased several parcels of goods, at sundry times, and on various terms of credit, as by the following statement. What is the average time for the payment of the whole?

Jan. 1,	1856, a bill amounting to		$175.80,	on 4 months'	cr.
" 16,	"	do. do.	96.46,	on 90 days'	"
Feb. 11,	"	do. do.	78.39,	on 3 months'	"
" 23,	"	do. do.	49.63,	on 60 days'	"
Mar. 19,	"	do. do.	114.92,	on 6 months'	"

Ans. May 30, or in 45 da.

3. Sold S. Dana several parcels of goods, at sundry times, and on various terms of credit, as by the following statement:

Jan. 7,	1854, a bill amounting to		$375.60,	on 4 months'	cr.
April 18,	"	do. do.	687.25,	on 4 months'	"
June 7,	"	do. do.	568.50,	on 6 months'	"
Sept. 25,	"	do. do.	300.00,	on 6 months'	"
Nov. 5,	"	do. do.	675.75,	on 9 months'	"
Dec. 1,	"	do. do.	100.00,	on 3 months'	"

What is the average time for the payment of all the bills?

Ans. Dec. 24, or in 231 da.

4. The following is my account against G. M. Holbrook, and I wish to ascertain the average time of payment.

Jan. 1,	1857,	97 yards of broadcloth,	at $4.50,	on 3 mos.'	cr.
Feb. 10,	"	7 bales of cotton cloth,	" 18.50,	on 60 days'	"
May 1,	"	9 tons of iron,	" 45.00,	on 4 mos.'	"
June 15,	"	11 hhds. of molasses,	" 12.00,	on 30 days'	"
July 5,	"	8 doz. shovels,	" 9.00,	on 2 mos.'	"
Sept. 25,	"	14cwt. of sugar,	" 6.50,	on 1 mo.'s	"
Dec. 1,	"	8 chests of tea,	" 15.00,	on 90 days'	"

Ans. July 16, or in 106 da.

5. The following is an account of my bills against J. Crowell:

Jan. 1,	1854, a bill amounting to		$300,	on 6 months'	credit.
June 1,	"	do. do.	500,	on 5 months'	"
Sept. 1,	"	do. do.	200,	on 6 months'	"
Feb. 1,	1855,	do. do.	800,	on 8 months'	"
July 1,	1856,	do. do.	400,	on 9 months'	"
Dec. 1,	"	do. do.	900,	on 7 months'	"
May 1,	1857,	do. do.	100,	on 3 months'	"

What is the average time of payment on the above bills?

Ans. March 9, 1856, or in 20 mo. 8 da.

ART. **232.** To find the average or mean time of paying the balance of a debt, when partial payments have been made before the debt is due.

Ex. 1. I have purchased goods to the amount of $800, on a credit of 6 months. At the end of 2 months I pay $100, and at the end of another month I pay $200 more. How long, in equity, after the expiration of the 6 months, ought the balance to remain unpaid? Ans. 2 months.

OPERATION.

$$\begin{array}{rr} \$100 \times 4 = & 400 \\ \$200 \times 3 = & 600 \\ \hline 300 & 500)\overline{1000} \\ \$800 - \$300 = \$500. & 2\text{ mo.} \end{array}$$

The interest on the $100 for 4 months is equal to the interest of $1 for 400 months; and the interest of the $200 for 3 months, to that of $1 for 600 months; and thus the interest on both partial payments, at the expiration of the 6 months, is equal to the interest of $1 for 400 + 600, or 1000 months. To equal this credit of interest, the balance of the debt, which we find to be $500, should remain unpaid, after the 6 months, $\frac{1}{500}$ of 1000 months, or 2 months.

RULE. — *Multiply each payment by the time, in months or days, it was made before it became due, and divide the sum of the products by the balance remaining unpaid. The quotient will be the average time required.*

EXAMPLES FOR PRACTICE.

2. Sold, March 11, 1855, James Stone goods to the amount of $1850, on a credit of 4 months. I received from him, April 7, $400; May 15, $270; and June 20, $350. When in equity should I receive the balance? Ans. Sept. 22, 1855.

3. Bought, June 12, 1855, of William Jones, goods to the amount of $1200, on a credit of 8 months. I paid him, September 1, $400; November 1, $200; and December 1, $100. When in equity can he require the balance of me?
Ans. Aug. 17, 1856.

4. I sold, September 25, 1855, John Eckles 144 barrels of flour, at $12 per barrel, and 370 bushels of wheat, at $3 per bushel, on 6 months' credit. I received of him, September 25, $1000; November 1, $800; and December 21, $600. When ought I to be paid the remainder? Ans. June 14, 1858.

QUESTION. — Art. 232. What is the rule for finding the average time of paying the balance of a debt, when partial payments have been made?

5. Wilson Seymour bought March 20, 1855, of a merchant in Troy, merchandise to the amount of $2000, on 6 months' credit. He pays down $500; May 10, $350, and June 7, $400. When ought he to pay the balance? Ans. May 18, 1856.

ART. **233.** To equate an account containing items of both debit and credit.

Ex. 1. At what time did the balance of the following account become due, allowing that each item drew interest from its date?

Dr. *Martin Jordan in account with David Hill & Co.* *Cr.*

1856.				1856.			
Jan. 22,	To merchandise,	$89	00	Jan. 4,	By merchandise,	$77	00
" 24,	" "	76	00	Apr. 16,	" "	40	00
Feb. 20,	" "	25	00	May 14,	" "	143	00
" 23,	" "	210	00				
April 4,	" "	189	00				
May 21,	" "	30	00				

Ans. February 9, 1856.

OPERATION.

Debits.

Jan. 22, $89
" 24, 76 × 2 = 152
Feb. 20, 25 × 29 = 725
" 23, 210 × 32 = 6720
April 4, 189 × 73 = 13797
May 21, 30 × 120 = 3600

619)24994(40$\frac{234}{619}$ days.
2476
234

Average date of purchase, 40 days from Jan. 22, or on March 2.

Credits.

Jan. 4, $77
April 16, 40 × 103 = 4120
May 14, 143 × 131 = 18733

260)22853(87$\frac{233}{260}$ days.
2080
2053
1820
233

Average of credits, 88 days from Jan. 4, or on April 1.

Difference between March 2 and April 1 = 30 days,

260
30

$619 — $260 = $359, or balance,

359) 7800 (21$\frac{261}{359}$ days.
718
620
359
261

22 days back from March 2 = February 9.

QUESTION. — Art. 233. How do you equate an account having items of debit and credit?

On equating each side of the account (Art. 230), we find the debits $619, became due 40 days from January 22, or on March 2; and the credits, $260, became due 88 days from January 4, or on April 1.

If the account had been settled on March 2, it is evident the credits, $260, would have been paid 30 days, or the time from March 2 to April 1, before having become due. This would have been a loss of the use or interest of that sum to the credit side of the account, and a corresponding gain to the debit side. Now, as the settlement is required to be one of equity, we find how long it will take the balance of the account, $359, to gain the same interest that $260 would gain in the 30 days. If it takes $260 to gain a certain interest in 30 days, it would take $1 to gain the same interest 260 times 30 days, or 7800 days; and $359 to gain the same amount of interest $\frac{1}{359}$ of 7800 days, or 22 days nearly. Hence, the balance became due 22 days back of March 2, or on February 9, which is the answer sought.

In this example, the time was counted back from the average date of the larger amount, since it became due *first;* but when that amount becomes due *last,* the time is counted forward from its average time.

RULE. — *Find the average time of each side becoming due.*

Multiply the amount of the smaller side by the number of days between the two average dates, and divide the product by the balance of the account.

The quotient will be the time of the balance becoming due, counted from the average date of the larger side, BACK *when the amount of that side is due* FIRST, *but* FORWARD *when it is due* LAST.

NOTE. — Having the average time of a balance becoming due, its CASH VALUE can be ascertained *when the balance is due before the time of settling the account, by adding to it the interest up to the time of settlement; and when due after that time, by finding the present worth* (Art. 213), *from the time of settlement to the time of the balance becoming due.*

EXAMPLES FOR PRACTICE.

2. In settling the following account, when did the balance become due, the merchandise items being on 6 months' credit?

Dr. *Hiram Lewis in account with Joseph Warren.* *Cr.*

1854.				1854.			
Feb. 16,	To merchandise,	$375	80	Mar. 20,	By cash,	$300	00
April 8,	" "	432	18	June 17,	" merchandise,	371	50
May 17,	" "	320	15	July 4,	" cash,	200	00
July 13,	" "	158	12	Sept. 25,	" merchandise,	85	20

Ans. March 3, 1855.

QUESTIONS. — What is the rule? How can the cash value of the balance of an account be found?

3. Edward Doton owes Daniel Stetson, 1855, May 1, for merchandise, $500; May 15, for timber, $400; June 14, for a horse, $300; July 24, for bill of labor, $100. Stetson owes Doton, 1855, March 7, for a pleasure-boat, $400; April 2, for merchandise, $200; May 6, for merchandise, $300; June 13, for a carriage, $120. Allowing all the items to be on 6 months' credit, when will the balance of the account become due?

Ans. April 27, 1856.

---

§ XXX. RATIO.

ART. 234. RATIO is the relation, in respect to magnitude or value, which one quantity or number has to another of the same kind, or the quotient arising from the division of one number by another. Thus, the ratio of 6 to 3 is 2.

Of the two numbers necessary to form a ratio, the first is called the *antecedent*, and the last the *consequent*. Thus, in the example given, 6 is the *antecedent*, and 3 the *consequent*.

When there is but one antecedent and one consequent, the ratio is called a *simple ratio*. The antecedent and consequent are also called the *terms* of the ratio.

ART. 235. A ratio may be expressed in two ways. The ratio of 6 to 3 may be expressed by two dots between the terms, thus, 6 : 3; or in the form of a fraction, by making the antecedent the numerator and the consequent the denominator, thus, $\frac{6}{3}$.

The terms of a ratio must be of the *same kind*, or such as may be reduced to the same denomination, in order that they may have a ratio to each other. Thus, shillings have a ratio to shillings, and shillings to pounds, &c.; but shillings have not a ratio to gallons, nor pounds to days, because they are not commensurable.

ART. 236. A ratio may be either *direct* or *inverse*. A *direct* ratio is when the antecedent is divided by the consequent; an *inverse* ratio is when the consequent is divided by the antecedent. Thus, the *direct* ratio of 6 to 3 is $\frac{6}{3}$, and the *inverse* ratio of 6 to 3 is $\frac{3}{6}$, or $\frac{1}{2}$.

The direct ratio of one quantity or number to another is found by dividing the number whose ratio is required, which is

---

QUESTIONS. — Art. 234. What is ratio? How many numbers are necessary to form a ratio? What are the antecedent and consequent called? — Art. 235. What two ways are there of expressing a ratio? — Art. 236. What is a direct ratio? What an inverse ratio?

the antecedent, by the number with which it is compared, which is the consequent. The inverse ratio is found by reversing this process.

Examples for Practice.

1. What is the direct ratio of 9 to 3? Ans. 3. Of 18 to 6? Of 16 to 4? Of 24 to 12? Of 20 to 5? Of 15 to 3?
2. What is the direct ratio of 7 to 21? Ans. $\frac{1}{3}$. Of 4 to 28? Of 6 to 30? Of 9 to 11? Of 9 to 99? Of 30 to 90?
3. What is the direct ratio of 60 to 12? Of 132 to 11? Of 40 to 120? Of 32 to 96? Of 200 to 50? Of 144 to 1728? Of 360 to 60?
4. What is the inverse ratio of 10 to 5? Ans. $\frac{1}{2}$. Of 27 to 81? Of 16 to 48? Of 72 to 9? Of 11 to 88?
5. What is the direct ratio of 2£. 5s. to 9s.? Ans. 5. Of 9in. to 1ft. 6in.?

Art. **237.** A *compound* ratio consists of two or more simple ratios, whose corresponding terms are to be multiplied together. Thus,

The simple ratio of	8 : 4	is 2
And " " of	12 : 3	is 4
The compound ratio of	8 × 12 : 4 × 3	is 2 × 4
Or " " of	96 : 12	is 8

When a compound ratio is composed of *two* equal ratios, it is called a *duplicate* ratio; when of *three*, it is called a *triplicate* ratio, &c.

The simple ratio of	4 : 2	is 2
" " " of	6 : 3	is 2
" " " of	8 : 4	is 2
The *triplicate* ratio of	4 × 6 × 8 : 2 × 3 × 4	is × 2 × 2 × 2
Or " " of	192 : 24	is 8

If the terms of a ratio are both multiplied or divided by the same number, the ratio is not altered. Thus, the ratio of 8 : 2 is 4; the ratio 8 × 2 : 2 × 2 is 4; and the ratio of 8 ÷ 2 : 2 ÷ 2 is 4.

Questions. — Art. 237. What is a compound ratio? What a duplicate ratio? What a triplicate ratio? What is the effect of multiplying or dividing the terms of a ratio?

§ XXXI. PROPORTION.

ART. **238.** PROPORTION is an equality of ratios. Thus the ratios 9 : 3 and 12 : 4 are equal, and when united form a proportion.

Proportion is usually expressed by four dots between the two ratios; thus, the proportion in the preceding example is written 9 : 3 : : 12 : 4, and is read, 9 is to 3 as 12 to 4.

The numbers which form a proportion are called proportionals. The *first* and *third* are called *antecedents*, the *second* and *fourth* are called *consequents;* also, the *first* and *last* are called *extremes*, and the remaining two the *means*.

ART. **239.** Any four numbers are said to be proportional to each other when the first contains the second as many times as the third contains the fourth; or when the second contains the first as many times as the fourth contains the third. Thus, 9 has the same ratio to 3 that 12 has to 4, because 9 contains 3 as many times as 12 contains 4.

ART. **240.** *If the antecedents or consequents of a proportion, or both, are divided by the same number, they are still proportionals.* Thus, dividing the antecedents of the proportion 4 : 8 : : 10 : 20 by 2, we have 2 : 8 : : 5 : 20; dividing the consequents by 2, we have 4 : 4 : : 10 : 10; and dividing both the consequents and antecedents by 2, we have 2 : 4 : : 5 : 10; each of which is a proportion, since if we divide the second term of each by the first, and the fourth by the third, the two quotients will be equal. The effect is the same when the terms are *multiplied* by the same number.

ART. **241.** *The product of the extremes of a proportion is equal to the product of the means.* Thus, the proportion 14 : 7 : : 18 : 9 may be expressed fractionally, $\frac{14}{7} = \frac{18}{9}$. Now, if we reduce these fractions to a common denominator, we have $\frac{126}{63} = \frac{126}{63}$; but in this operation we multiplied together the two *extremes* of the proportion, 14 and 9, and the two *means*, 18 and 7; thus, $14 \times 9 = 18 \times 7$.

QUESTIONS. — Art. 238. What is proportion? How is proportion expressed? What are the numbers called that form a proportion? Which are the extremes? Which the means? — Art. 239. When are numbers said to be in proportion to each other? — Art. 240. What is the effect of dividing the antecedents or consequents of a proportion? Of multiplying them? — Art. 241. How does the product of the extremes compare with that of the means?

ART. **242.** *If the extremes and one of the means are given the other mean may be found by dividing the product of the extremes by the given mean.* Thus, if the extremes are 3 and 24, and the given mean 6, the other mean is 12; because 24 × 3 = 72; and 72 ÷ 6 = 12.

ART. **243.** *If the means and one of the extremes are given the other extreme may be found by dividing the product of the means by the given extreme.* Thus, if the means are 8 and 16, and the given extreme 4, the other extreme is 32; because 16 × 8 = 128; and 128 ÷ 4 = 32.

SIMPLE PROPORTION.

ART. **244.** SIMPLE PROPORTION is an expression of the equality between two simple ratios.

NOTE. — Simple Proportion is sometimes called the Rule of Three.

ART. **245.** Method of stating and solving questions in Simple Proportion.

Ex. 1. If 7lb. of sugar cost 56 cents, what will 36lb. cost?
Ans. $2.88.

OPERATION.

```
Extreme.   Mean.       Mean.
 7 lb. : 3 6 lb. : : 5 6 cts.
                     3 6
                    3 3 6
                  1 6 8
              7 ) 2 0.1 6
                  $2.8 8 Extreme.
```

Since 7lb. have the same ratio to 36lb. as 56 cents, the cost of the former, have to the cost of the latter, we have the first three terms of a proportion given, namely, one of the extremes and the two means. Now, to ascertain which of these terms are the means, and which the extreme, we arrange them in the order of a proportion, or *state the question*, by making 56 cents the *third* term, because it is of the same kind, and has the same proportion to the required answer, or fourth term, as the first has to the second. From the nature of the question, since the answer will be more than 56 cents, or the third term, the *second* term must be greater than the *first;* we make 36lb. the *second* term, and 7lb. the *first*, and then proceed as in Art. 246.

QUESTIONS. — Art. 242. If the extremes and one of the means are given, how can the other mean be found? — Art. 243. When the means and one of the extremes are given, how can the other extreme be found? — Art. 244. What is simple proportion? How many terms are given in questions in simple proportion?

BY ANALYSIS. — If 7lb. cost 56 cents, 1 pound will cost $\frac{1}{7}$ of 56 cents, which is 8 cents. Then, if 1lb. cost 8 cents, 36lb. will cost 36 times as much; that is, 36 times 8 cents, which are $2.88, Ans. as before.

Ex. 2. If 76 barrels of flour cost $456, what will 12 barrels cost? Ans. $72.

OPERATION.

```
bar. bar.    $.
76 : 12 : : 456
             12
  76 ) 5472 ( $72
       532
        152
        152
```

We state this question by making $456 the third term, because it is of the same kind of the required answer. Then, since the answer must be less than $456, because 12 barrels will cost less than 76 barrels, we make 12 barrels, the smaller of the other two terms, the *second* term, and 76 barrels the *first* term, and proceed as before.

BY ANALYSIS. — If 76 barrels cost $456, 1 barrel will cost $\frac{1}{76}$ of $456, which is $6. Then, if 1 barrel cost $6, 12 barrels will cost 12 times as much; that is, $72, Ans. as before.

Ex. 3. If 3 men can dig a well in 20 days, how long will it take 12 men? Ans. 5 days.

OPERATION.

```
men. men. days.
12 : 3 : : 20
             3
     12 ) 60
           5 days.
```

Since the required answer is days, we make 20 days the *third* term. And as 12 men will dig the well in less time than 3 men, the answer must be less than 20 days. Therefore we make 3 men the second term, and 12 men the first, and proceed as in the other examples.

BY ANALYSIS. — If 3 men dig the well in 20 days, it will take one man 3 times as long, that is, 60 days. Again, we say, If one man dig the well in 60 days, 12 men would dig it in $\frac{1}{12}$ of 60 days, that is, 5 days, Ans. as before.

From the preceding examples we deduce the following

RULE. — *Write the given number that is of the same kind as the required fourth term, or answer, for the third term of the proportion.*

Of the other two numbers, write the larger for the second term, and the less for the first, when the answer should exceed the third term; but write the less for the second term, and the larger for the first, when the answer should be less than the third term.

Multiply the second and third terms together, and divide their product by the first.

QUESTIONS. — What is meant by stating the question? Which of the terms given in the example do you make the third? Why? Which the second? Why? Which the first? Why? After the question is stated, how do you obtain the answer?

NOTE 1. — When the first and second terms consist of different denominations, they must be reduced to the same denomination; and when the third term is a compound number, it must be reduced to the lowest denomination mentioned in it. The answer will be the same denomination as the third term.

NOTE 2. — To shorten the operations, factors common to the dividend and divisor may be cancelled.

NOTE 3. — The pupil should perform these questions by analysis as well as by proportion, and introduce cancellation when it will abbreviate the operation.

Ex. 4. If 16 bushels of wheat are worth $24, what are 96 bushels worth? Ans. $144.

OPERATION BY CANCELLATION.

bu. bu. $.
16 : 96 : : 24

$$\frac{\overset{6}{\cancel{96}} \times 24}{\cancel{16}} = \$144$$

BY ANALYSIS AND CANCELLATION.

$$\frac{\$24 \times \overset{6}{\cancel{96}}}{\cancel{16}} = \$144$$

We first state the question as directed in the rule, and then write the second and third terms above a horizontal line, with the sign of multiplication between them, for a dividend, and the first term below the line for a divisor, and cancel the common factors.

By this method of analysis we first place the $24, which is of the same kind of the required answer, above a line for a dividend; and then say, Since $24 is the price of 16 bushels, 1 bushel will cost $\frac{1}{16}$ of $24, and express the division by placing the 16 below the line for a divisor. Now, since we have an expression for the price of 1 bushel, we next express the multiplication of it by 96 bushels, the price of which is required, and then cancel as before.

EXAMPLES FOR PRACTICE.

5. What cost 9 gallons of molasses, if 63 gallons cost $14.49? Ans. $2.07.

6. What cost 97 acres of land, if 19 acres can be obtained for $337.25? Ans. $1721.75.

7. If a man travel 319 miles in 11 days, how far will he travel in 47 days? Ans. 1363 miles.

8. When $120 are paid for 15 barrels of mackerel, what will be the cost of 79 barrels? Ans. $632.

QUESTIONS. — What is the rule for simple proportion? How should the pupil perform the questions? How do you state the question and arrange the terms for cancellation? What do you cancel? How do you arrange the terms for cancellation by analysis?

9. If 9 horses eat a load of hay in 12 days, how many horses would it require to eat the hay in 3 days? Ans. 36 horses.

10. When $5.88 are paid for 7 gallons of oil, what cost 27 gallons? Ans. $22.68.

11. When $10.80 are paid for 9lb. of tea, what cost 147lb Ans. $176.40.

12. What cost 27 tons of coal, when 9 tons can be purchased for $85.95? Ans. $257.85.

13. If 15 tons of lead cost $105, what cost 765 tons? Ans. $5355.00.

14. If 16hhd. of molasses cost $320, what cost 176hhd.? Ans. $3520.00.

15. If 15cwt. 3qr. 17lb. of sugar cost $124.67, what cost 76cwt. 2qr. 19lb.? Ans. $600.56+.

16. If the cars on the Boston and Portland Railroad go one mile in 2 minutes and 8 seconds, how long will they be in passing from Haverhill to Boston, the distance being 32 miles? Ans. 1h. 8min. 16sec.

17. If a man travels 3m. 7fur. 18rd. in one hour, how far will he travel in 9h. 45min. 19sec.? Ans. 38m. 2fur. 32rd+

18. A fox is 96 rods before a greyhound, and while the fox is running 15 rods the greyhound will run 21 rods; how far will the dog run before he can catch the fox? Ans. 336 rods.

19. If 5 men can reap a field in 12 hours, how long would it take them if 4 men were added to their number? Ans. $6\frac{2}{3}$ hours.

20. Ten men engage to build a house in 63 days, but 3 of their number being taken sick, how long will it take the rest to complete the house? Ans. 90 days.

21. If a 4 cent loaf weighs 5oz. when flour is $5 per barrel, what should it weigh when flour is $7.50 per barrel? Ans. $3\frac{1}{3}$oz.

22. If 7 men can mow a field in 10 days when the days are 14 hours long, how long would it take the same men to mow the field when the days are 13 hours long? Ans. $10\frac{10}{13}$ days.

23. If 29lb. of butter will purchase 40lb. of cheese, how many pounds of butter will buy 79lb. of cheese? Ans. $57\frac{11}{40}$lb.

24. If $\frac{3}{4}$ of a yard cost $\frac{5}{8}$ of a dollar, what will $\frac{11}{12}$ of a yard cost? Ans. 0.76\frac{7}{18}$.

$$\overset{yd.}{\tfrac{3}{4}} : \overset{yd.}{\tfrac{11}{12}} :: \overset{\$.}{\tfrac{5}{8}};\ \tfrac{4}{3} \times \tfrac{11}{12} \times \tfrac{5}{8} = \tfrac{220}{288} = \$0.76\tfrac{7}{18}, \text{ Ans.}$$

25. If $4\frac{7}{8}$ yards of cloth cost $\$2\frac{7}{8}$, what will $19\frac{1}{2}$ yards cost?
Ans. $11.50.

$$\overset{\text{yd.}}{4\tfrac{7}{8}} : \overset{\text{yd.}}{19\tfrac{1}{2}} : : \overset{\$.}{2\tfrac{7}{8}} ; \ \frac{\not{8}}{\not{39}} \times \frac{\not{39}}{2} \times \frac{23}{\not{8}} = \frac{23}{2} = \$.11.50, \text{ Ans.}$$

26. If for $4\frac{7}{11}$ yards of velvet there be received $11\frac{4}{5}$ yards of calico, how many yards of velvet will be sufficient to purchase 100 yards of calico? Ans. $39\frac{189}{649}$ yards.

27. A certain piece of labor was to have been performed by 144 men in 36 days, but a number of them having been sent away, the work was performed in 48 days; required the number of men discharged. Ans. 36 men.

28. James can mow a certain field in 6 days, John can mow it in 8 days; how long will it take John and James both to mow it? Ans. $3\frac{3}{7}$ days.

29. A. Atwood can hoe a certain field in 10 days, but with the assistance of his son Jerry he can hoe it in 7 days, and he and his son Jacob can hoe it in 6 days; how long would it take Jerry and Jacob to hoe it together? Ans. $9\frac{3}{23}$ days.

30. Bought a horse for $75; for what must I sell him to gain 10 per cent.?

1.00 : 1.10 : : $75 : $82.50, Ans.

31. Bought 40 yards of cloth at $5.00 per yard; for what must I sell the whole amount to gain 15 per cent.?
Ans. $230.00.

32. My chaise cost $175.00, but, having been injured, I am willing to sell it at a loss of 30 per cent.; what should I receive? Ans. $122.50.

33. Bought a cargo of flour on speculation at $5.00 per barrel, and sold it at $6.00 per barrel; what did I gain per cent.?
Ans. 20 per cent.

34. Bought a hogshead of molasses for $15.00, but, it not proving so good as I expected, I sell it for $12; what do I lose per cent.? Ans. 20 per cent.

35. Bought a hogshead of molasses for $27.50, at 25 cents per gallon; how much did it contain? Ans. 110 gallons.

36. A certain farm was sold for $1728, it being $15.75 per acre; what was the quantity of land?
Ans. 109A. 2R. $34\frac{2}{7}$p.

37. A certain cistern has 3 cocks; the first will empty it in 2 hours, the second in 3 hours, and the third in 4 hours; in what time would they all empty the cistern together?
Ans. $55\frac{5}{13}$ minutes.

COMPOUND PROPORTION.

ART. **246.** COMPOUND PROPORTION is an expression of the equality between a *compound* and a simple *ratio.*

It is employed in performing such questions as require two or more operations in Simple Proportion.

ART. **247.** Method of stating and solving questions in Compound Proportion, sometimes called the Double Rule of Three.

Ex. 1. If $100 will gain $8 in 12 months, what will $600 gain in 10 months? Ans. $40.

OPERATION.

$$\left.\begin{array}{l}\overset{\text{Extreme.}}{\$100} : \overset{\text{Mean.}}{\$600} \\ 12\text{ mo.} : 10\text{ mo.}\end{array}\right\} :: \overset{\text{Mean.}}{\$8}$$

$$\frac{600\times10\times8}{100\times12}=\frac{48000}{1200}=\$40, \text{ Extreme.}$$

In stating this question, we make $8, the gain, which is the same name of the required answer, the third term. Then, taking $100 and $600, two of the remaining terms of the same kind, we inquire if the answer, depending on these alone, must be greater or less than the third term; and since it must be greater, because $600 will gain more than $100 in the same time, we make $600 the second term, and $100 the first. Again, we take the two remaining terms, and make 10 mo. the second term, and 12 mo. the first, since the same sum would gain less in 10 mo. than in 12 mo. We then find the continued products of the second and third terms, and divide it by the continued product of the first terms, for the answer. Hence the following

RULE. — *Make that number which is of the same kind as the answer required the third term of a proportion. Of the remaining numbers, take any two, that are of the same kind, and consider whether an answer, depending upon these alone, would be greater or less than the third term, and place them as directed in Simple Proportion.*

Then take any other two, and consider whether an answer, depending only upon them, would be greater or less than the third term, and arrange them accordingly; and so on until all are used.

QUESTIONS. — Art. 246. What is compound proportion? For what is it employed? — Art. 247. By what other name is it sometimes called? In stating the question, which of the numbers do you make the third term? Why? What do you do with the remaining terms? How do you know which of the two to take for the second term? Which for the first? After all the terms have been arranged, how do you find the answer? What is the rule for compound proportion?

Multiply the product of the second terms by the third, and divide the result by the product of the first terms. The quotient will be the fourth term, or answer.

Note. — The following questions should be performed not only by the *rule*, but by an analysis and cancellation.

Ex. 2. If $100 will gain $6 in 12 months, what will $800 gain in 8 months?

OPERATION BY CANCELLATION.

$$\left.\begin{array}{l}\$100 : \$800 \\ 12 \text{ mo.} : 8 \text{ mo.}\end{array}\right\} :: \$6$$

$$\frac{\$\overset{8}{\cancel{800}} \times \overset{4}{\cancel{8}} \times \cancel{6}}{\cancel{100} \times \underset{2}{\cancel{12}}} = \$32$$

We state the question according to the rule, and then write the second and third terms for a dividend and the first terms for a divisor, and cancel the common factors.

BY ANALYSIS AND CANCELLATION.

$$\frac{\$\overset{4}{\cancel{6}} \times \cancel{8} \text{ mo.} \times \$\overset{8}{\cancel{800}}}{\underset{2}{\cancel{12}} \text{ mo.} \times \$\cancel{100}} = \$32$$

By this method of analysis we say, if $6 are the gain of $100 in 12 mo., in 1 mo. the gain of $100 will be $\frac{1}{12}$ as much, or $\frac{6}{12}$, and in 8 mo. 8 times as much, or $\$\frac{6 \times 8}{12}$. Again, if $100 gain $\$\frac{6 \times 8}{12}$ in 8 mo., $1 will gain $\frac{1}{100}$ of it, or $\$\frac{6 \times 8}{12 \times 100}$, and $800 will gain 800 times as much, or $\$\frac{6 \times 8 \times 800}{12 \times 100}$, the same as in the operation. Cancelling the common factors, we obtain $32 for the answer.

Examples for Practice.

3. If $100 gain $6 in 12 months, in how many months will $800 gain $32? Ans. 8 months.

4. If $100 gain $6 in 12 months, how large a sum will it require to gain $32 in 8 months? Ans. $800.

5. If $800 gain $32 in 8 months, what is the per cent.? Ans. 6 per cent.

6. If 15 carpenters can build a bridge in 60 days when the days are 15 hours long, how long will it take 20 men to build the bridge when the days are 10 hours long? Ans. $67\frac{1}{2}$ days.

Questions. — How does the author say the questions under this rule should be performed? How are questions stated for cancellation? Which terms are taken for the dividend? Which for the divisor? What are cancelled?

7. If a regiment of soldiers, consisting of 939 men, can eat 351 bushels of wheat in 3 weeks, how many soldiers will it require to eat 1404 bushels in 2 weeks?

Ans. 5634 soldiers.

8. If 8 men spend $64 in 13 weeks, what will 12 men spend in 52 weeks? Ans. $384.

9. If 8 horses consume 42 bushels of grain in 24 days, how many bushels will suffice 32 horses 48 days?

Ans. 336 bushels.

10. If 6 men in 16 days of 9 hours each build a wall 20 feet long, 6 feet high, and 4 feet thick, in how many days of 16 hours each will 24 men build a wall 200 feet long, 16 feet high, and 6 feet thick? Ans. 90 days.

11. If a man travel 117 miles in 15 days, employing only 9 hours a day, how far would he go in 20 days, travelling 12 hours a day? Ans. 208 miles.

12. If 12 men in 15 days can build a wall 30 feet long, 6 feet high, and 3 feet thick, when the days are 12 hours long, in what time will 30 men build a wall 300 feet long, 8 feet high, and 6 feet thick, when they work 8 hours a day?

Ans. 240 days.

13. If the carriage of 5cwt. 3qr. 150 miles cost $24.58, what must be paid for the carriage of 7cwt. 2qr. 15lb. 32 miles, at the same rate? Ans. $6.97+.

14. A received of B $9 for the use of $600 for 6 months; now B wishes to hire of A $1800 until the interest shall amount to the same sum. How long may he keep it? Ans. 2 months.

15. If 15 oxen or 20 cows will eat 3 tons of hay in 8 weeks, how much hay will be sufficient for 15 oxen and 8 cows 12 weeks? Ans. $6\frac{3}{10}$ tons.

16. If 5 men, by laboring 10 hours a day, can mow a field of 30 acres in 10 days, how long will it require 8 men and 7 boys, provided each boy can do $\frac{7}{11}$ as much as a man, to mow a field containing 54 acres? Ans. $7\frac{31}{107}$ days.

17. If 2 men can build $12\frac{3}{4}$ rods of wall in $6\frac{1}{2}$ days, how long will it take 18 men to build $247\frac{2}{13}$ rods?

Ans. 14 days.

18. If 248 men, in $5\frac{1}{2}$ days of 11 hours each, dig a trench of 7 degrees of hardness, and $232\frac{1}{2}$ feet long, $3\frac{2}{3}$ feet wide, and $2\frac{1}{3}$ feet deep, in how many days of 9 hours each will 24 men dig a trench of 4 degrees of hardness, and $337\frac{1}{2}$ feet long, $5\frac{3}{5}$ feet wide, and $3\frac{1}{2}$ feet deep? Ans. 132 days.

§ XXXII. PROFIT AND LOSS.

ART. **248.** PROFIT and Loss is the process by which merchants and other traders estimate their gain or loss in buying and selling goods.

The following questions may be performed either by analysis or by proportion.

ART. **249.** To find the profit or loss per cent. when the cost and selling price are given.

Ex. 1. If I buy flour at $4 per barrel, and sell it at $5 per barrel, what is the gain per cent.? Ans. 25 per cent.

OPERATION.

$$\$5 - \$4 = \$1;\ \tfrac{1}{4} = 1.00 \div 4 = .25, \text{ or } 25 \text{ per cent.}$$

By subtracting the cost from the selling price, we find the gain per barrel to be $1. Now, if the gain is $1 on $4, on $1 it will be $\frac{1}{4}$ of $1 = $\frac{1}{4}$ of a dollar, or 25 per cent.

OPERATION BY PROPORTION.

$$\$5 - \$4 = \$1;\ \$4 : \$1 :: 1.00 : .25, \text{ that is, } 25 \text{ per cent.}$$

2. If I buy flour at $5 per barrel, and sell it at $4 per barrel, what is the loss per cent.? Ans. 20 per cent.

OPERATION.

$$\$5 - \$4 = \$1;\ \tfrac{1}{5} = 1.00 \div 5 = .20, \text{ or } 20 \text{ per cent.}$$

By subtracting the selling price from the cost, we find the loss per barrel to be $1. Now, if the gain is $1 on $4, on $1 it will be $\frac{1}{5}$ of $1 = $\frac{1}{5}$ of a dollar, or 20 per cent. From this analysis and that of the preceding example, it is seen that the operation is equivalent to making the gain or loss the numerator of a fraction, and the cost the denominator, and then reducing this fraction to a decimal; or, in short, to simply dividing the gain or loss by the cost.

OPERATION BY PROPORTION.

$$\$5 - \$4 = \$1;\ \$5 : \$1 :: 100 \text{ per cent} : 20 \text{ per cent.}$$

RULE 1. — *Divide the gain or loss by the cost, and the quotient will be the gain or loss per cent.* Or,

RULE 2. — *As the cost is to the gain or loss, so is* 100 *per cent. to the gain or loss per cent.*

QUESTIONS. — Art. 248. What is profit and loss? What is the first rule for finding the profit or loss in buying or selling goods? What is the second rule?

EXAMPLES FOR PRACTICE.

3. Bought 40 yards of broadcloth at \$5.40 per yard, and I sell $\frac{3}{4}$ of it at \$6 per yard, and the remainder at \$7 per yard; what do I gain per cent.? Ans. $15\frac{20}{27}$ per cent.

4. A merchant purchased for cash 50 barrels of flour, at \$5 per barrel, and immediately sold the same on 8 months' credit, at \$5.98 per barrel; what does he gain per cent.?

Ans. 15 per cent.

5. A grocer bought a hogshead of molasses, containing 100 gallons, at 30 cents per gallon; but 30 gallons having leaked out, he disposed of the remainder at 40 cents per gallon. Did he gain or lose, and how much per cent.?

Ans. Lost $6\frac{2}{3}$ per cent.

6. A gentleman in Rochester, N. Y., purchased 3000 bushels of wheat, at $\$1.12\frac{1}{2}$ per bushel. He paid 5 cents per bushel for its transportation to N. Y. city, and then sold it at $\$1.37\frac{1}{2}$ per bushel; what did he gain per cent.? Ans. $17\frac{1}{47}$ per cent.

7. J. Morse bought, in Lawrence, a lot of land $7\frac{3}{11}$ rods square, for \$5 per square rod. He sold the land at 5 cents per square foot; what did he gain per cent.?

Ans. $172\frac{1}{4}$ per cent.

ART. **250.** To find the selling price when the cost and the gain or loss per cent. are given.

Ex. 1. If I buy flour at \$4 per barrel, for how much must I sell it per barrel to gain 25 per cent.? Ans. \$5.

OPERATION.

$$\$4 \times .25 = \$1.00\,;\ \text{then}\ \$4 + \$1 = \$5,\ \text{Ans.}$$

It is evident, if I sell the flour for 25 per cent. gain, I sell it for .25 more than it cost. Therefore, if I add to the cost .25 of the cost, the sum will be the price per barrel for which the flour must be sold; as seen in the operation.

OPERATION BY PROPORTION.

$$1.00 + .25 = 1.25\ ;\ 1.00 : 1.25 :: \$4 : \$5,\ \text{Ans.}$$

2. If I buy flour at \$5 per barrel, for what must I sell it per barrel to lose 20 per cent.?

QUESTION. — Art. 250. Explain how you find the selling price when the cost and the gain or loss per cent. are given.

OPERATION.

$$\$5 \times .20 = \$1.00;\ \$5 - \$1 = \$4, \text{ Ans.}$$

It is evident, if I sell the flour for 20 per cent. loss, I sell it for .20 less than it cost. Therefore, if I subtract from the cost .20 of the cost, the remainder will be the price per barrel for which the flour must be sold; as seen in the operation.

OPERATION BY PROPORTION.

$$1.00 - .20 = .80;\ 1.00 : .80 :: \$5 : \$4, \text{ Ans.}$$

RULE 1. — *Find the percentage on the cost at the given rate per cent., and add it to the cost, or subtract it from the same, according as the selling price is to be that of profit or loss.* Or,

RULE 2. — *As* 1 *is to* 1 *with the profit per cent. added, or loss per cent. subtracted, expressed decimally, so is the given price to the price required.*

EXAMPLES FOR PRACTICE.

3. Bought a hogshead of molasses, containing 120 gallons, for 30 cents per gallon, but it not proving so good as was expected, I am willing to lose 10 per cent. on the cost; what shall I receive for it? Ans. $32.40.

4. A grocer bought a hogshead of sugar, weighing net 8cwt. 3qr. 5lb., for $88; for what must he sell it per pound to gain 20 per cent.? Ans. 12 cents per pound.

5. J. Simpson bought a farm for $1728; for what must it be sold to gain 12 per cent., provided he is to wait 8 months, without interest, for his pay? Ans. $2012.77+.

6. J. Fox purchased a barrel of vinegar, containing 32 gallons, for $4; but 8 gallons having leaked out, for how much must he sell the remainder per gallon to gain 10 per cent. on the cost? Ans. $0.18⅓ per gallon.

7. Bought a horse for $90, and gave my note to be paid in 6 months, without interest; what must be my cash price to gain 20 per cent. on my bargain? Ans. $104.84+.

8. H. Tilton bought 7cwt. of coffee at $11.50 per cwt., but finding it injured, he is willing to lose 15 per cent.; for how much must he sell the 7cwt.? Ans. $68.42+.

ART. **251.** To find the cost when the selling price and the gain or loss per cent. are given.

Ex. 1. If I sell flour at $5 per barrel, and by so doing make 25 per cent., what was the cost of the flour?

Ans. $4 per barrel.

QUESTIONS. — Art. 250. What is the first rule for finding at what price goods must be sold to gain or lose a given per cent.? What is the second rule?

OPERATION.

$$\$5.00 \div 1.25 = \$4, \text{ Ans.}$$

Since the gain evidently is 25 per cent. of the cost, the selling price, $5, is equal to the cost increased by 25 per cent. of the cost; or, as it may be expressed, equal to 1.25 of the cost. Hence, the cost must be as many dollars as the $5 contains times 1.25; and, dividing, we obtain $4, the answer required,

OPERATION BY PROPORTION.

$$1.00 + .25 = 1.25;\ 1.25 : 1.00 :: \$5 : \$4, \text{ Ans.}$$

2. If I sell flour at $4 per barrel, and by so doing lose 20 per cent., what was the cost of the flour? Ans. $5 per barrel.

OPERATION.

$$\$4.00 \div .80 = \$5, \text{ Ans.}$$

Since the loss evidently is 20 per cent. of the cost, the selling price, $4, is equal to the cost decreased by 20 per cent. of the cost; or, as it may be expressed, equal to .80 of the cost. Hence, the cost must be as many dollars as the $5 contains times .80; and, dividing, we obtain $5, the answer required.

OPERATION BY PROPORTION.

$$1.00 - .20 = .80;\ .80 : 1.00 :: \$4 : \$5, \text{ Ans.}$$

RULE 1. — *Divide the selling price by* 1 *increased by the gain per cent., or by* 1 *decreased by the loss per cent., expressed decimally, and the quotient will be the cost.* Or,

RULE 2. — *As* 1 *with the gain per cent. added, or loss per cent. subtracted, expressed decimally, is to* 1, *so is the selling price to the cost.*

EXAMPLES FOR PRACTICE.

3. Having used my chaise 16 years, I am willing to sell it for $80; but by so doing I lose 62½ per cent.; what was the cost of the chaise? Ans. $213.33⅓.

4. If I sell wood at $7.20 per cord, and gain 20 per cent., what did the wood cost me per cord? Ans. $6 per cord.

5. J. Adams sold 40 cases of shoes for $1600, and gained 18 per cent.; what was the first cost of the shoes?
Ans. $1355.93+.

QUESTIONS. — Art. 251. What is the first rule for finding the cost, when the selling price and the gain or loss per cent. are given? What is the second rule?

6. Sold 17 barrels of flour at $8 per barrel, for which I received a note payable in 3 months. This note I had discounted at the Granite Bank, but, on examining my account, I find I have lost 10 per cent. on the flour; what was the cost of it?

Ans. $148.76+.

ART. **252.** The selling price of goods and the rate per cent. being given, to find what the gain or loss per cent. would be, if sold at another price.

Ex. 1. If I sell flour at $5 per barrel, and gain 25 per cent., what should I gain if I were to sell it for $7 per barrel?

OPERATION.

The solution of this question involves two principles: First, to find the cost of the flour per barrel. (Art. 251.)

Thus, $5.00 ÷ 1.25 = $4.00, the cost per barrel. Second, to find the gain per cent. on the cost when sold at $7 per barrel. (Art. 248.)

Thus, $7 — $4 = $3; 3.00 ÷ 4 = .75, or 75 per cent.

OPERATION BY PROPORTION.

1.00 + .25 = 1.25; $5 : $7 : : 1.25 : 1.75;
1.75 — 1.00 = .75, that is, 75 per cent.

RULE 1. — *Find the cost* (Art. 251), *and then the gain or loss per cent. on this cost at the proposed selling price.* (Art. 248.) Or,

RULE 2. — *As the first price is to the proposed price, so is* 1 *with the gain per cent. of the first price added, or the loss per cent. of the first price subtracted, to* 1 *with the gain per cent. of the proposed price added, or with the loss per cent. of the proposed price subtracted.*

NOTE. — If the result by the last rule exceeds 1.00, the excess is the gain per cent.; but, if it is less than 1.00, the deficiency is the loss per cent.

EXAMPLES FOR PRACTICE.

2. Sold a quantity of oats at 28 cents per bushel, and gained 12 per cent.; what per cent. should I gain or lose, if I were to sell them at 24 cents per bushel? Ans. Lose 4 per cent.

3. S. Rice sold a horse for $37.50, and lost 25 per cent.; what would have been his gain per cent. if he had sold him for $75? Ans. 50 per cent.

4. S. Phelps sold a quantity of wheat for $1728, and took

QUESTIONS. — Art. 252. What is the first rule for finding what gain or loss is made by selling goods at another price when the selling price and rate per cent. are given? What is the second rule? If the answer exceeds $100 what is the excess? If it is less than $100, what is the deficiency?

a note payable in 9 months without interest, and made 10 per cent. on his purchase; what would have been his gain per cent. if he had sold it to James Wilson for $2000 cash?

Ans. 33+ per cent.

MISCELLANEOUS EXERCISES IN PROFIT AND LOSS.

1. A horse that cost $84, having been injured, was sold for $75.60; what was the loss per cent.? Ans. 10 per cent.

2. Sold a horse for $75.60, and lost 10 per cent. on the cost; but, if I had sold him for $97.44, what per cent. should I have gained on the cost of the horse? Ans. 16 per cent.

3. M. Star sold a horse for $97.44, and gained 16 per cent.; what would have been his loss per cent. if he had sold the horse for $75.60, and what his actual loss?

Ans. Loss 10 per cent. $8.40 loss.

4. If I buy cloth at $5 per yard, on 9 months' credit, for what must I sell it per yard for cash to gain 12 per cent.?

Ans. $5.35+.

5. A. Pemberton bought a hogshead of molasses, containing 120 gallons, for $40; but 20 gallons having leaked out, for what must he sell the remainder per gallon to gain 10 per cent. on his purchase? Ans. $0.44.

6. H. Jones sells flour, which cost him $5 per barrel, for $7.50 per barrel; and J. B. Crosby sells coffee for 14 cents per pound, which cost him 10 cents per pound; which makes the greater per cent.?

Ans. H. Jones makes 10 per cent. most.

7. J. Gordon bought 160 gallons of molasses, but having sold 40 gallons, at 30 cents per gallon, to a man who proved a bankrupt, and could pay only 30 cents on the dollar, he disposed of the remainder at 35 cents per gallon, and gained 10 per cent. on his purchase; what was the cost of the molasses?

Ans. $41.45+.

8. D. Bugbee bought a horse for $75.60, which was 10 per cent. less than his real value, and sold him for 16 per cent. more than his real value; what did he receive for the horse, and what per cent. did he make on his purchase?

Ans. Received $97.44, and made $28\frac{8}{9}$ per cent.

9. A merchant bought 70 yards of broadcloth that was $1\frac{3}{4}$ yards wide, for $4.50 per yard, but the cloth having been wet, it shrunk 5 per cent. in length, and 5 in width; for what must the cloth be sold per square yard to gain 12 per cent.?

Ans. $3.19+.

§ XXXIII. PARTNERSHIP, OR COMPANY BUSINESS.

ART. **253.** PARTNERSHIP is the association of two or more persons in business, with an agreement to share the profits and losses in proportion to the amount of capital stock, or the value of the labor and experience of each.

The association is called a *Firm*, or *Company;* the money or property invested is called the *Joint Stock*, or *Capital;* each of the owners is called a *Partner*, and the profit or gain the *Dividend.*

ART. **254.** To find each partner's share of the profit or loss when the stock is employed for the same time.

Ex. 1. John Smith and Henry Gray enter into partnership for three years; Smith puts in \$4000, and Gray \$2000. They gain \$570. What is each man's share of the gain?

Ans. Smith's gain, \$380; Gray's gain, \$190.

OPERATION.

\$4000, Smith's stock, $\frac{4000}{6000} = \frac{2}{3}$, Smith's part of the stock.
\$2000, Gray's " $\frac{2000}{6000} = \frac{1}{3}$, Gray's part of the stock.
\$6000, Whole stock.

Then $\frac{2}{3}$ of \$570, the whole gain, = \$380, is Smith's share of the gain.
And $\frac{1}{3}$ of \$570, " " = \$190, is Gray's share of the gain.
Proof, \$570

Since the sum of \$4000 and \$2000, equal to \$6000, is the whole stock, it is evident that Smith's part of the stock is $\frac{4000}{6000} = \frac{2}{3}$; and that Gray's part is $\frac{2000}{6000} = \frac{1}{3}$. Then, since each man's gain must be in proportion to his stock, $\frac{2}{3}$ of \$570, = \$380, is Smith's share of the gain; and $\frac{1}{3}$ of \$570, = \$190, is Gray's share of the gain.

OPERATION BY PROPORTION.

\$6000 : \$570 : : \$4000 : \$380, Smith's gain.
\$6000 : \$570 : : \$2000 : \$190, Gray's gain.

QUESTIONS. — Art. 253. What is partnership? What is the association called? What the property invested? What are the owners called? What the profit or loss?

RULE 1. — *Multiply the whole gain or loss by each partner's fractional part of the stock, and the product will be the gain or loss of each.* Or,

RULE 2. — *As the whole stock is to each partner's stock, so is the whole gain or loss to each partner's gain or loss.*

EXAMPLES FOR PRACTICE.

2. Three merchants, A, B, and C, engaged in trade. A put in $6000, B put in $9000, and C put in $5000. They gain $840. What is each man's share of the gain?

Ans. A's gain $252, B's gain $378, C's gain $210.

3. A bankrupt owes Peter Parker $8750, James Dole $3610, and James Gage $7000. His effects, sold at auction, amount to $6875; of this sum $375 are to be deducted for expenses, &c. What will each receive of the dividend?

Ans. Parker, 2937.75\frac{100}{121}$; Dole, 1212.03\frac{62}{121}$; Gage, 2350.20\frac{80}{121}$.

4. A merchant, failing in trade, owes A $500, B $386, C $988, and D $126. His effects are sold for $100. What will each man receive?

Ans. A receives $25.00, B $19.30, C $49.40, D $6.30.

5. A, B, and C, engaged in trade. A put in $700, B put in $300, and C put in 100 barrels of flour. They gained $90; of which sum C took $30 for his part; what will A and B receive, and what was C's flour valued per barrel?

Ans. A receives $42, B $18, C's flour $5 per barrel.

ART. **255.** To find each partner's share of the profit or loss, when the stock is employed for unequal times.

Ex. 1. Josiah Brown and George Dole trade in company Brown put in $600 for 8 months, and Dole put in $400 for 6 months. They gain $60. What is each man's share of the gain?

OPERATION.

$600 × 8 = $4800, Brown's money for 1 month.
$\frac{4800}{7200} = \frac{2}{3}$, Brown's part of stock.
$400 × 6 = $2400, Dole's money for 1 month.
$\frac{2400}{7200} = \frac{1}{3}$, Dole's part of stock.
$7200, Whole stock for 1 month.

Then $\frac{2}{3}$ of $60, the whole gain, = $40, is Brown's share of gain.
And $\frac{1}{3}$ of $60, " " " = $20, is Dole's " "

QUESTION. — Art. 254. What is the rule for finding the shares of profit or loss when the stock is employed for the same time?

It is evident that \$600 for 8 months is the same as $\$600 \times 8 = \$4800$ for 1 month, because \$4800 would gain as much in 1 month as \$600 in 8 months. And for the same reason, \$400 for 6 months is the same as $\$400 \times 6 = \$2400$ for 1 month. The question is, therefore, the same as if Brown had put in \$4800 and Dole \$2400 for 1 month each. The whole stock would then be $\$4800 + \$2400 = \$7200$, and Brown's share of the gain would be $\frac{4800}{7200} = \frac{2}{3}$ of $\$60 = \$40$. Dole's share will be $\frac{2400}{7200} = \frac{1}{3}$ of $\$60 = 20$

OPERATION BY PROPORTION.

\$4800	\$7200 : \$4800 : : \$60 : \$40, B's share.
\$2400	7200 : \$2400 : : \$60 : \$20, D's share.
\$7200	

RULE 1. — *Multiply each partner's stock by the time it was in trade, and consider each product a numerator, to be written over the sum of the products, as a common denominator. Then, multiply the whole gain or loss by each of these fractions, and the product will be the gain or loss of each partner.* Or,

RULE 2. — *Multiply each partner's stock by the time it was in trade; then, as the sum of these products is to each product, so is the whole gain or loss to each partner's gain or loss.*

EXAMPLES FOR PRACTICE.

2. A, B, and C, trade in company. A put in \$700 for 5 months; B put in \$800 for 6 months; and C put in \$500 for 10 months. They gain \$399. What is each man's share of the gain? Ans. A's gain \$105, B's gain \$144, C's gain \$150.

3. Leverett Johnson, William Hyde, and William Tyler, formed a connection in business, under the firm of Johnson, Hyde & Co. Johnson at first put in \$1000, and at the end of 6 months he put in \$500 more. Hyde at first put in \$800, and at the end of 4 months he put in \$400 more; but, at the end of 10 months, he withdrew \$500 from the firm. Tyler at first put in \$1200, and at the end of 7 months he put in \$300 more, and at the end of 10 months he put in \$200. At the end of the year they found their net gain to be \$1000. What is each man's share?

Ans. Johnson's gain $\$348.02\frac{338}{431}$, Hyde's $\$273.78\frac{82}{431}$, Tyler's $\$378.19\frac{11}{431}$.

4. George Morse hired of William Hale, of Haverhill, his best horse and chaise for a ride to Newburyport, for \$3.00, with the privilege of one person's having a seat with him. Having

QUESTIONS. — Art. 255. What are the rules for finding the shares of profit or loss when the stock is employed for unequal times? Why do you multiply each man's stock by the time it was in trade?

rode 4 miles, he took in John Jones, and carried him to Newburyport, and brought him back to the place from which he took him. What share of the expense should each pay, the distance from Haverhill to Newburyport being 15 miles?

Ans. Morse pays $1.90, Jones pays $1.10.

5. J. Jones and L. Cotton enter into partnership for one year. January 1, Jones put in $1000, but Cotton did not put in any until the first of April. What did he then put in, to have an equal share with Jones at the end of the year?

Ans. 1333.33\frac{1}{3}$.

6. S, C, and D, engage in partnership, with a capital of $4700. S's stock was in trade 8 months, and his share of the profits was $96; C's stock was in the firm 6 months, and his share of the gain was $90; D's stock was in the firm 4 months, and his gain was $80. Required the amount of stock which each had in the firm.

Ans. { S's stock $1200. C's stock $1500. D's stock $2000.

7. A, B, and C, engage in trade. A put in $300 for 7 months, B put in $500 for 8 months, and C put in $200 for 12 months; they gain $85; what share of the gain does each receive?

Ans. A $21, B $40, and C $24.

8. A and B engage in trade, with $500. A put in his stock for 5 months, and B put in his for 4 months. A gained $10, and B gained $12; what sum did each put in?

Ans. A $200, B $300.

9. A and B trade in company. A put in $3000, and at the end of 6 months put in $2000 more; B put in $6000, and at the end of 8 months took out $3000; they trade one year, and gain $1080; what is each man's share of the gain?

Ans. A's share is $480, B's $600.

10. Four men hired a pasture for $50. A put in 5 horses for 4 weeks; B put in 6 horses for 8 weeks; C put in 12 oxen for 5 weeks, calling 3 oxen equal to 2 horses; and D put in 3 horses for 14 weeks. How much ought each man to pay?

Ans. A 6.66\frac{2}{3}$, B $16.00, C 13.33\frac{1}{3}$, and D $14.00.

11. A, B, and C contract to build a piece of railroad for $7500. A employs 30 men 50 days; B employs 50 men 36 days; and C employs 48 men and 10 horses 45 days, each horse to be reckoned equal to one man, and he is also to have $112.50 for overseeing the work. How much is each man to receive? Ans. A receives $1875; B $2250; C $3375.

§ XXXIV. CURRENCIES.

ART. **256.** The currency of a state or country is its money or circulating medium of trade. In the United States, the gold silver, and copper coins of the country, foreign coins whose value has been fixed by law, and bank notes, redeemable in specie, pass as money. The *legal tender*, however, in payment of debts, is gold and silver.

The *intrinsic* value of foreign coins is their mint value, or that depending upon the weight and purity of the metal of which they are made; their *commercial* value is the price they will bring in the market, and their *legal* value is that fixed by law.

The value of foreign coins, as fixed by present laws of the United States, is shown in the following

TABLE OF FOREIGN CURRENCIES.

Pound Ster. of G. Britain,	$4.84
Pound Ster. of Br. Prov., Nova Scotia, N. Bruns., Newfoundland, and Can.,	4.00
Dollar of Mexico, Peru, Chili, and Cen. Amer.,	1.00
Specie Dollar of Sweden and Norway,	1.06
Specie Dol. of Denmark,	1.05
Rix Dollar of Bremen,	.78¾
Rix Dol., or Thaler, of Prussia and Northern States of Germany,	.69
Ruble, silver, of Russia,	.75
Guilder of Netherlands,	.40
Florin of Netherlands,	.40
Florin of South of Ger.,	.40

Ounce of Sicily,	$2.40
Pagoda of India,	1.84
Tael of China,	1.48
Milrea of Portugal,	1.12
Milrea of Azores,	.83⅓
Ducat of Naples,	.80
Rupee of British India,	.44½
Marco Banco of Hamburg,	.35
Franc of France and Bel.,	$.18\frac{6}{10}$
Livre Tournois of France,	.18½
Leghorn Livre,	.16
Lira of Lombardy, Venetian Kingdom,	.16
Lira of Tuscany,	.16
Lira of Sardinia,	$.18\frac{6}{10}$
Real Plate of Spain,	.10
Real Vellon of Spain,	.05

The legal currency of this countr , previous to 1786, was sterling money, or that of pounds, shillings, and pence. On the adoption of the currency of dollars and cents, there were in circulation colonial notes, or bills of credit, which had depreciated in value. This depreciation being greater in some sections

QUESTIONS. — Art. 256. What is currency? What pass for money in the United States? What is the intrinsic value of foreign coins? What is the commercial value? What is the legal value? — Art. 257. Mention some of the foreign coins whose value has been fixed by law. What was the currency of this country previous to 1786?

than in others, gave rise to the variation, in the states, as to the number of shillings equivalent to a dollar, as shown in the following

TABLE.

$1 in	New Eng. States, Virginia, Kentucky, Tennessee,	= 6s. = $\frac{3}{10}$£., called New Eng. currency, of which 1£. = \$3$\frac{1}{3}$; 1s. = 16$\frac{2}{3}$cts.
$1 in	New York, Ohio, Michigan, North Carolina,	= 8s. = $\frac{2}{5}$£., called New York currency, of which 1£. = \$2$\frac{1}{2}$; 1s. = 12$\frac{1}{2}$cts.
$1 in	Pennsylvania, New Jersey, Delaware, Maryland,	= 7s. 6d. = $\frac{3}{8}$£., called Pennsylvania currency; of which 1£. = \$2$\frac{2}{3}$; 1s. = 13$\frac{1}{3}$cts.
$1 in	Georgia, South Carolina,	= 4s. 8d. = $\frac{7}{30}$£., called Georgia currency; of which 1£. = \$4$\frac{2}{7}$; 1s. = 21$\frac{3}{7}$cts.
$1 in	Canada, Nova Scotia, New Brunswick, Newfoundland,	= 5s. = $\frac{1}{4}$£., called Canada currency; of which 1£. = \$4; 1s. = 20cts.
$1 in	Great Britain,	= 4$\frac{16}{121}$s. = $\frac{25}{121}$£., called English or Sterling money; of which 1£. = \$4.84; 1s. = 24$\frac{1}{5}$cts.

NOTE. — The old currencies of the states are no longer used in keeping accounts, yet the price of articles is still named by some traders in the old currency of their state.

REDUCTION OF CURRENCIES.

ART. **257.** REDUCTION of Currencies is the process of finding the value of the denominations of one currency in the denominations of another.

ART. **258.** To reduce pounds, shillings, pence, and farthings, of the different currencies, to United States money.

Ex. 1. Reduce 18£. 15s. 6d. New England currency to United States money. Ans. \$62.58$\frac{1}{3}$.

OPERATION.

18£. 15s. 6d. = 18.775£.
18.775 ÷ $\frac{3}{10}$ = \$62.58$\frac{1}{3}$

We first reduce the shillings and pence to the decimal of a pound (Art. 188), and then annexing it to the pounds, we divide

QUESTIONS. — What gave rise to the variation in the old currency of this country? Repeat the table. How are the old currencies of the states now used? — Art. 260. What is reduction of currencies?

the sum by $\frac{3}{10}$, because 6s., or a dollar in this currency, is $\frac{3}{10}$ of a pound, and thus obtain the answer in dollars and the decimal of a dollar. Hence the following

RULE. — *Divide the given sum expressed in pounds and decimals of a pound, by the value of $1 expressed in a fraction of a pound. The quotient will be the value in dollars.*

EXAMPLES FOR PRACTICE.

2. Change 144£. 7s. 6d. of the old New England currency to United States money. Ans. $481.25.
3. Change 74£. 1s. 6d. of the old currency of New York to United States money. Ans. $185.18¾.
4. Change 129£. of the old currency of Pennsylvania to United States money. Ans. $344.
5. Change 84£. of the old currency of South Carolina to United States money. Ans. $360.
6. Change 144£. 4s. of Canada and Nova Scotia currency to United States money. Ans. $576.80.
7. Change 257£. 8s. 6d. English or sterling money to United States money. Ans. $1245.937.

ART. **259.** To reduce United States money to pounds, shillings, pence, and farthings, of the different currencies.

Ex. 1. Reduce $152.625 to old New England currency. Ans. 45£. 15s. 9d.

OPERATION.

$\$152.625 \times \frac{3}{10} = 45.7875£.$
45.7875£. = 45£. 15s. 9d.

Since 6s., or a dollar, in this currency, is $\frac{3}{10}$ of a pound, we multiply the given sum by the fraction $\frac{3}{10}$, and reduce the decimal to shillings and pence. (Art. 189.)

RULE. — *Multiply the given sum expressed in dollars by the value of $1 expressed in a fraction of a pound. The quotient will be the value in pounds.*

EXAMPLES FOR PRACTICE.

2. Change $481.25 to the old currency of New England. Ans. 144£. 7s. 6d.

QUESTIONS. — Art. 260. How do you reduce United States money to pounds, shillings, pence, and farthings, New England currency? Why multiply by $\frac{3}{10}$£.? How would you reduce United States money to pounds, &c., Ohio currency? How, to Pennsylvania currency? What is the general rule? — Art. 261. What is the rule for reducing United States money to pounds, shillings, pence, and farthings, of the different currencies?

3. Change 185.18\frac{3}{4}$ to the old currency of New York.
Ans. 74£. 1s. 6d.

4. Change $344 to the old currency of Pennsylvania.
Ans. 129£.

5. Change $360 to the old currency of South Carolina.
Ans. 84£.

6. Change $576.50 to Canada and Nova Scotia currency.
Ans. 144£. 2s. 6d.

7. Change $1245.93,7 to English or sterling money.
Ans. 257£. 8s. 6d.

ART. **260.** To reduce any foreign currency to United States money, and United States money to any foreign currency, when the value of a unit of the foreign currency is known (Art. 256), simply

Multiply or divide, as the case may require, by the value of the unit of the given currency expressed in United States money.

EXAMPLES FOR PRACTICE.

Ex. 1. Reduce 123 rubles, silver, of Russia, to United States money. Ans. $92.25.

2. Reduce $27.90 to francs. Ans. 150 francs.

3. What is the value of 121 thalers of Prussia in United States money? Ans. $83.49.

4. What is the value of $165.20 in florins?
Ans. 413 florins.

5. A merchant purchased tea in China to the amount of 216 taels. What did it cost in United States money?
Ans. $319.68.

6 How many reals, plate of Spain, are equal to $5137.90?
Ans. 51379.

---

§ XXXV. EXCHANGE.

ART. **261.** EXCHANGE, in commerce, is the paying or receiving of money in one place for an equivalent sum in another, by means of *drafts*, or *bills of exchange*.

QUESTIONS. — Art. 260. How do you reduce any foreign currency to United States money, and United States money to any foreign currency? Art. 261. What is exchange?

A bill of exchange is a written order, to some person at a distance, to pay a certain sum, at an appointed time, to another person, or to his order.

The person who draws a bill is termed the *maker*, or *drawer*; the person for whom it is drawn, the *buyer*, *taker*, or *remitter*; the person on whom it is drawn, the *drawee*, and after he has accepted, the *acceptor*. The person who endorses it is termed the *endorser*; and the person in whose legal possession the bill may be at any time is termed the *holder*, or *possessor*.

Exchange is at *par* when a certain sum, at the place from which it is remitted, will pay an equal sum at the place to which it is remitted. It is said to be at a *premium*, or *above par*, when the balance of trade is against the place from which the bill is remitted; and *below par* when the balance of trade is in favor of the place from which the bill is remitted.

INLAND BILLS.

ART. **262.** An INLAND BILL of exchange, or draft, is one of which the drawer and drawee are both residents of the same country.

ART. **263.** To find the value of an inland bill, or draft,

Add to the face of the bill, or draft, the amount of premium, or subtract from the face of the bill, or draft, the amount of discount.

EXAMPLES FOR PRACTICE.

Ex. 1. What is the value of the following bill of exchange, or draft, at $1\frac{1}{2}$ per cent. discount? Ans. $445.22.

$452. *Boston, March* 6, 1856.

At sight, pay to William Dura, or order, four hundred and fifty-two dollars, value received, and charge the same to my account.

EDWIN DANTON.

To LEWIS FONTENAY,
Merchant, New Orleans.

2. A merchant in Chicago purchased a bill on New York for $1164, at 1 per cent. premium; what did he pay? Ans. $1175.64.

QUESTIONS. — What is a bill of exchange? Who is the maker, or drawer, of a bill? Who is the buyer, taker, or remitter? Who is the drawee? Who is the endorser? Who is the holder, or possessor? When is exchange at par? When at a premium? When at a discount? What is an inland bill, or draft? How do you find the value of an inland bill, or draft?

3. What costs a bill on Burlington, Iowa, for \$4000, at $2\frac{1}{2}$ per cent. discount? Ans. \$3900.

4. What costs a bill on Buffalo for \$450, at $\frac{5}{8}$ of 1 per cent. discount? Ans. \$447.18$\frac{3}{4}$.

5. What costs a draft from the Girard Bank, Philadelphia, on the Bank of Commerce, Boston, for \$2517.70, at $\frac{1}{8}$ of 1 per cent. premium? Ans. \$2520.84+.

FOREIGN BILLS.

ART. **264.** A FOREIGN BILL of exchange is one of which the drawer and drawee are residents of different countries.

Foreign bills are usually drawn in sets; that is, at the same time there are drawn two or more bills of the same tenor and date, each containing a condition that it shall continue payable only while the others remain unpaid.

NOTE. — Each bill of a set is remitted in a different manner, in order to guard against loss or delay; and when one of the set has been accepted and paid, the others become worthless.

EXCHANGE ON ENGLAND.

ART. **265.** The exchange value, in the United States, of the pound sterling of Great Britain, is that of its former legal value, or $\$4\frac{4}{9} = \$4.44\frac{4}{9}$, which is considerably below either its intrinsic or commercial value. The commercial value is generally about 9 per cent. more than this exchange, or nominal par value.

Thus, nominal par value being	=	\$4.44$\frac{4}{9}$
To which we add 9 per cent. premium,	=	.40
The commercial par value will be	=	\$4.84$\frac{4}{9}$.

Therefore, when the nominal exchange between the United States and Great Britain exceeds 9 per cent. premium, it is above true par; when less, it is below true par.

ART. **266.** To find the value in United States currency of a bill on England.

Ex. 1. What should be paid for the following bill at $9\frac{1}{2}$ per cent. premium? Ans. \$4866.66$\frac{2}{3}$.

QUESTIONS. — Art. 264. What is a foreign bill of exchange? How are foreign bills usually drawn? Why? — Art. 265. What is the exchange value of the pound sterling of Great Britain, in United States money? How does this differ from the commercial or true par value?

Exchange for £1000. *New York, May* 16, 1856.

Thirty days after sight of this first of exchange (second and third of the same tenor and date unpaid), pay J. W. Hathaway & Co., or order, in London, one thousand pounds sterling, value received, and place the same to my account.

To BATES, BARING & Co., *London.* RUFUS W. KING.

OPERATION.

$$1£. + .09\tfrac{1}{2}£. = 1.095£.;\ 1.095 \times \$\tfrac{40}{9} = \$4.866\tfrac{2}{3};\ 1000 \times \$4.866\tfrac{2}{3} = \$4866.66\tfrac{2}{3}.$$

We add to 1£. the premium on 1£., and obtain 1.095£., which, multiplied by $\$\frac{40}{9}$, or $\$4\frac{4}{9}$, the nominal value of a pound, gives $\$4.866\frac{2}{3}$ as the value of a pound at the given rate of exchange; and 1000 multiplied by this value of a pound gives $\$4866.66\frac{2}{3}$ as the value of the bill. If there had been with the pounds shillings, pence, or farthings, they would have been reduced to a decimal of a pound, and as such annexed to the pounds in the operation.

RULE. — *Multiply the amount of the bill, expressed in pounds and decimals of a pound, by the value of one pound at the given rate of exchange, and the product will be the value in dollars.*

EXAMPLES FOR PRACTICE.

2. A merchant in Boston wishes to purchase a bill of 572£. 10s., on Liverpool, the premium being $8\frac{1}{2}$ per cent.; what will it cost him in dollars and cents? Ans. $\$2760.72\frac{2}{9}$.

3. If J. C. Sherman, of Chicago, should remit to London 1200£., exchange being at $9\frac{1}{4}$ per cent., what will be the cost of the bill in United States money? Ans. $\$5826.66\frac{2}{3}$.

ART. **267.** To find the amount of a bill on England, which can be purchased for a given sum of United States currency.

Ex. 1. When exchange is at $9\frac{1}{2}$ per cent. premium, what will be the amount of a bill on London which I can purchase for $\$4866.66\frac{2}{3}$? Ans. 1000£.

OPERATION.

$$1£. + .09\tfrac{1}{2}£. = 1.095£.;\ 1.095 \times \$\tfrac{40}{9} = \$4.866\tfrac{2}{3};\ \$4866.66\tfrac{2}{3} \div \$4.866\tfrac{2}{3} = 1000£.$$

QUESTIONS. — Art. 266. What is the rule for finding the value of a bill on England in United States currency? — Art. 267. What is the rule for finding the amount of a bill on England, which can be purchased for a given sum of United States currency?

We find, as by Art. 266, the value of one pound at the given rate of exchange. The given sum, $4866.66⅔, we divide by the value of a pound, and obtain 1000£. as the required amount of the bill.

RULE. — *Divide the given sum by the value of one pound at the given rate of exchange, and the quotient will be the amount in pounds and decimals of a pound.*

EXAMPLES FOR PRACTICE.

2. J. Reed, of Cincinnati, proposes to make a remittance to Liverpool of $1640, exchange being at 8½ per cent. premium; what will be the amount of the bill he can remit for that sum?
Ans. 340£. 1s. 10d.

3. A merchant wishes to remit $500 to England, exchange being at 10 per cent. premium; what will be the amount of the bill he can purchase for that sum? Ans. 102£. 5s. 5d.+.

EXCHANGE ON FRANCE.

ART. **268.** In France accounts are kept in francs and centimes. The centimes are hundredths of a franc. All bills of exchange on France are drawn in francs, and are bought, sold, and quoted, as at a certain number of francs to the dollar.

ART. **269.** To find the value in United States currency of a bill on France,

Divide the amount of the bill by the value of one dollar in francs, and the quotient will be the value in dollars.

EXAMPLES FOR PRACTICE.

Ex. 1. What must be paid, in United States currency, for a bill on Paris of 2380 francs, exchange being 5.15 francs per dollar? Ans. $462.13+.

2. How many dollars will purchase a bill on Havre of 30000 francs, exchange being 5.17½ francs per dollar?
Ans. $5797.10+.

3. What is the value of a bill on Paris of 62500 francs, exchange being 5.12 francs per dollar? Ans. $12207.03+.

ART. **270.** To find the amount of a bill on France, which can be purchased for a given sum of United States currency,

QUESTIONS. — Art. 268. How are accounts kept in France? How are all bills of exchange on France drawn? — Art. 269. How do you find the value in United States currency of a bill on France? — Art. 270. How do you find the amount of a bill on France, which can be purchased for a given sum of United States money?

Multiply the given sum by the value of one dollar in francs and the product will be the amount of the bill in francs.

Ex. 1. Alfred Walker, of New York, pays $2500 for a bill on Paris, exchange being 5.12 francs per dollar. What was the amount of the bill in francs? Ans. 12800.

2. When exchange on France is at 5.13 francs per dollar, a bill of how many francs should $700 purchase? Ans. 3591.

3. Morton and Blanchard, of Boston, wish to remit $675 to Paris, exchange being 5.16 francs per dollar; what will be the amount of the bill of exchange they can purchase with the money? Ans. 3483 francs

---

§ XXXVI. DUODECIMALS.

ART. **271.** DUODECIMALS are a kind of compound numbers in which the unit, or foot, is divided into 12 equal parts, and each of these parts into 12 other equal parts, and so on indefinitely; thus, $\frac{1}{12}$, $\frac{1}{144}$, &c.

Duodecimals decrease from left to right in *a twelve-fold* ratio; and the different orders, or denominations, are distinguished from each other by accents, called *indices*, placed at the right of the numerators. Hence the denominators are not expressed. Thus,

1 inch or prime, equal to $\frac{1}{12}$ of a foot, is written 1 in. or 1′.
1 second " $\frac{1}{144}$ " " 1″.
1 third " $\frac{1}{1728}$ " " 1‴.
1 fourth " $\frac{1}{20736}$ " " 1⁗.

Hence the following

TABLE.

12 fourths	make 1‴.	12 seconds	make 1′.
12 thirds	" 1″.	12 inches or primes	" 1ft.

ADDITION AND SUBTRACTION OF DUODECIMALS.

ART. **272.** Duodecimals are added and subtracted in the same manner as compound numbers.

QUESTIONS. — Art. 271. What are duodecimals? In what ratio do duodecimals decrease from left to right? How are the different denominations distinguished from each other? — Art. 272. How are duodecimals added and subtracted?

EXAMPLES FOR PRACTICE.

1. Add together 12ft. 6′ 9″, 14ft. 7′ 8″, 165ft. 11′ 10″.
Ans. 193ft. 2′ 3″.

2. Add together 182ft. 11′ 2″ 4‴, 127ft. 7′ 8″ 11‴, 291ft. 5′ 11″ 10‴. Ans. 602ft. 0′ 11″ 1‴.

3. From 204ft. 7′ 9″ take 114ft. 10′ 6″.
Ans. 89ft. 9′ 3″.

4. From 397ft. 9′ 6″ 11‴ 7⁗ take 201ft. 11′ 7″ 8‴ 10⁗.
Ans. 195ft. 9′ 11″ 2‴ 9⁗.

MULTIPLICATION AND DIVISION OF DUODECIMALS.

ART. **273.** To find the denomination of the product of any two numbers in duodecimals, when multiplied together.

Ex. 1. What is the product of 9ft. multiplied by 3ft.?
Ans. 27ft.

OPERATION.

9ft. × 3ft. = 27ft.

2. What is the product of 7ft. multiplied by 6′? Ans. 3ft. 6′.

OPERATION.

$6' = \frac{6}{12}$ of a foot; then 7ft. × $\frac{6}{12}$ ft. = $\frac{42}{12}$ = 42′; 42′ ÷ 12 = 3ft. 6′.

3. What is the product of 5′ multiplied by 4′? Ans. 1′ 8″.

OPERATION.

$5' = \frac{5}{12}$, and $4' = \frac{4}{12}$; then $\frac{5}{12} \times \frac{4}{12} = \frac{20}{144} = 20''$; 20″ ÷ 12 = 1′ 8″.

4. What is the product of 9′ multiplied by 11‴?
Ans. 8‴ 3⁗.

OPERATION.

$9' = \frac{9}{12}$, and $11''' = \frac{11}{1728}$; then $\frac{9}{12} \times \frac{11}{1728} = \frac{99}{20736} = 99''''$; 99⁗ ÷ 12 = 8‴ 3⁗.

It will be observed in the examples above, that feet multiplied by feet produce feet; feet multiplied by primes produce primes; primes multiplied by primes produce seconds, &c.; and that the several products are of the same denomination as denoted by the sum of the indices of the numbers multiplied together. Hence,

When two numbers are multiplied together, the sum of their indices annexed to their product denotes its denomination.

ART. **274.** To multiply duodecimals together.

Ex. 1. Multiply 8ft. 6in. by 3ft. 7in. Ans. 30ft. 5′ 6″.

QUESTION. — Art. 273. How is the denomination of the product denoted when duodecimals are multiplied together?

OPERATION.
8ft. 6′
3ft. 7′
4ft. 11′ 6″
25ft. 6′
30ft. 5′ 6″

We first multiply each of the terms in the multiplicand by the 7′ in the multiplier, thus, 7′ into 6′ = 42″ = 3′ and 6″. Placing the 6″ under its multiplier, we add the 3′ to the product of 7′ into 8ft. = 59′ = 4ft. and 11′, which we write down. We then multiply by the 3ft., thus: 3ft into 6′ = 18′ = 1ft. and 6′. We write the 6′ under its multiplier, and add the 1ft. to the product of the 3ft. into 8ft., making 25ft., which we write down. The two products being added together, we obtain 30ft. 5′ 6″ for the answer.

RULE. — *Write the multiplier under the multiplicand, so that the same denominations shall stand in the same column.*

Beginning at the right hand, multiply each term in the multiplicand by each term of the multiplier, and write the first term of each partial product directly under its multiplier, observing to carry a unit for every twelve from each lower denomination to the next higher.

The sum of the several partial products will be the product required.

EXAMPLES FOR PRACTICE.

2. Multiply 8ft. 3in. by 7ft. 9in. Ans. 63ft. 11′ 3″.

3. Multiply 12ft. 9′ by 9ft. 11′. Ans. 126ft. 5′ 3″.

4. My garden is 18 rods long and 10 rods wide; a ditch is dug round it 2 feet wide and 3 feet deep; but the ditch not being of a sufficient breadth and depth, I have caused it to be dug 1 foot deeper, and, outside, 1 ft. 6 in. wider. How many solid feet will it be necessary to remove? Ans. 7540.

5. I have a room 12 feet long, 11 feet wide, and $7\frac{1}{2}$ feet high. In it are two doors, 6 feet 6 inches high, and 30 inches wide, and the mop-boards are 8 inches high. There are 3 windows, 3 feet 6 inches wide, and 5 feet 6 inches high; how many square yards of paper will it require to cover the walls?

Ans. $25\frac{29}{108}$ square yards.

ART. **275.** To divide one duodecimal by another.

Ex. 1. A certain aisle contains 68ft. 10′ 8″ of floor. The width of the floor being 2ft. 8′, what is its length? Ans. 25ft. 10′.

OPERATION.
2ft. 8′) 68ft. 10′ 8″ (25ft. 10′
66ft. 8′
2ft. 2′ 8″
2ft. 2′ 8″

We first divide the 68ft. by the divisor, and obtain 25ft. for the quotient. We multiply the entire divisor by the 25ft., and subtract the product, 66ft. 8′, from the corresponding portion of the

QUESTION. — Art. 274. What is the rule for the multiplication of duodecimals?

dividend, and obtain 2ft. 2′, to which remainder we bring down the 8″, and dividing, we obtain 10′ for the quotient. Multiplying the entire divisor by the 10′, we obtain 2ft. 2′ 8″, which subtract, as before, leaves no remainder. Therefore, 25ft. 10′ is the length of the aisle.

RULE. — *Find how many times the highest term of the dividend will contain the divisor. By this quotient multiply the entire divisor, and subtract the product from the corresponding terms of the dividend. To the remainder annex the next denomination of the dividend, and divide as before, and so continue till the division is complete.*

EXAMPLES FOR PRACTICE.

2. What must be the length of a board, that is 1ft. 9in. wide, to contain 22ft. 2in.? Ans. 12ft. 8in

3. I have engaged E. Holmes to cut me a quantity of wood. It is to be cut 4ft. 6in. in length, and to be "corded" in a range 256 ft. long. Required the height of the range to contain 75 cords. Ans. 8ft. 4in.

---

§ XXXVII. INVOLUTION.

ART. **276.** INVOLUTION is the method of finding any power of a given quantity.

A *power* is a quantity produced by taking any given number, a certain number of times, as a factor. The factor, thus taken, is called the *root* of the power.

The number denoting the power is called the *index* or *exponent* of the power, and is a small figure placed at the right of the root. Thus, the second power of 6 is written 6^2; the third power of 4 is written 4^3, and the fourth power of $\frac{2}{3}$ is written $(\frac{2}{3})^4$.

ART. **277.** To raise a number to any required power.

$3 =$ 3, the first power of 3, is written 3^1 or 3.
$3 \times 3 =$ 9, the second power of 3, is written 3^2.
$3 \times 3 \times 3 =$ 27, the third power of 3, " " 3^3.
$3 \times 3 \times 3 \times 3 =$ 81, the fourth power of 3, " " 3^4.
$3 \times 3 \times 3 \times 3 \times 3 =$ 243, the fifth power of 3, " " 3^5.

QUESTIONS. — Art. 275. What is the rule? — Art. 276. What is Involution? What is a power? What is the number called that denotes the power? Where is it placed? — Art. 277. To what is the index in each power equal?

By examining the several powers of 3 in the examples given, we see that the index of each power is equal to the number of times 3 is used as a factor in the multiplications producing the power, and that the number of times the number is multiplied into itself is one less than the power denoted by the index. Hence the

RULE. — *Multiply the given number continually by itself, till the number of multiplications is one less than the index of the power to be found, and the last product will be the power required.*

NOTE. — 1. A fraction may be raised to any power by this rule, by multiplying its terms continually together. Thus, the second power of $\frac{2}{5}$ is $\frac{2}{5} \times \frac{2}{5} = \frac{4}{25}$.

NOTE. — 2. A mixed number may be either reduced to an improper fraction, or the fractional part reduced to a decimal, and then raised to the required power.

EXAMPLES FOR PRACTICE.

1. What is the 2d power of 6? Ans. 36.
2. What is the 3d power of 5? Ans. 125.
3. What is the 6th power of 4? Ans. 4096.
4. What is the 4th power of $\frac{1}{3}$? Ans. $\frac{1}{81}$.
5. What is the 5th power of $3\frac{2}{3}$? Ans. $662\frac{185}{243}$.
6. What is the 3d power of .25? Ans. .015625.
7. What is the 1st power of 17? Ans. 17.

ART. **278.** To raise a number to any required power without producing all the intermediate powers.

Ex. 1. What is the 8th power of 4? Ans. 65536.

OPERATION.

$$\overset{1}{4},\quad \overset{2}{16},\quad \overset{3}{64};\quad \overset{3}{64} \overset{+}{\times} \overset{3}{64} \overset{+}{\times} \overset{2}{16} \overset{=}{=} \overset{8.}{65536}.$$

We raise the 4 to the 2d and to the 3d power, and write above each power its exponent. We then add the exponent 3 to itself, and, increasing the sum by the exponent 2, obtain 8, a number equal to the power required. We next multiply 64, the power belonging to the exponent 3, into itself, and this product by 16, the power belonging to the exponent 2, and obtain 65536 for the 8th power. Hence the following

RULE. — *Raise the given number to any convenient number of powers, and write above each of the respective powers its exponent.*

QUESTIONS. — What is the rule for raising a number to any required power? How may a vulgar fraction be raised to a required power? How a mixed number? — Art. 278. What are the numbers placed over the several powers of 4 called, and what do they denote?

Then add together such exponents as will make a number equal to the required power, repeating any one when it is more convenient, and the product of the powers belonging to these exponents will be the required answer.

EXAMPLES FOR PRACTICE.

2. What is the 7th power of 5? Ans. 78125.
3. What is the 9th power of 6? Ans. 10077696.
4. What is the 12th power of 7? Ans. 13841287201.
5. What is the 8th power of 8? Ans. 16777216.
6. What is the 20th power of 4? Ans. 1099511627776.
7. What is the 30th power of 3? Ans. 205891132094649.

---

§ XXXVIII. EVOLUTION.

ART. **279.** EVOLUTION is the method of finding the root of a given power or number, and is therefore the reverse of Involution.

A *root* of any power is a number which, being multiplied into itself a certain number of times, produces the given power. Thus 4 is the second or *square* root of 16, because $4 \times 4 = 16$; and 3 is the third or *cube* root of 27, because $3 \times 3 \times 3 = 27$.

The root takes the name of the power of which it is the root. Thus, if the number is a second power, the root is called the second or *square* root; if it is a third power, the root is called the third or *cube* root; and if it is a fourth power, its root is called the fourth or *biquadrate* root.

Those roots which can be exactly found are called *rational* roots; those which cannot be exactly found, but approximate towards the true root, are called *surd* roots.

Numbers that have exact roots are called *perfect* powers, and all other numbers are called *imperfect* powers.

QUESTIONS. — What is the rule for involving a number without producing all the intermediate powers? — Art. 279. What is Evolution? What is a root? From what does the root take its name? What are rational roots? What surd roots?

Roots are denoted by writing the character $\sqrt{}$, called the radical sign, before the power, with the index of the root over it, or by a fractional index or exponent. The third or *cube* root of 27 is expressed thus, $\sqrt[3]{27}$, or $27^{\frac{1}{3}}$; and the second or *square* root of 25 is expressed thus, $\sqrt{25}$, or $25^{\frac{1}{2}}$.

NOTE. — The index 2 over $\sqrt{}$ is usually omitted when the square root is required. Thus, $\sqrt{64}$ denotes the square root of 64.

EXTRACTION OF THE SQUARE ROOT.

ART. **280.** The Square Root is the root of any second power, and is so called because the square or second power of any number represents the contents of a square surface, of which the root is the length of one side.

ART. **281.** To extract the square root of any number is to find a number which, being multiplied by itself, will produce the given number.

The following numbers in the upper line represent roots, and those in the lower line their second powers, or squares.

Roots,	1	2	3	4	5	6	7	8	9	10
Squares,	1	4	9	16	25	36	49	64	81	100

It will be observed that the second power or square of each of the numbers above contains twice as many figures as the root, or twice as many wanting one. Hence,

To ascertain the number of figures in the square root of any given number, it must be divided into periods, beginning at the right, each of which, excepting the last, must always contain two figures; and the number of periods will denote the number of figures of which the root will consist.

Ex. 1. I wish to arrange 625 tiles, each of which is 1 foot square, into a square pavement; what will be the length of one of the sides? Ans. 25 feet.

OPERATION.

```
     6̇ 2 5̇ ( 2 5, Ans.
     4
     -----
4 5 ) 2 2 5
      2 2 5
      -----
```

It is evident, if we extract the square root of 625, we shall obtain one side of the pavement, in feet. (Art. 280.)

Beginning at the right hand, we divide the number into periods, by placing a point over the right-hand figure of each period; and then find the greatest square number in

QUESTIONS. — How are roots denoted? What is said of the index 2? — Art. 280. What is meant by the square root, and why is it so called? — Art. 281. What is meant by extracting the square root? How do you ascertain the number of figures in the square root of any number?

the left-hand period, 6 (hundreds) to be 4 (hundreds), and that its root is 2, which we write in the quotient. As this 2 is in the place of tens, its value is 20, and represents the side of a square, the area or superficial contents of which are 400 square feet, as seen in Fig. 1.

Fig. 1.

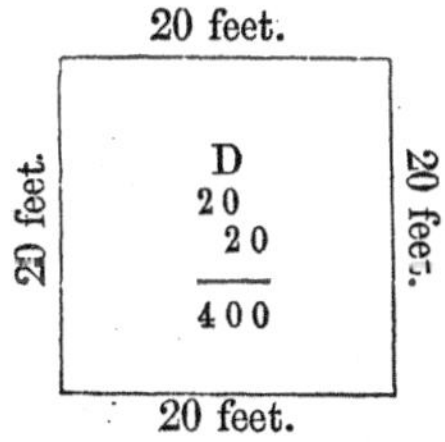

20 feet.

D
20
20
400

20 feet. 20 feet. 20 feet.

We now subtract 400 feet from 625 feet, and have 225 feet remaining, which must be added on two sides of Fig. 1, in order that it may remain a square. We therefore double the root 2 (tens) or 20, one side of the square, to obtain the length of the two sides to be enlarged, making 40 feet; and then inquire how many times 40, as a divisor, is contained in the dividend 225, and find it to be 5 times. This 5 we write in the quotient or root, and also on the right of the divisor, and it represents the width of the additions to the square, as seen in Fig. 2.

The width of the additions being multiplied by 40, the length of the two additions, makes 200 square feet, the contents of the two additions E and F, which are 100 feet for each. The space G now remains to be filled, to complete the square, each side of which is 5 feet, or equal to the width of E and F. If, therefore, we square 5, we have the contents of the last addition, G, equal to 25 square feet. It is on account of this last addition that the last figure of the root is placed in the divisor; for we thus obtain 45 feet for the length of all the additions made, which, being multiplied by the width (5ft.), the last figure in the root, the product, 225 square feet, will be the contents of the three additions, E, F, and G, and equal to the feet remaining after we had found the first square. Hence, we obtain 25 feet for the length of one side of the pavement, since $25 \times 25 = 625$, the number of tiles to be arranged, and equal to the sum of the several parts of Fig. 2; thus, $400 + 100 + 100 + 25 = 625$.

Fig. 2.

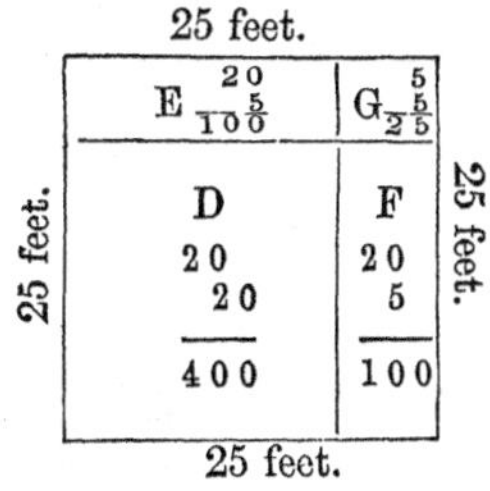

25 feet.

E 20 / 5 / 100	G 5 / 5 / 25
D 20 / 20 / 400	F 20 / 5 / 100

25 feet. 25 feet. 25 feet.

This illustration and explanation is founded upon the principle,

---

QUESTIONS. — What is first done after dividing the number into periods? What part of Fig. 1 does this greatest square number represent? What place does the figure of the root occupy, and what part of the figure does it represent? Why do you double the root for a divisor? What part of Fig. 2 does the divisor represent? What part does the last figure of the root represent? Why do you multiply the divisor by the last figure of the root? What parts of the figure does the product represent? Why do you square the last figure of the root? What part of the figure does this square represent? What other way of finding the contents of the additions without multiplying the parts separately by the width?

That *the square of the sum of two numbers is equal to the squares of the numbers, plus twice their product.* Thus, 25 being equal to $20 + 5$, its square is equal to the squares of 20 and of 5, plus twice the product of 20 and 5, or to $400 + 2 \times 20 \times 5 + 25 = 625$.

RULE. — *Separate the given number into periods of two figures each, by putting a point over the place of units, another over the place of hundreds, and so on.*

Find the greatest square number in the left-hand period, writing the root of it at the right hand of the given number, after the manner of a quotient in division, for the first figure of the root. Subtract this square number from the first period, and to the remainder bring down the next period for a dividend.

Double the root already found for a divisor, and find how often the divisor is contained in the dividend, omitting the right-hand figure, and annex the result to the root for the second figure of it, and likewise to the divisor. Multiply the divisor with the figure last annexed by the figure annexed to the root, and subtract the product from the dividend. To the remainder bring down the next period for a new dividend.*

Double the root already found for a new divisor, and continue the operation as before, till all the periods have been brought down.

NOTE 1. — It is evident, when the given number contains an odd number of figures, the left-hand period can contain but one figure.

2. If the dividend does not contain the divisor, a cipher must be placed in the root, and also at the right of the divisor; then, after bringing down the next period, this last divisor must be used as the divisor of the new dividend.

3. When there is a remainder after extracting the root of a number, periods of ciphers may be annexed, and the figures of the root thus obtained will be decimals.

4. If the given number is a decimal, or a whole number and a decimal, the root is extracted in the same manner as in whole numbers, except, in pointing off the decimals either alone or in connection with the whole number, we place a point over every second figure toward the right, from the separatrix, and fill the last period, if incomplete, with a cipher.

5. The square root of any number ending with 2, 3, 7, or 8, cannot be exactly found.

EXAMPLES FOR PRACTICE.

2. What is the square root of 148996?

* The figure of the root must generally be diminished by one or two units, on account of the deficiency in enlarging the square.

QUESTIONS. —What is the rule for extracting the square root? What is to be done if the dividend does not contain the divisor? What must be done if there is a remainder after extracting the root? What do you do if the given number is a decimal? Of what numbers can the square root not be found?

OPERATION.

```
14̇89̇96̇(386
9
———
68)589
   544
   ———
766)4596
    4596
    ————
```

3. What is the square root of 516961? Ans. 719.
4. What is the square root of 182329? Ans. 427.
5. What is the square root of 23804641? Ans. 4879.
6. What is the square root of 10673289? Ans. 3267.
7. What is the square root of 20894041? Ans. 4571.
8. What is the square root of 42025? Ans. 205.
9. What is the square root of 1014049? Ans. 1007.
10. What is the square root of 538? Ans. 23.194+.
11. What is the square root of 71? Ans. 8.426+.
12. What is the square root of 7? Ans. 2.645+.
13. What is the square root of .1024? Ans. .32.
14. What is the square root of .3364? Ans. .58.
15. What is the square root of .895? Ans. .946+.
16. What is the square root of .120409? Ans. .347.
17. What is the square root of 61723020.96? Ans. 7856.4.
18. What is the square root of 9754.60423716? Ans. 98.7654.

ART. 282. If it is required to extract the square root of a common fraction, or of a mixed number, the mixed number must be reduced to an improper fraction; and in both cases the fractions must be reduced to their lowest terms, and the root of the numerator and denominator extracted.

NOTE. — When the exact root of the terms of a fraction cannot be found, it must be reduced to a decimal, and the root of the decimal extracted.

EXAMPLES FOR PRACTICE.

1. What is the square root of $\frac{49}{529}$? Ans. $\frac{7}{23}$.
2. What is the square root of $\frac{196}{625}$? Ans. $\frac{14}{25}$.
3. What is the square root of $\frac{3721}{7569}$? Ans. $\frac{61}{87}$.
4. What is the square root of $\frac{1849}{12769}$? Ans. $\frac{43}{113}$.
5. What is the square root of $60\frac{1}{16}$? Ans. $7\frac{3}{4}$.
6. What is the square root of $28\frac{57}{64}$? Ans. $5\frac{3}{8}$.

QUESTION. — Art. 282. What do you do when it is required to extract the square root of a common fraction, or of a mixed number?

7. What is the square root of $47\frac{17}{64}$? Ans. $6\frac{7}{8}$.
8. What is the square root of $\frac{42}{57}$? Ans. .858+.
9. What is the square root of $83\frac{2}{3}$? Ans. 9.14+.
10. What is the square root of $121\frac{17}{18}$? Ans. 11.042+.
11. What is the square root of $\frac{339\frac{3}{7}}{462}$? Ans. $\frac{6}{7}$.
12. What is the square root of $\frac{76\frac{12}{13}}{1557\frac{9}{13}}$? Ans. $\frac{2}{9}$.

APPLICATION OF THE SQUARE-ROOT.

ART. **283.** The square root may be applied to finding the dimensions and areas of squares, triangles, circles, and other surfaces.

1. A general has an army of 226576 men; how many must he place rank and file to form them into a square? Ans. 476.

2. A gentleman purchased a lot of land in the form of a square, containing 640 acres; how many rods square is his lot? Ans. 320 rods.

3. I have three pieces of land; the first is 125 rods long, and 53 wide; the second is $62\frac{1}{2}$ rods long, and 34 wide; and the third contains 37 acres; what will be the length of the side of a square field whose area will be equal to the three pieces? Ans. 121.11+ rods.

4. W. Scott has 2 house-lots; the first is 242 feet square, and the second contains 9 times the area of the first; how many feet square is the second? Ans. 726 feet.

5. There are two pastures, one of which contains 124 acres, and the area of the other is to the former as 5 to 4; how many rods square is the latter? Ans. 157.48+ rods.

6. I wish to set out an orchard containing 216 fruit-trees, so that the length shall be to the breadth as 3 to 2, and the distance of the trees from each other 25 feet; how many trees will there be in a row each way, and how many square feet of ground will the orchard cover?

Ans. 18 in length; 12 in breadth; 116875sq. ft.

ART. **284.** A TRIANGLE is a figure having three sides and three angles.

A *right-angled triangle* is a figure having three sides and three angles, one of which is a right angle.

QUESTIONS. — Art. 283. To what may the square root be applied? — Art. 284. What is a triangle? What is a right-angled triangle? What is the longest side called? What the other two?

The side A B is called the *base* of the triangle A B C, the side B C the *perpendicular*, the side A C the *hypothenuse*, and the angle at B is a *right angle*.

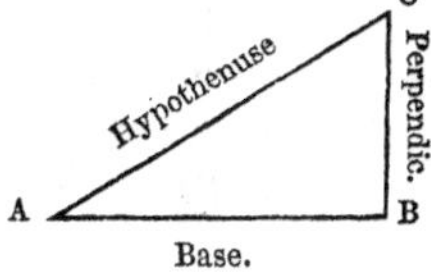

ART. **285.** In every right-angled triangle, the square of the hypothenuse is equal to the sum of the squares of the base and perpendicular, as shown by the following diagram.

It will be seen, by examining this diagram, that the large square, formed on the hypothenuse A C, contains the same number of small squares as the other two counted together. Hence, the propriety of the following rules.

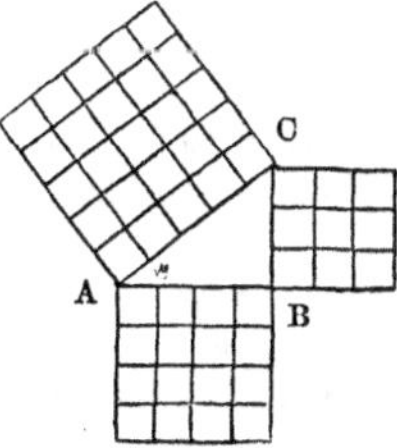

ART. **286.** To find the hypothenuse, the base and perpendicular being given.

RULE. — *Add the square of the base to the square of the perpendicular, and extract the square root of their sum.*

ART. **287.** To find the perpendicular, the base and hypothenuse being given.

RULE. — *Subtract the square of the base from the square of the hypothenuse, and extract the square root of the remainder.*

ART. **288.** To find the base, the hypothenuse and perpendicular being given.

RULE. — *Subtract the square of the perpendicular from the square of the hypothenuse, and extract the square root of the remainder.*

EXAMPLES FOR PRACTICE.

1. What must be the length of a ladder to reach to the top of a house 40 feet in height, the bottom of the ladder being placed 9 feet from the sill? Ans. 41 feet.

QUESTIONS. — Art. 285. How does the square of the hypothenuse compare with the base and perpendicular? How does this fact appear from Fig. 2? — Art. 286. What is the rule for finding the hypothenuse? — Art. 287. What for finding the perpendicular? — Art. 288. What for finding the base?

2. Two vessels sail from the same port; one sails due north 360 miles, and the other due east 450 miles; what is their distance from each other? Ans. 576.2+ miles.

3. The hypothenuse of a certain right-angled triangle is 60 feet, and the perpendicular is 36 feet; what is the length of the base? Ans. 48 feet.

4. A line drawn from the top of the steeple of a certain meeting-house to a point at the distance of 50 feet on a level from the base of the steeple, is 120 feet in length; what is the height of the steeple? Ans. 109.08+ feet.

5. The height of a tree on an island in a certain river is 160 feet. The base of the tree is 100 feet on a horizontal line from the river, and is elevated 20 feet above its surface. A line extending from the top of the tree to the further shore of the river is 500 feet. Required the width of the river.
Ans. 366.47+ feet.

6. On the edge of a perpendicular rock, whose base is 90 feet, on a level, from a certain road that is 110 feet wide, there is a tower 160 feet high; the length of a line extending from the top of the tower to a point on the opposite side of the road is 300 feet. What is the elevation of the base of the tower above the road? Ans. 63.6+ feet.

7. John Snow's dwelling is 60 rods north of the meeting-house, James Briggs's is 80 rods east of the meeting-house, Samuel Jenkins's is 70 rods south, and James Emerson's 90 rods west of the meeting-house; how far will Snow have to travel to visit his three neighbors, and then return home?
Ans. 428.47+ rods.

8. A certain room is 24 feet long, 18 feet wide, and 12 feet high; required the distance from one of the lower corners to an opposite upper corner. Ans. 32.3+ feet.

ART. **289.** A CIRCLE is a plane figure bounded by a curved line, every part of which is equally distant from a point called the centre.

The *circumference* or *periphery* of a circle is the line which bounds it.

The *diameter* of a circle is a line drawn through the centre, and terminated by the circumference; as A B.

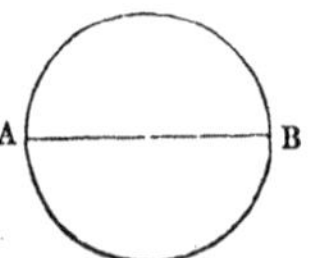

QUESTIONS. — Art. 289. What is a circle? What is the circumference of a circle? What the diameter?

ART. **290.** All circles are to each other as the squares of their diameters, semi-diameters, or circumferences.

All similar triangles and other rectilineal figures are to each other as the squares of their homologous or corresponding sides.

ART. **291.** To find the side, diameter, or circumference, of any surface, which is similar to a given surface.

RULE. — *State the question as in Proportion, and square the given sides, diameters, or circumferences, and the square root of the fourth term of the proportion will be the required answer.*

ART. **292.** To find the area of any surface which is similar to a given surface.

RULE. — *State the question as in Proportion, and square the given sides, diameters, or circumferences, and the fourth term of the proportion is the required answer.*

EXAMPLES FOR PRACTICE.

Ex. 1. I have a triangular piece of land containing 65 acres, one side of which is 100 rods in length; what is the length of the corresponding side of a similar triangle containing $32\frac{1}{2}$ acres?
Ans. 70.71+ rods.

OPERATION.

$$65 : 32\frac{1}{2} :: 100^2 : 5000; \sqrt{5000} = 70.71 + \text{ rods.}$$

2. I have a board in the form of a triangle; the length of one of its sides is 16 feet. My neighbor wishes to purchase one half the board; at what distance from the smaller end must it be divided parallel to the base or larger end? Ans. 11.31+ feet.

3. There is a triangular piece of land, the length of one side of which is 11 rods; required the length of the corresponding side of a similar triangle containing three times as much.
Ans. 19.05+ rods.

4. The diameter of a circle is 6 feet, and its area is 28.3 feet; what is the diameter of a circle whose area is 42.5 feet?
Ans. 7.35+ feet.

5. If an anchor, which weighs 2000lb., requires a cable 3 inches in diameter, what should be the diameter of the cable, when the anchor weighs 4000lb.? Ans. 4.24+ inches.

6. A rope 4 inches in circumference will sustain a weight of 1000lb.; what must be the circumference of a rope that will sustain 5000lb.? Ans. 8.94+ inches.

7. There is a triangle containing 72 square rods, and one of its

---

QUESTIONS. — Art. 290. What proportion do circles have to each other? — Art. 291. What is the rule for finding the side, diameter, &c., of a surface similar to a given surface? — Art. 292. What is the rule for finding the area of a surface similar to a given surface?

sides measures 12 rods; what is the area of a similar triangle whose corresponding side measures 8 rods? Ans. 32 rods.

8. A gentleman has a park, in the form of a right-angled triangle, containing 950 square rods, the longest side or hypothenuse of which is 45 rods. He wishes to lay out another in the same form, with an hypothenuse $\frac{1}{3}$ the length of the first; required the area. Ans. 105.55+ square rods.

9. If a cylinder 6 inches in diameter contain 1.178+ cubic feet, how many cubic feet will a cylinder of the same length contain that is 9 inches in diameter? Ans. 2.65+ feet.

10. If a pipe 2 inches in diameter will fill a cistern in $20\frac{1}{4}$ minutes, how long would it take a pipe that is 3 inches in diameter? Ans. 9 minutes.

11. A tube $\frac{3}{4}$ of an inch in diameter will empty a cistern in 50 minutes; required the time it will empty the cistern, when there is another pipe running into it $\frac{1}{3}$ of an inch in diameter. Ans. $62\frac{4}{13}$ minutes.

ART. **293.** To find the side of a square that can be inscribed in a circle of a given diameter.

A square is said to be inscribed in a circle when each of its angles or corners touches the circumference. It may be conceived to be composed of two right-angled triangles, the base and perpendicular of each being equal, and their hypothenuse the diameter of the circle, as seen in the diagram. Hence the

RULE. — *Extract the square root of half the square of the diameter, and it is the side of the inscribed square.*

EXAMPLES FOR PRACTICE.

1. What is the length of one side of a square that can be inscribed in a circle, whose diameter is 12 feet? Ans. 8.48+ feet.

2. How large a square stick may be hewn from a round one, which is 30 inches in diameter? Ans. 21.2+ inches square.

3. A has a cylinder of lignum-vitæ, $19\frac{4}{5}$ inches long and $1\frac{1}{2}$ inches in diameter; how large a square ruler may be made from it? Ans. 1.06+ inches square.

QUESTIONS. — Art. 293. When is a square said to be inscribed in a circle? Of what may the inscribed square be conceived to be composed? What part of the circle is the hypothenuse of the two triangles? What is the rule for finding the side of the inscribed square?

EXTRACTION OF THE CUBE ROOT.

ART. **294.** The CUBE ROOT is the root of any third power. It is so called, because the cube or third power of any number represents the contents of a cubic body, of which the cube root is one of its sides.

ART. **295.** To extract the cube root is to find a number which, being multiplied into its square, will produce the given number. The following numbers in the upper line represent roots, and those in the lower line their third powers, or cubes.

Roots,	1	2	3	4	5	6	7	8	9	10
Cubes,	1	8	27	64	125	216	343	512	729	1000

It will be observed that the cube or third power of each of the numbers above contains three times as many figures as the root, or three times as many wanting *one*, or *two* at most. Hence, to determine the number of figures in the cube root of a given number,

Divide it into periods, beginning at the right, each of which, excepting the last, must always contain three figures; and the number of periods will denote the number of figures which the root will contain.

Ex. 1. I have 17576 cubical blocks of marble, which measure one foot on each side; what will be the length of one of the sides of a cubical pile, which may be formed of them?

Ans. 26 feet.

OPERATION.

$$
\begin{array}{rl}
 & 1\dot{7}\,5\,7\,\dot{6}\,(\,2\,6,\ \text{Root.} \\
 & 8 \\
2^2 \times 300 = 1200\,) & 9\,5\,7\,6,\ \text{1st dividend.} \\
 & 7\,2\,0\,0,\ \text{1st addition.} \\
6^2 \times 2 \times 30 \ = & 2\,1\,6\,0,\ \text{2d addition.} \\
6^3 \ = & 2\,1\,6,\ \text{3d addition.} \\
 & 9\,5\,7\,6,\ \text{Subtrahend.}
\end{array}
$$

It is evident that the number of blocks or feet on a side will be equal to the cube root of 17576. (Art. 294.)

Beginning at the right hand, we divide the number into periods, by placing a point over the right-hand figure of each period. We then find the greatest cube number in the left-hand period, 17 (thousands), to be 8 (thousands),

QUESTIONS. — Art. 294. What is the cube root, and why so called? — Art. 295. What is meant by extracting the cube root? How many more figures in the cube of any number than in the root? How do you ascertain the number of figures in the cube root of any number? What is found by extracting the cube root of the number in the example? What is first done after separating the number into periods?

and its root 2, which we place in the quotient or root. As 2 is in the place of tens, because there is to be another figure in the root, its value is 20, and it represents the side of a cube (Fig. 1) the contents of which are 8000 cubic feet; thus, $20 \times 20 \times 20 = 8000$.

We now subtract the cube of 2 (tens) $= 8$ (thousands) from the first period, 17 (thousands), and have 9 (thousand) feet remaining, which, being increased by the next period, makes 9576 cubic feet. This must be added to three sides of the cube, Fig. 1, in order that it may remain a cube. To do this, we must find the superficial contents of the three sides of the cube, to which the additions are to be made. Now, since one side is 2 (tens) or 20 feet square, its superficial contents will be $20 \times 20 = 400$ square feet, and this multiplied by 3 will be the superficial contents of three sides; thus, $20 \times 20 \times 3 = 1200$, or, which is the same thing, we multiply the square of the quotient figure, or root, by 300; thus, $2^2 \times 300 = 1200$ square feet. Making this number a divisor, we divide the dividend 9576 by it, and obtain 6, which we place in the root. This 6 represents the thickness of each of the three additions to be made to the cube, and their superficial contents being multiplied by it, we have $1200 \times 6 = 7200$ cubic feet for the contents of the three additions, A, B, and C, as seen in Fig. 2.

Fig. 1.

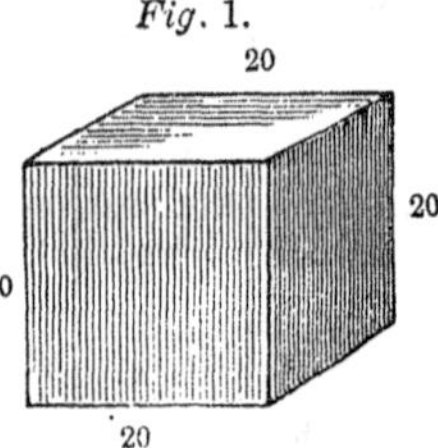

Fig. 2.

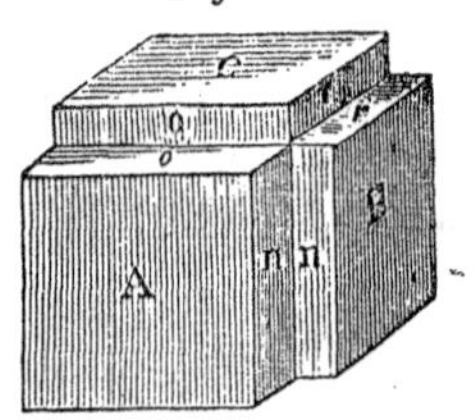

Having made these additions to the cube, we find that there are three other deficiencies, *n n*, *o o*, and *r r*, the length of which is equal to one side of the additions, 2 (tens), or 20 feet; and their breadth and thickness, 6 feet, equal to the thickness of the additions. Therefore, to find the solid contents of the additions, necessary to supply these deficiencies, we multiply the product of their length, breadth, and thickness, by the number of additions; thus, $6 \times 6 \times 20 \times 3 = 2160$, or, which is the same thing, we multiply the square of the last quotient figure by the former figure of the root, and that product by 30; thus, $6^2 \times 2$

QUESTIONS. — What is done with this greatest cube number, and what part of Fig. 1 does it represent? What is done with the root? What is its value, and what part of the figure does it represent? How are the cubical contents of the figure found? What constitutes the remainder after subtracting the cube number from the left-hand period? To how many sides of the cube must this remainder be added? How do you find the divisor? What parts of the figure does it represent? How do you obtain the last figure of the root? What part of Fig. 2 does it represent? What parts of the figure does the product represent? What three other deficiencies in the figure?

$\times$ 30 = 2160 cubic feet for the contents of the additions *s s*, *u u*, and *v v*, as seen in Fig. 3.

These additions being made to the cube, we still observe another deficiency of the cubical space *x x x*, the length, breadth, and thickness of which are each equal to the thickness of the other additions, which is 6 feet. Therefore, we find the contents of the addition necessary to supply this deficiency by multiplying its length, breadth, and thickness together, or cubing the last figure of the root; thus, 6 $\times$ 6 $\times$ 6 = 216 cubic feet for the contents of the addition *z z z*, as seen in Fig. 4.

Fig. 3.

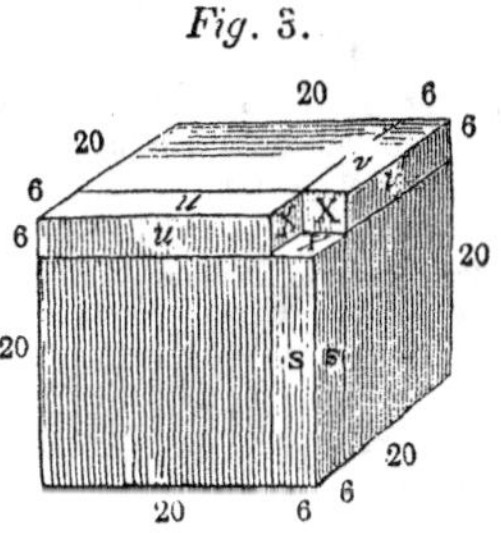

The cube is now complete, and, if we add together the several additions that have been made to it, thus, 7200 + 2160 + 216 = 9576, we obtain the number of cubic feet remaining after subtracting the first cube, which, being subtracted from the dividend in the operation, leaves no remainder. Hence, the cubical pile formed is 26 feet on each side; since 26 $\times$ 26 $\times$ 26 = 17576, the given number of blocks, and the sum of the several parts of Fig. 4. Thus, 8000 + 7200 + 2160 + 216 = 17576. Hence the following

Fig. 4.

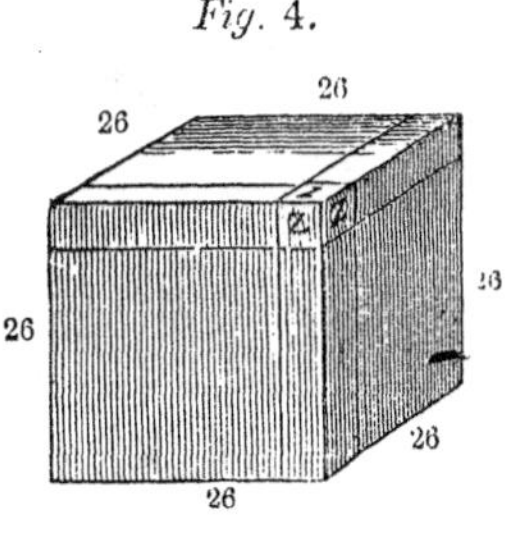

RULE. — *Separate the given number into as many periods as possible of three figures each, by placing a point over the unit figure, and every third figure beyond the place of units.*

Find the greatest cube in the left-hand period, and place its root on the right. Subtract the cube, thus found, from this period, and to the remainder bring down the next period for a dividend.

Multiply the square of the root already found by 300 *for a divisor, by which divide the dividend, and place the quotient, usually diminished by one or two units, for the next figure of the root.*

Multiply the divisor by the last figure of the root, and write the product under the dividend; then multiply the square of the last figure of the root by its former figure or figures, and this product by 30, *and place the product under the last; under all set the cube of the last figure of the root, and call their sum the subtrahend.*

QUESTIONS. — How do you find their contents? What parts of Fig. 3 does the product represent? What other deficiency do you observe? To what are its length, breadth, and thickness, equal? How do you find its contents? What part of Fig. 4 does it represent? What is the rule for extracting the cube root?

Subtract the subtrahend from the dividend, and to the remainder bring down the next period for a new dividend, with which proceed as before; and so on, till the whole is completed.

NOTE 1. — In separating the given number into periods, when the number of the figures is not divisible exactly by 3, the left-hand period will contain less than 3 figures.

NOTE 2. The observations made in Notes 2, 3, and 4, under square root, are equally applicable to the cube root, except in pointing off decimals each period must contain *three* figures, and *two* ciphers must be placed at the right of the divisor when it is not contained in the dividend.

EXAMPLES FOR PRACTICE.

1. What is the cube root of 78402752? Ans. 428

OPERATION.

```
           7 8 4 0 2 7 5 2 ( 4 2 8 Root
           6 4
         -----------
4 8 0 0 )  1 4 4 0 2 = 1st dividend.
         -----------
             9 6 0 0
               4 8 0
                   8
           -----------
           1 0 0 8 8 = 1st subtrahend.
           ---------------
5 2 9 2 0 0 ) 4 3 1 4 7 5 2 = 2d dividend.
              ---------------
              4 2 3 3 6 0 0
                  8 0 6 4 0
                      5 1 2
              ---------------
              4 3 1 4 7 5 2 = 2d subtrahend.
```

2. What is the cube root of 74088? Ans. 42.
3. What is the cube root of 185193? Ans. 57.
4. What is the cube root of 80621568? Ans. 432.
5. What is the cube root of 176558481? Ans. 561.
6. What is the cube root of 257259456? Ans. 636.
7. What is the cube root of 1860867? Ans. 123.
8. What is the cube root of 1879080904? Ans. 1234.
9. What is the cube root of 41673648.563? Ans. 346.7.
10. What is the cube root of 483921.516051? Ans. 78.51.
11. What is the cube root of 8.144865728? Ans. 2.012.
12. What is the cube root of .075686967? Ans. .423.

QUESTION. — How many ciphers must be placed at the right of the divisor when it is not contained in the dividend?

ART. **296.** When it is required to extract the cube root of a common fraction, or a mixed number, it is prepared in the same manner as directed in square root. (Art. 272.)

EXAMPLES FOR PRACTICE.

1. What is the cube root of $81\frac{5}{11}$? Ans. 4.334+
2. What is the cube root of $\frac{729}{4096}$? Ans. $\frac{9}{16}$.
3. What is the cube root of $49\frac{8}{27}$? Ans. $3\frac{2}{3}$.
4. What is the cube root of $166\frac{3}{8}$? Ans. $5\frac{1}{2}$.
5. What is the cube root of $85\frac{23}{125}$? Ans. $4\frac{2}{5}$.

APPLICATION OF THE CUBE ROOT.

ART. **297.** THE cube root may be applied in finding the dimensions and contents of cubes and other solids.

1. A carpenter wishes to make a cubical cistern that shall contain 2744 cubic feet of water; what must be the length of one of its sides? Ans. 14 feet.

2. A farmer has a cubical box that will hold 400 bushels of grain; what is the depth of the box? Ans. 7.92+ feet.

3. There is a cellar, the length of which is 18 feet, the width 15 feet, and the depth 10 feet; what would be the depth of another cellar of the same size, having the length, width, and depth equal? Ans. 13.92+ feet.

ART. **298.** A SPHERE is a solid bounded by one continued convex surface, every part of which is equally distant from a point within, called the centre.

The *diameter* of a sphere is a straight line passing through the centre, and terminated by the surface; as A B.

ART. **299.** A CONE is a solid having a circle for its base, and its top terminated in a point, called the vertex.

QUESTIONS. — Art. 296. How is a common fraction or a mixed number prepared for extracting the square root? — Art. 297. To what may the cube root be applied? — Art. 298. What is a sphere? What is the diameter of a sphere? — Art. 299. What is a cone?

The *altitude* of a cone is its perpendicular height, or a line drawn from the vertex perpendicular to the plane of the base; as B C.

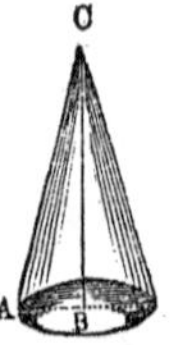

ART. **300.** Spheres are to each other as the cubes of their diameters, or of their circumferences.

Similar cones are to each other as the cubes of their altitudes, or the diameters of their bases.

All similar solids are to each other as the cubes of their homologous or corresponding sides, or of their diameters.

ART. **301.** To find the contents of any solid which is similar to a given solid.

RULE. — *State the question as in Proportion, and cube the given sides, diameters, altitudes, or circumferences, and the fourth term of the proportion is the required answer.*

ART. **302.** To find the side, diameter, circumference, or altitude, of any solid, which is similar to a given solid.

RULE. — *State the question as in Proportion, and cube the given sides, diameters, circumferences, or altitudes, and the cube root of the fourth term of the proportion is the required answer.*

EXAMPLES FOR PRACTICE.

1. If a cone 2 feet in height contains 456 cubic feet, what are the contents of a similar cone, the altitude of which is 3 feet? Ans. 1539 cubic feet.

OPERATION.

$$2^3 : 3^3 :: 456 : 1539.$$

2. If a cubic piece of metal, the side of which is 2 feet, is worth \$6.25, what is another cubical piece of the same kind worth, one side of which is 12 feet? Ans. \$1350.

3. If a ball, 4 inches in diameter, weighs 50lb., what is the weight of a ball 6 inches in diameter? Ans. 168.7+lb.

QUESTIONS. — What is the altitude of a cone? Art. 300. What proportion do spheres have to each other? What proportion do cones have to each other? What proportion do all similar solids have to each other? — Art. 301. What is the rule for finding the contents of a solid similar to a given solid? — Art. 302. What is the rule for finding the side, diameter, &c., of a solid similar to a given solid?

4. If a sugar loaf, which is 12 inches in height, weighs 16lb., how many inches may be broken from the base, that the residue may weigh 8lb.? Ans. 2.5+ in.

5. If an ox, that weighs 800lb., girts 6 feet, what is the weight of an ox that girts 7 feet? Ans. 1270.3lb.

6. If a tree, that is one foot in diameter, make one cord, how many cords are there in a *similar* tree, whose diameter is two feet? Ans. 8 cords.

7. If a bell, 30 inches high, weighs 1000lb., what is the weight of a bell 40 inches high? Ans. 2370.3lb.

8. If an apple, 6 inches in circumference, weighs 16 ounces, what is the weight of an apple 12 inches in circumference? Ans. 128 ounces.

9. A and B own a stack of hay in a conical form. It is 15 feet high, and A owns $\frac{2}{3}$ of the stack; it is required to know how many feet he must take from the top of it for his share. Ans. 13.1+ feet.

---

§ XXXIX. ARITHMETICAL PROGRESSION.

ART. **303.** WHEN a series of numbers increases or decreases by a constant difference, it is called Arithmetical Progression, or Progression by Difference. Thus,

2, 5, 8, 11, 14, 17, 20, 23, 26, 29.
29, 26, 23, 20, 17, 14, 11, 8, 5, 2.

The first is called an *ascending* series or progression. The second is called a *descending* series or progression. The numbers which form the series are called the *terms* of the progression. The *first* and *last* terms are called the *extremes*, and the other terms the *means*. The constant difference is called the *common difference* of the progression.

Any *three* of the *five* following things being given, the other *two* may be found:

QUESTIONS. — Art. 303. What is arithmetical progression? What is an ascending series? What a descending series? What are the terms of a progression? What the extremes? What the means?

1st. The first term, or first extreme;
2d. The last term, or last extreme;
3d. The number of terms;
4th. The common difference;
5th. The sum of the terms.

ART. **304.** To find the common difference, the first term, last term, and number of terms, being given.

ILLUSTRATION. — In the following series,

2, 5, 8, 11, 14, 17, 20, 23, 26, 29,

2 and 29 are the extremes, 3 the common difference, 10 the number of terms, and the sum of the series 155.

It is evident that the number of common differences in any series must be 1 less than the number of terms. Therefore, since the number of terms in this series is 10, the number of common differences will be 9, and their sum will be equal to the difference of the extremes; hence, if the difference of the extremes (29 — 2 = 27) be divided by the number of common differences, the quotient will be the *common difference*. Thus, 27 ÷ 9 = 3, the common difference. Hence the following

RULE. — *Divide the difference of the extremes by the number of terms less one, and the quotient is the common difference.*

EXAMPLES FOR PRACTICE.

1. The extremes of a series are 3 and 35, and the number of terms is 9; what is the common difference? Ans. 4.

OPERATION.

$$\frac{35-3}{9-1} = 4 \text{ common difference.}$$

2. If the first term is 7, the last term 55, and the number of terms 17, required the common difference. Ans. 3.

3. If the first term is 4, the last term 14, and the number of terms 15, what is the common difference? Ans. $\frac{5}{7}$.

4. If a man travels 10 days, and the first day goes 9 miles,

QUESTIONS. — Art. 304. What is the common difference? What five things are named, any three of which being given the other two can be found. What is the rule for finding the common difference, the first term, last term, and number of terms, being given?

and the last 17 miles, and increases each day's travel by an equal difference, what is the daily increase? Ans. $\frac{8}{9}$ miles.

ART. **305.** To find the sum of all the terms, the first term, last term, and number of terms, being given.

ILLUSTRATION. — Let the two following series be arranged as follows:

2,	5,	8,	11,	14,	17,	20,	= 77, sum of first series.
20,	17,	14,	11,	8,	5,	2,	= 77, sum of inverted series.
22,	22,	22,	22,	22,	22,	22,	= 154, sum of both series.

From the arrangement of the above series, we see that, by adding the two as they stand, we have the same number for the sum of the successive terms, and that the sum of both series is double the sum of either series.

It is evident that, if 22 in the above series be multiplied by 7, the number of terms, the product will be the sum of *both* series; thus, $22 \times 7 = 154$; and, therefore, the sum of either series will be $154 \div 2 = 77$. But 22 is the sum of the *extremes* in each series; thus, $20 + 2 = 22$. Therefore, if the sum of the extremes be multiplied by the number of terms, the product will be double the sum of either series. Hence,

RULE 1. — *Multiply the sum of the extremes by the number of terms and half the product will be the sum of the series.* Or,

RULE 2. — *Multiply the sum of the extremes by half the number of terms, and the product is the sum required.*

EXAMPLES FOR PRACTICE.

1. If the extremes of a series are 5 and 45, and the number of terms 9, what is the sum of the series? Ans. 225.

OPERATION.

$$\frac{(45 + 5) \times 9}{2} = 225, \text{ sum of the series.}$$

2. John Oaks engaged to labor for me 12 months. For the first month I was to pay him \$7, and for the last month \$51. In each successive month he was to have an equal addition to his wages; what sum did he receive for his year's labor? Ans. \$348.

QUESTION. — Art. 305. What is the rule for finding the sum of all the terms, the first term, last term, and number of terms, being given?

3. I have purchased from W. Hall's nursery 100 fruit-trees of various kinds, to be set around a circular lot of land, at the distance of one rod from each other. Having deposited them on one side of the lot, how far shall I have travelled when I have set out my last tree, provided I take only one tree at a time, and travel on the same line each way? Ans. 9801 rods.

ART. **306.** To find the number of terms, the extremes and common difference being given.

ILLUSTRATION. — Let the extremes of a series be 2 and 29, and the common difference 3. The difference of the extremes will be $29 - 2 = 27$. Now, it is evident that, if the difference of the extremes be divided by the common difference, the quotient will be the number of common differences; thus, $27 \div 3 = 9$. It has been shown (Art. 303) that the number of terms is 1 more than the number of differences; therefore, $9 + 1 = 10$ is the number of terms in this series. Hence the following

RULE. — *Divide the difference of the extremes by the common difference, and the quotient, increased by 1, will be the number of terms required.*

EXAMPLES FOR PRACTICE.

1. If the extremes of a series are 4 and 44, and the common difference 5, what is the number of terms? Ans. 9.

OPERATION.

$$\frac{44 - 4}{5} + 1 = 9, \text{ number of terms.}$$

2. A man going a journey travelled the first day 8 miles, and the last day 47 miles, and each day increased his journey by 3 miles. How many days did he travel? Ans. 14 days.

ART. **307.** To find the sum of the series, the extremes and common difference being given.

ILLUSTRATION. — Let the extremes be 2 and 29, and the common difference 3. The difference of the extremes will be $29 - 2 = 27$; and it has been shown (Art. 306) that if the difference of the extremes be divided by the common difference, the

---

QUESTION. — Art. 306. What is the rule for finding the number of terms, the extremes and common difference being given?

quotient will be the number of terms less *one*. Therefore, the number of terms less one will be $27 \div 3 = 9$, and the number of terms $9 + 1 = 10$. It was also shown (Art. 305) that, if the number of terms be multiplied by the sum of the extremes, and the product divided by 2, the quotient will be the sum of the series. Hence the

RULE. — *Divide the difference of the extremes by the common difference, and to the quotient add* 1; *multiply this sum by the sum of the extremes, and half the product is the sum of the series.*

EXAMPLES FOR PRACTICE.

1. If the two extremes are 11 and 74, and the common difference 7, what is the sum of the series? Ans. 425.

OPERATION.

$$\frac{74-11}{7} + 1 = 10; \quad \frac{(74+11) \times 10}{2} = 425, \text{ sum of series.}$$

2. A pupil commenced Virgil by reading 12 lines the first day, 17 lines the second day, and thus increased every day by 5 lines, until he read 137 lines in a day. How many lines did he read in all? Ans. 1937 lines.

ART. 308. To find the last term, the first term, the number of terms, and the common difference, being given.

ILLUSTRATION. — Let the first term of a series be 2, the number of terms 10, and the common difference 3. It has been shown (Art. 304) that the number of common differences is always 1 less than the number of terms; and that the sum of the common differences is equal to the difference of the extremes; therefore, since the number of terms is 10, and the common difference 3, the difference of the extremes will be $(10 - 1) \times 3 = 27$; and this difference, added to the first term, must give the last term; thus, $2 + 27 = 29$. Hence the following

RULE. — *Multiply the number of terms less* 1 *by the common difference, and add this product to the first term for the last term.*

NOTE. — If the series is descending, the product must be *subtracted* from the first term.

QUESTIONS. — Art. 307. What is the rule for finding the sum of the series, the extremes and common difference being given? — Art. 308. What is the rule for finding the last term, the first term, the number of terms, and common difference, being given?

EXAMPLES FOR PRACTICE.

1. If the first term is 1, the number of terms 7, and the common difference 6, what is the last term? Ans. 37.

OPERATION.

$$1 + (7 - 1) \times 6 = 37, \text{ last term.}$$

2. If a man travel 7 miles the first day of his journey, and 9 miles the second, and shall each day travel 2 miles further than the preceding, how far will he travel the twelfth day?

Ans. 29 miles.

3. If A set out from Portland for Boston, and travel $20\frac{1}{4}$ miles the first day, and on each succeeding day $1\frac{1}{2}$ miles less than on the preceding, how far will he travel the tenth day?

Ans. $6\frac{3}{4}$ miles.

ANNUITIES AT SIMPLE INTEREST BY ARITHMETICAL PROGRESSION.

ART. **309.** AN ANNUITY is a sum of money to be paid annually, or at any other regular period, either for a limited time or forever.

The *present worth* of an annuity is that sum which being put at interest will be sufficient to pay the annuity.

The *amount* of an annuity is the interest of all the payments added to their sum.

Annuities are said to be *in arrears* when they remain unpaid after they have become due.

ART. **310.** To find the amount of an annuity at simple interest.

Ex. 1. A man purchased a farm for $2000, and agreed to pay for it in 5 years, paying $400 annually; but, finding himself unable to make the annual payments, he agreed to pay the whole amount at the end of the 5 years, with the simple interest, at 6 per cent., on each payment, from the time it became due till the time of settlement; what did the farm cost him?

Ans. $2240.

ILLUSTRATION. — It is evident the *fifth* payment will be $400, without interest; the *fourth* will be on interest 1 year, and will amount to $424; the *third* will be on interest 2 years, and will amount to $448; the *second* will be on interest 3

QUESTIONS. — Art. 309. What is an annuity? What is meant by the present worth of an annuity? By the amount? When are annuities said to be in arrears?

years, and will amount to $472; and the *first* will be on interest 4 years, and will amount to $496. Therefore, these several sums form an arithmetical series; thus, 400, 424, 448, 472, 496; of which the fifth payment, or the *annuity*, is the *first term*, the *interest* on the annuity for one year the *common difference*, the *time* in years, the *number* of terms, and the *amount* of the annuity, the *sum* of the series. The sum of this series is found by Art. 305; thus, $\frac{(400 + 496) \times 5}{2} = \$2240$. Hence the

RULE. — *First find the last term of the series* (Art. 308), *and then the sum of the series* (Art. 305).

NOTE. — If the payments are to be made semi-annually, quarterly, &c., these periods will be the number of terms, and the interest of the annuity for each period the common difference.

EXAMPLES FOR PRACTICE.

2. What will an annuity of $250 amount to in 6 years, at 6 per cent. simple interest? Ans. $1725.

3. What will an annuity of $380 amount to in 10 years, at 5 per cent. simple interest? Ans. $4655.

4. An annuity of $825 was settled on a gentleman, January 1, 1840, to be paid annually. It was not paid until January 1, 1848; how much did he receive, allowing 6 per cent. simple interest? Ans. $7986.

5. A gentleman let a house for 3 years, at $200 a year, the rent to be paid semi-annually, at 8 per cent. per annum, simple interest. The rent, however, remained unpaid until the end of the three years; what did he then receive? Ans. $660.

6. A certain clergyman was to receive a salary of $700, to be paid annually; but, for certain reasons, which we fear were not very good, his parishioners neglected to pay him for 8 years; but he agreed to settle with them, and allow them $100 if they would pay him his just due with interest; required the sum received. Ans. $6676.

7. A certain gentleman in Boston has a very fine house, which he rents at $50 per month. Now, if his tenant shall omit payment until the end of the year, what sum should the owner receive, reckoning interest at 12 per cent.? Ans. $633.

QUESTIONS. — Art. 310. What forms the first term of a progression in an annuity? What the common difference? What the number of terms? What the sum of the series? What is the rule for finding the amount of an annuity at simple interest? If the payments are made semi-annually, quarterly, &c., what constitute the terms? What the common difference?

§ XL. GEOMETRICAL PROGRESSION.

ART. **311.** WHEN there are three or more numbers, and the same quotient is obtained by dividing the second by the first, the third by the second, and the fourth by the third, &c., these numbers are in *Geometrical Progression*, and may be called a *Geometrical Series.* Thus,

2, 4, 8, 16, 32, 64.
64, 32, 16, 8, 4, 2.

The former is called an *ascending* series, and the latter a *descending* series.

In the first series the quotient is 2, and is called the *ratio;* in the second, it is $\frac{1}{2}$. Hence, if the series is *ascending*, the quotient is more than unity; if it is *descending*, it is less than unity.

The first and last terms of a series are called *extremes*, and the other terms *means.*

Any *three* of the *five* following things being given, the other *two* may be found:

1st. The first term, or first extreme;
2d. The last term, or last extreme;
3d. The number of terms;
4th. The ratio;
5th. The sum of the terms, or series.

ART. **312.** One of the extremes, the ratio, and the number of terms, being given, to find the other extreme.

ILLUSTRATION. — Let the first term be 2, the ratio 3, and the number of terms 7. It is evident that, if we multiply the *first* term by the ratio, the product will be the second term in the series; and if we multiply the *second* term by the ratio, the product will be the third term; and, in this manner, we may carry the series to any desirable extent. By examining the following series, we find that 2 carried to the 7th term is 1458; thus,

---

QUESTIONS. — Art. 311. When are numbers in geometrical progression? What is an ascending series? What a descending series? What is the ratio of a progression? Is the ratio greater or less than unity in an ascending series? In a descending series? What are the extremes of a series? What the means? What five things are mentioned, any three of which being given, the other two may be found?

1	2	3	4	5	6	7
2	6	18	54	162	486	1458

The factors of 1458 are 3, 3, 3, 3, 3, 3, and 2, the last of which is the first term of the series, and the others the ratio repeated a number of times one less than the number of terms. But multiplying these factors together is the same as raising the ratio to the sixth power, and then multiplying that power by the first term. Hence the following

RULE. — *Raise the ratio to a power whose index is equal to the number of terms less one; then multiply this power by the first term, and the product is the last term, or other extreme.*

NOTE. — This rule may be applied in computing compound interest, the principal being the first term, the amount of one dollar for one year the ratio, the time, in years, one less than the number of terms, and the amount the last term.

EXAMPLES FOR PRACTICE.

1. The first term of a series is 1458, the number of terms 7, and the ratio $\frac{1}{3}$; what is the last term? Ans. 2.

OPERATION.

Ratio $(\frac{1}{3})^6 = \frac{1}{729}$; $\frac{1}{729} \times 1458 = \frac{1458}{729} = 2$, the last term.

2. If the first term of a series is 4, the ratio 5, and the numbers of terms 7, what is the last term? Ans. 62500.

3. If the first term of a series is 28672, the ratio $\frac{1}{4}$, and the number of terms 7, what is the last term? Ans. 7.

4. The first term of a series is 5, the ratio 4, and the number of terms is 8; required the last term. Ans. 81920.

5. If the first term of a series is 10, the ratio 20, and the number of terms 5, what is the last term? Ans. 1600000.

6. If the first term of a series is 30, the ratio 1.06, and the number of terms 6, what is the last term? Ans. 40.146767328.

7. What is the amount of $1728 for 5 years, at 6 per cent., compound interest? Ans. $2312.453798+.

8. What is the amount of $328.90 for 4 years, at 5 per cent., compound interest? Ans. $399.78+.

9. A gentleman purchased a lot of land containing 15 acres, agreeing to pay for the whole what the last acre would come

---

QUESTIONS. — Art. 312. What is the rule for finding the other extreme, one of the extremes, the ratio, and number of terms, being given? To what may this rule be applied?

to, reckoning 5 cents for the first acre, 15 cents for the second and so on, in a three-fold ratio. What did the lot cost him?

Ans. $239148.45.

ART. **313.** To find the sum of all the terms, the first term the ratio, and the number of terms, being given.

ILLUSTRATION. — Let it be required to find the sum of the following series:

2, 6, 18, 54.

If we multiply each term of this series by the ratio 3, the products will be 6, 18, 54, 162, forming a second series, whose sum is three times the sum of the first series; and the *difference* between these two series is *twice* the sum of the first series. Thus,

	6,	18,	54,	162, the second series.
2,	6,	18,	54,	the first series.
2,	0,	0,	0,	162 — 2 = 160, difference of the two series

Now, since this difference is *twice* the sum of the first series, one half this difference will be the *sum* of the first series; thus 160 ÷ 2 = 80.

It will be observed, by examining the operation above, that if we had simply multiplied 54, the last term of the first series, by the ratio 3, and subtracted 2, the first term, from it, we should have obtained 160; and this being divided by the ratio, 3 less 1, would have given 80, the same number as before, for the sum of the first series. Hence the

RULE. 1. — *Find the last term as in the preceding article, multiply it by the ratio, and from the product subtract the first term. Then divide this remainder by the ratio, less* 1, *and the quotient will be the sum of the series.* Or,

RULE. 2. — *Raise the ratio to a power whose index is equal to the number of terms, from which subtract* 1; *divide the remainder by the ratio, less* 1, *and the quotient, multiplied by the given extreme, will be the sum of the series required.*

NOTE 1. — If the ratio is less than a unit, the product of the last term, multiplied by the ratio, must be subtracted from the first term; and, to obtain the divisor, the ratio must be subtracted from unity, or 1.

QUESTIONS. — Art. 313. What is the rule for finding the sum of all the terms, the first term, ratio, and number of terms, being given? If the ratio is less than a unit, what must be done with the product of the last term multiplied by the ratio? How is the divisor obtained when the ratio is less than 1?

NOTE 2. — If the second rule is employed, when the ratio is less than 1, its power, denoted by the number of terms, must be subtracted from 1, and the remainder divided by the difference between 1 and the ratio.

EXAMPLES FOR PRACTICE.

1. If the first term of a series is 12, the ratio 3, and the number of terms 8, what is the sum of the series? Ans. 39360.

OPERATION.

Ratio $3^7 \times 12 = 26244$, the last term; $26244 \times 3 = 78732$; $78732 - 12 = 78720$; $78720 \div (3 - 1) = 39360$, the sum of the series.

2. The first term of a series is 5, the ratio $\frac{2}{3}$, and the number of terms 6; required the sum of the series. Ans. $13\frac{166}{243}$.

OPERATION.

Ratio $(\frac{2}{3})^5 \times 5 = \frac{160}{243}$, the last term; $\frac{160}{243} \times \frac{2}{3} = \frac{320}{729}$; $5 - \frac{320}{729} = \frac{3325}{729}$; $\frac{3325}{729} \div (1 - \frac{2}{3}) = \frac{3325}{243} = 13\frac{166}{243}$, the sum of the series.

3. If the first term of a series is 8, the ratio 4, and the number of terms 7, required the sum of the series. Ans. 43688.

4. If the first term is 10, the ratio $\frac{3}{4}$, and the number of terms 5, what is the sum of the series? Ans. $30\frac{65}{128}$.

5. If the first term is 18, the ratio 1.06, and the number of terms 4, what is the sum of the series? Ans. 78.743+.

6. When the first term is $144, the ratio $1.05, and the number of terms 5, what is the sum of the series? Ans. $795.6909.

7. D. Baldwin agreed to labor for E. Thayer for 6 months. For the first month he was to receive $3, and each succeeding month's wages were to be increased by $\frac{2}{3}$ of his wages for the month next preceding; required the sum he received for his 6 months' labor. Ans. $$91\frac{77}{81}$.

8. If the first term of a series is 2, the ratio 6, and the number of terms 4, what is the sum of the series? Ans. 518.

9. A lady, wishing to purchase 10 yards of silk for a new dress, thought $1.00 per yard too high a price; she, however, agreed to give 1 cent for the first yard, 4 for the second, 16 for the third, and so on, in a four-fold ratio; what was the cost of the dress? Ans. $3495.25.

QUESTION. — When the second rule is employed, if the ratio is less than 1, what must be done?

ANNUITIES AT COMPOUND INTEREST BY GEOMETRICAL PROGRESSION

ART. **314.** WHEN compound interest is reckoned on an annuity in arrears, the annuity is said to be at compound interest; and the amounts of the several payments form a geometrical series, of which the annuity is the first term, the amount of $1.00 for one year the ratio, the years the number of terms, and the amount of the annuity the sum of the series. Hence,

ART. **315.** To find the amount of an annuity at compound interest, we have the following

RULE 1. — *Find the sum of the series by either of the preceding rules.* (Art. 313.) Or,

RULE 2. — *Multiply the amount of $1.00, for the given time, found in the table, by the annuity, and the product will be the required amount*

TABLE,

Showing the amount of $1 *annuity* from 1 year to 40.

Years.	5 per cent.	6 per cent.	Years.	5 per cent.	6 per cent.
1	1.000000	1.000000	21	35.719252	39.992727
2	2.050000	2.060000	22	38.505214	43.392290
3	3.152500	3.183600	23	41.430475	46.995828
4	4.310125	4.374616	24	44.501999	50.815577
5	5.525631	5.637093	25	47.727099	54.864512
6	6.801913	6.975319	26	51.113454	59.156383
7	8.142008	8.393838	27	54.669126	63.705766
8	9.549109	9.897468	28	58.402583	68.528112
9	11.026564	11.491316	29	62.322712	73.639798
10	12.577893	13.180795	30	66.438847	79.058186
11	14.206787	14.971643	31	70.760790	84.801677
12	15.917127	16.869941	32	75.298829	90.889778
13	17.712983	18.882138	33	80.063771	97.343165
14	19.598632	21.015066	34	85.066959	104.183755
15	21.578564	23.275970	35	90.220307	111.434780
16	23.657492	25.672528	36	95.836323	119.120867
17	25.840366	28.212880	37	101.628139	127.268119
18	28.132385	30.905653	38	107.709546	135.904206
19	30.539004	33.759992	39	114.095023	145.058458
20	33.065954	36.785591	40	120.799774	154.761966

QUESTIONS. — Art. 314. When is an annuity said to be at compound interest? What do the amounts of the several payments form? What is the first term of the series? What the ratio? What the number of terms? What the sum of the series? — Art. 315. What is the first rule for finding the amount of an annuity? What the second? What does the table show?

EXAMPLES FOR PRACTICE.

1. What will an annuity of $378 amount to in 5 years, at 6 per cent. compound interest? Ans. $2130.821+.

OPERATION BY RULE FIRST.

$$\frac{1.06^5 - 1}{1.06 - 1} \times 378 = \$2130.821+.$$

OPERATION BY RULE SECOND.

$$5.637093 \times 378 = \$2130.821+.$$

2. What will an annuity of $1728 amount to in 4 years, at 5 per cent. compound interest? Ans. $7447.896+.

3. What will an annuity of $87 amount to in 7 years, at 6 per cent. compound interest? Ans. $730.263+.

4. What will an annuity of $500 amount to in 6 years, at 6 per cent. compound interest? Ans. $3487.659+.

5. What will an annuity of $96 amount to in 10 years, at 6 per cent. compound interest? Ans. $1265.356+.

6. What will an annuity of $1000 amount to in 3 years, at 6 per cent. compound interest? Ans. $3183.60.

7. July 4, 1842, H. Piper deposited in an annuity office, for his daughter, the sum of $56, and continued his deposits each year, until July 4, 1848. Required the sum in the office July 4, 1848, allowing 6 per cent. compound interest.

Ans. $470.054+.

8. C. Greenleaf has two sons, Samuel and William. On Samuel's birth-day, when he was 15 years old, he deposited for him, in an annuity office, which paid 5 per cent. compound interest, the sum of $25, and this he continued yearly, until he was 21 years of age. On William's birth-day, when he was 12 years old, he deposited for him, in an office which paid 6 per cent. compound interest, the sum of $20, and continued this until he was 21 years of age. Which will receive the larger sum, when 21 years of age?

Ans. $60.065+ William receives more than Samuel.

9. I gave my daughter Lydia $10 when she was 8 years old, and the same sum on her birth-day each year, until she was 21 years old. This sum was deposited in the savings bank, which pays 5 per cent. annually. Now, supposing each deposit to remain on interest until she is 21 years of age, required the amount in the bank. Ans. $195.986+.

§ XLI. ALLIGATION.

ART. **316.** ALLIGATION is a rule employed in the solution of questions relating to the compounding or mixing of several ingredients. The term signifies the act of connecting or tying together. It is of two kinds: *Alligation Medial* and *Alligation Alternate.*

ALLIGATION MEDIAL.

ART. **317.** *Alligation Medial* is the method of finding the mean price of a mixture composed of articles of different values, the quantity and price of each being given.

ART. **318.** To find the mean price of several articles or ingredients, at different prices, or of different qualities.

RULE. — *Find the value of each of the ingredients, and divide the amount of their values by the sum of the ingredients. The quotient will be the price of the mixture.*

EXAMPLES FOR PRACTICE.

EX. 1. A grocer mixed 20lb. of tea worth $0.50 a pound, with 30lb. worth $0.75 a pound, and 50lb. worth $0.45 a pound; what is 1 pound of the mixture worth? Ans. $0.55.

OPERATION.

$0.50 × 20 = $10.00
$0.75 × 30 = $22.50
$0.45 × 50 = $22.50
Sum of ingredients, 100 $55.00, value.
Then, $55.00 ÷ 100 = $0.55 per pound.
Proof, $0.55 × 20 lb. = $11.00
$0.55 × 30 lb. = $16.50
$0.55 × 50 lb. = $27.50
} = $55.00.

2. I have four kinds of molasses, and a different quantity of each, as follows: 30 gal. at 20 cents, 40 gal. at 25 cents, 70 gal. at 30 cents, and 80 gal. at 40 cents; what is a gallon of the mixture worth? Ans. 0.31\frac{4}{11}$.

3. A farmer mixed 4 bush. of oats at 40 cents, 8 bush. of

QUESTIONS. — Art. 316. What is alligation? What two kinds are there? — Art. 317. What is alligation medial? — Art. 318. What is the rule for finding the mean price of several articles at different prices? How does it appear that this process will give the mean price of the mixture?

corn at 85 cents, 12 bush. rye at $1.00, and 10 bush. of wheat at $1.50 per bushel. What will one bushel of the mixture be worth? Ans. $\$1.04\frac{2}{17}$.

ALLIGATION ALTERNATE.

ART. **319.** *Alligation Alternate* is the method of finding what quantity of ingredients or articles, whose prices or qualities are given, must be taken, to compose a mixture of any given price or quality.

ART. **320.** To find what quantity of each ingredient must be taken to form a mixture of a given price.

Ex. 1. I wish to mix spice, at 20 cents, 23 cents, 26 cents, and 28 cents per pound, so that the mixture may be worth 25 cents per pound. How many pounds of each must I take?

FIRST OPERATION.

Mean price 25 cts.	1lb. at 20cts.	gain 5cts.
	1lb. at 23cts.	gain 2cts.
	1lb. at 26cts.	loss 1ct.
	1lb. at 28cts.	loss 3cts.
	1lb. at 28cts.	loss 3cts.

PROOF.

Ans.	1lb. at 20cts.	= 20cts.
	1lb. at 23cts.	= 23cts.
	1lb. at 26cts.	= 26cts.
	2lb. at 28cts.	= 56cts.
	5lbs. whole val.	$1.25

$\$1.25 \div 5 = 25$cts. per lb.

Compared with the mean or average price given, by taking 1lb. at 20 cents there is a gain of 5 cents, by taking 1lb. at 23 cents a gain of 2 cents, by taking 1lb. at 26 cents a loss of 1 cent, and by taking 1lb. at 28 cents a loss of 3 cents; making an excess of gain over loss of 3 cents. Now, it is evident that the mixture, to be of the average value named, should have the several items of gain and loss in the aggregate exactly offset one another. This balance we can effect, in the present case, either by taking 3lb. more of the spice at 26 cents, or 1lb. more of spice at 28 cents. We take the 1lb. at 28 cents, and thus have a mixture of the required average value, by having taken, in all, 1lb. at 20 cents, 1lb. at 23 cents, 1lb. at 26 cents, and 2lb. at 28 cents. We prove the correctness of the result by dividing the value of the whole mixture, or $1.25, by the number of pounds taken, or 5, which gives 25 cents, or the given mean price per pound.

SECOND OPERATION.

25cts.	20cts.	3lb.	Ans.
	23cts.	1lb.	
	26cts.	2lb.	
	28cts.	5lb.	

Having arranged in a column the prices of the ingredients, with the given mean price on the left, we connect together the terms denoting the price of each ingredient, so that a price less than the given

QUESTIONS. — Art. 319. What is alligation alternate? How do you connect the prices? Explain the first operation. How is it proved to be correct?

mean is united with one that is greater. We then proceed to find what quantity of each of the two kinds, whose prices have been connected, can be taken, in making a mixture, so that what shall be gained on the one kind shall be balanced by the loss on the other. By taking 1lb. of spice at 20 cents, the gain will be 5 cents; and by taking 1lb. at 28 cents, the loss will be 3 cents. To equalize the gain and loss in this case, it is evident we should take as many more pounds of that at 28 cents as the loss on 1lb. of it is less than the gain on 1lb. of that at 20 cents; or, in other words, *the ingredients taken should be in the inverse ratio* (Art. 236) *of the difference between their respective prices and the given mean price.* Therefore, we take 5lbs. at 28 cents, and 3lbs. of that at 20 cents, and the loss, 3cts. $\times$ 5 = 15 cents, on the former, exactly offsets the gain, 5cts. $\times$ 3 = 15 cents, on the latter. We write the 3lb. against its price, 20 cents; and the 5lb. against its price, 28 cents. In like manner we determine the quantity that may be taken of the other two ingredients, whose prices are connected, by finding the difference between each price and the mean price; and, as before, write the quantity taken against its price.

We obtain, as a result, 3lb. at 20 cents, 1lb. at 23 cents, 2lb. at 26 cents, 5lb. at 28 cents; this, in the same manner as the other answer, may be proved to satisfy the conditions of the question, since examples of this kind admit of several answers.

RULE. — *Write the prices of the ingredients in a column, with the mean price on the left, and connect the price of each ingredient which is less than the given mean price with one that is greater.*

Write the difference between the mean price and that of each of the ingredients opposite to the price with which it is connected; and the number set against each price is the quantity of the ingredient to be taken at that price.

NOTE. — There will be as many different answers as there are different ways of connecting the prices, and by multiplying and dividing these answers they may be varied indefinitely.

EXAMPLES FOR PRACTICE.

2. A farmer wishes to mix corn at 75 cents a bushel, with rye at 60 cents a bushel, and oats at 40 cents a bushel, and wheat at 95 cents a bushel; what quantity of each must he take to make a mixture worth 70 cents a bushel?

	FIRST OPERATION.	Ans.		SECOND OPERATION.	Ans.		THIRD OPERATION.	Ans.
70	40	25	70	40	5	70	40	25 + 5 = 30
	60	5		60	25		60	5 + 25 = 30
	75	10		75	30		75	10 + 30 = 40
	95	30		95	10		95	30 + 10 = 40

QUESTIONS. — What is the rule for alligation alternate? How can you obtain different answers? Are they all true?

3. I have 4 kinds of salt, worth 25, 30, 40, and 50 cents per bushel; how much of each kind must be taken, that a mixture might be sold at 42 cents per bushel?

Ans. 8 bushels at 25, 30, and 40 cents, and 31 bushels at 50 cents.

ART. **321.** When the quantity of one ingredient is given to find the quantity of each of the others.

Ex. 1. How much sugar, that is worth 6, 10, and 13 cents a pound, must be mixed with 20lb. worth 15 cents a pound, so that the mixture will be worth 11 cents a pound?

OPERATION.

Mean	Prices	Differences.
11	6	4
	10	2
	13	1
	15	5

Then, 5 : 1 :: 20 : 4
5 : 2 :: 20 : 8 } Ans.
5 : 4 :: 20 : 16

By the conditions of the question we are to take 20lb. at 15 cents a pound; but by the operation we find the difference at 15 cents a pound to be only 5lb., which is but $\frac{1}{4}$ of the given quantity. Therefore, if we increase the 5lb. to 20, the other differences must be increased in the same proportion. Hence the propriety of the following

RULE. — *Find the difference between each price and the mean price; then say, As the difference of that ingredient whose quantity is given is to each of the differences separately, so is the quantity given to the several quantities required.*

EXAMPLES FOR PRACTICE.

2. A farmer has oats at 50 cents per bushel, peas at 60 cents, and beans at $1.50. These he wishes to mix with 30 bushels of corn at $1.70 per bushel, that he may sell the whole at $1.25 per bushel; how much of each kind must he take?

Ans. 18 bushels of oats, 10 bushels of peas, and 26 bushels of beans.

3. A merchant has two kinds of sugar, one of which cost him 10 cents per lb., and the other 12 cents per lb.; he has also 100lb. of an excellent quality, which cost him 15 cents per lb. Now, as he ought to make 25 per cent. on his cost, how much of each quantity must be taken that he may sell the mixture at 14 cents per lb.?

Ans. 383$\frac{1}{3}$lb. at 10 cents, and 100lb. at 12 cents.

QUESTION. — Art. 321. What is the rule for finding the quantity of each of the other ingredients when one is given?

ART. **322.** When the sum of the ingredients and their mean price are given, to find what quantity of each must be taken.

Ex. 1. I have teas at 25 cents, 35 cents, 50 cents, and 70 cents a pound, with which I wish to make a mixture of 180lb. that will be worth 45 cents a pound. How much of each kind must I take?

OPERATION.

45	25	25	Then, 60 : 25 : : 180 : 75	Ans.
	35	5	60 : 5 : : 180 : 15	
	50	10	60 : 10 : : 180 : 30	
	70	20	60 : 20 : : 180 : 60	
Sum of differences,		60	Proof, 180	

By the conditions of the question, the weight of the mixture is 180lb., but by the operation we find the sum of the differences to be only 60lb., which is but ⅓ of the quantity required. Therefore, if we increase 60lb. to 180, each of the differences must be increased in the same proportion, in order to make a mixture of 180lb., the quantity required. Hence the

RULE. — *Find the differences as before; then say, As the sum of the differences is to each of the differences separately, so is the given quantity to the required quantity of each ingredient.*

EXAMPLES FOR PRACTICE.

2. John Smith's "great box" will hold 100 bushels. He has wheat worth $2.50 per bushel, and rye worth $2.00 per bushel. How much chaff, of no value, must he mix with the wheat and rye, that, if he fill the box, a bushel of the mixture may be sold at $1.80?

Ans. 40bu. each of wheat and rye, and 20 bushels of chaff.

3. I have two kinds of molasses, which cost me 20 and 30 cents per gallon; I wish to fill a hogshead, that will hold 80 gallons, with these two kinds. How much of each kind must be taken, that I may sell a gallon of the mixture at 25 cents per gallon, and make 10 per cent. on my purchase?

Ans. $58\frac{2}{11}$ of 20 cents, and $21\frac{9}{11}$ of 30 cents.

4. I have sugars at 10 cents and 15 cents per pound. How much of each must be taken, that a mixture containing 60 pounds shall be worth $7.20?

Ans. 36 pounds at 10 cents, and 24 pounds at 15 cents.

QUESTION. — Art. 322. How do you find what quantity of each ingredient must be taken when the sum and mean price are given?

§ XLII. PERMUTATION.

ART. **323.** PERMUTATION is the process of finding the different orders in which may be arranged a given number of things.

ART. **324.** To find the number of different arrangements that can be made of any given number of things.

Ex. 1. How many different numbers may be formed from the figures of the following number, 432, making use of three figures in each number? Ans. 6.

FIRST OPERATION.

4 3 2, 4 2 3, 3 4 2, 3 2 4, 2 4 3, 2 3 4.

SECOND OPERATION.

$$1 \times 2 \times 3 = 6.$$

In the 1st operation, we have made all the different arrangements that can be made of the given figures, and find the number to be 6. In the second operation, the same result is obtained by simply multiplying together the first three of the digits, a number equal to the number of figures to be arranged. Hence the following

RULE. — *Multiply together all the terms of the natural series of numbers, from* 1 *up to the given number, and the last product will be the answer required.*

EXAMPLES FOR PRACTICE.

2. My family consists of nine persons, and each person has his particular seat around my table. Now, if their situations were to be changed once each day, for how many days could they be seated in a different position?
Ans. 362880 days, or 994 years 70 days.

3. On a certain shelf in my library there are 12 books. If a person should remove them without noticing their order, what would be the probability of his replacing them in the same position they were at first? Ans. 1 to 479001600.

4. How many words can be made from the letters in the word "Embargo," provided that any arrangement of them may be used, and that all the letters shall be taken each time?
Ans. 5040 words.

QUESTIONS. — Art. 323. What is permutation? — Art. 324. What is the rule for finding the number of arrangements that can be made of any given number of things?

§ XLIII. MENSURATION OF SURFACES.

ART. **325.** A SURFACE is a magnitude, which has length and breadth without thickness.

The surface or superficial contents of a figure are called its *area.*

ART. **326.** An ANGLE is the inclination or opening of two lines, which meet in a point.

A *right angle* is an angle formed by one line falling perpendicularly on another, and it contains 90 degrees; as A B C.

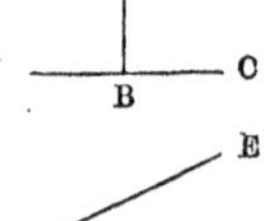

An *acute angle* is an angle less than a right angle, or less than 90 degrees; as E B C.

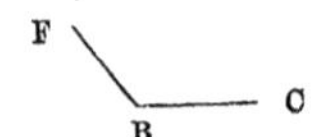

An *obtuse angle* is an angle greater than a right angle, or more than 90 degrees; as F B C.

F
B
C

THE TRIANGLE.

ART. **327.** A TRIANGLE is a figure having three sides and three angles. It receives the particular names of an *equilateral triangle*, *isosceles triangle*, and *scalene triangle.*

It is also called a *right-angled triangle* when it has one right angle; an *acute-angled triangle*, when it has all its angles acute; and an *obtuse-angled triangle*, when it has one obtuse angle.

The *base* of a triangle, or other plane figure, is the lowest side, or that which is parallel to the horizon; as C D.

The *altitude* of a triangle is a line drawn from one of its angles perpendicular to its opposite side or base; as A B.

An *equilateral triangle* is a figure which has its three sides equal.

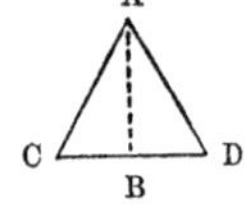

QUESTIONS. — Art. 325. What is a surface? What are the superficial contents of a figure called? — Art. 326. What is an angle? What is a right angle? An acute angle? An obtuse angle? — Art. 327. What is a triangle? What particular names does it receive? When is it called a right-angled triangle? When an acute-angled triangle? When an obtuse-angled triangle? What is the base of a triangle? What the altitude? What is an equilateral triangle?

An *isosceles triangle* is a figure which has two of its sides equal.

A *scalene triangle* is a figure which has its three sides unequal.

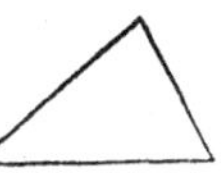

A *right-angled triangle* is a figure having three sides and three angles, one of which is a right angle.

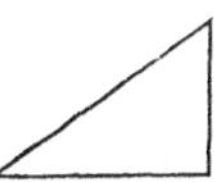

ART. **328.** To find the area of a triangle.

RULE 1. — *Multiply the base by half the altitude, and the product will be the area.* Or,

RULE 2. — *Add the three sides together, take half that sum, and from this subtract each side separately; then multiply the half of the sum and these remainders together, and the square root of this product will be the area.*

1. What are the contents of a triangle, whose base is 24 feet, and whose perpendicular height is 18 feet? Ans. 216 feet.

2. What are the contents of a triangular piece of land, whose sides are 50 rods, 60 rods, and 70 rods?

Ans. 1469.69+ rods.

THE QUADRILATERAL.

ART. **329.** A QUADRILATERAL is a figure having four sides, and consequently four angles. It comprehends the rectangle, square, rhombus, rhomboid, trapezium, and trapezoid.

ART. **330.** A PARALLELOGRAM is any quadrilateral whose opposite sides are parallel. It takes the particular names of *rectangle*, *square*, *rhombus*, and *rhomboid.*

The *altitude* of a parallelogram is a perpendicular line drawn between its opposite sides; as C D in the rhomboid.

QUESTIONS. — What is an isosceles triangle? A scalene triangle? A right-angled triangle? — Art. 328. What is the first rule for finding the area of a triangle? What the second? — Art. 329. What is a quadrilateral? What figures does it comprehend? — Art. 330. What is a parallelogram? What particular names does it take? What is the altitude of a parallelogram?

A *rectangle* is a right-angled parallelogram, whose opposite sides are equal.

A *square* is a parallelogram, having four equal sides and four right angles.

A *rhomboid* is an oblique-angled parallelogram, whose opposite sides are equal.

A *rhombus* is an oblique-angled parallelogram, having all its sides equal.

ART. **331.** To find the area of a parallelogram.

RULE. — *Multiply the base by the altitude, and the product will be the area.*

1. What are the contents of a board 25 feet long and 3 feet wide? Ans. 75 square feet.

2. What is the difference between the contents of two floors; one is 37 feet long and 27 feet wide, and the other is 40 feet long and 20 feet wide? Ans. 199 square feet.

3. The base of a rhombus is 15 feet, and its perpendicular height is 12 feet; what are its contents?
Ans. 180 square feet.

ART. **332.** A TRAPEZOID is a quadrilateral, which has only one pair of its opposite sides parallel.

ART. **333.** To find the area of a trapezoid.

RULE. —*Multiply half of the sum of the parallel sides by the altitude, and the product is the area.*

1. What is the area of a trapezoid, the longer parallel side

QUESTIONS. — What is a rectangle? A square? A rhomboid? A rhombus? — Art. 331. What is the rule for finding the area of a parallelogram? — Art. 332. What is a trapezoid? — Art. 333. What is the rule for finding the area of a trapezoid?

being 482 feet, the shorter 324 feet, and the altitude 216 feet? Ans. 87048 square feet.

2. What is the area of a plank, whose length is 22 feet, the width of the wider end being 28 inches, and of the narrower 20 inches? Ans. 44 square feet.

ART. **334.** A TRAPEZIUM is a quadrilateral, which has neither two of its opposite sides parallel.

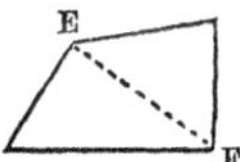

A *diagonal* is a line joining any two opposite angles of a quadrilateral; as E F.

ART. **335.** To find the area of a trapezium.

RULE. — *Divide the trapezium into two triangles by a diagonal, and hen find the areas of these triangles; their sum will be the area of the trapezium.*

1. What is the area of a trapezium, whose diagonal is 65 feet, and the length of the perpendiculars let fall upon it are 14 and 18 feet? Ans. 1040 square feet.

2. What is the area of a trapezium, whose diagonal is 125 rods, and the length of the perpendiculars let fall upon it are 70 and 85 rods? Ans. 9687.5 square rods.

THE POLYGON.

ART. **336.** A POLYGON is any rectilineal figure having more than four sides and four angles. It takes the particular names of *pentagon*, which is a polygon of five sides; *hexagon*, one of six sides; *heptagon*, one of seven sides; *octagon*, one of eight sides; *nonagon*, one of nine sides; *decagon*, one of ten sides; *undecagon*, one of eleven sides; and *dodecagon*, one of twelve sides.

ART. **337.** A REGULAR POLYGON is a plane rectilineal figure, which has all its sides and all its angles equal.

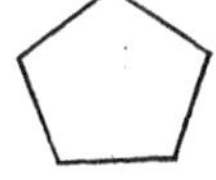

The *perimeter* of a polygon is the sum of all its sides.

ART. **338.** To find the area of a regular polygon.

QUESTIONS. — Art. 334. What is a trapezium? What is a diagonal? — Art. 335. What is the rule for finding the area of a trapezium? — Art. 336. What is a polygon? What particular names does it take? — Art. 337. What is a regular polygon?

RULE. — *Multiply the perimeter by half the perpendicular let fall from the centre on one of its sides, and the product will be the area.*

1. What is the area of a regular pentagon, whose sides are each 35 feet, and the perpendicular 24.08 feet?
Ans. 2107 square feet.

2. What is the area of a regular hexagon, whose sides are each 20 feet, and the perpendicular 17.32 feet?
Ans. 1039.20 square feet.

THE CIRCLE.

ART. **339.** A CIRCLE is a plane figure bounded by a curved line, every part of which is equally distant from a point, called its centre.

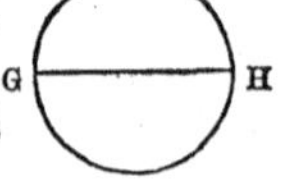

The *circumference* or *periphery* of a circle is the line which bounds it.

The *diameter* of a circle is a line drawn through the centre, and terminated by the circumference; as G H.

ART. **340.** To find the circumference of a circle, the diameter being given.

RULE. — *Multiply the diameter by* 3.141592, *and the product is the circumference.*

1. What is the circumference of a circle, whose diameter is 50 feet?
Ans. 157.0796+ feet.

2. A gentleman has a circular garden whose diameter is 100 rods; what is the length of the fence necessary to enclose it?
Ans. 314.15+ rods.

ART. **341.** To find the diameter of a circle, the circumference being given.

RULE. — *Multiply the circumference by* .318309, *and the product will be the diameter.*

1. What is the diameter of a circle, whose circumference is 80 miles?
Ans. 25.46+ miles.

2. If the circumference of a wheel is 62.84 feet, what is the diameter?
Ans. 20+ feet.

QUESTIONS. — What is the perimeter of a polygon? — Art. 338. What is the rule for finding the area of a regular polygon? — Art. 339. What is a circle? What is the circumference of a circle? The diameter of a circle? — Art. 340. What is the rule for finding the circumference of a circle, the diameter being given? — Art. 341. What is the rule for finding the diameter of a circle, the circumference being given?

ART. 342. To find the area of a circle, the diameter, the circumference, or both, being given.

RULE 1. — *Multiply the square of the diameter by* .785398, *and the product is the area.* Or,
RULE 2. — *Multiply the square of the circumference by* .079577, *and the product is the area.* Or,
RULE 3. — *Multiply half the diameter by half the circumference, and the product is the area.*

1. If the diameter of a circle be 200 feet, what is the area?
Ans. 31415.92 square feet.

2. There is a certain farm, in the form of a circle, whose circumference is 400 rods; how many acres does it contain?
Ans. 79A. 2R. 12+p.

ART. 343. To find the side of a square equal in area to a given circle.

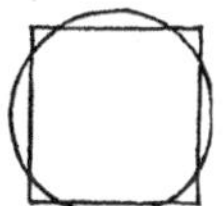

The square in the figure is supposed to have the same area as the circle.

RULE 1. — *Multiply the diameter by* .886227, *and the product is the side of an equal square.* Or,
RULE 2. — *Multiply the circumference by* .282094, *and the product is the side of an equal square.*

1. We have a round field 40 rods in diameter; what is the side of a square field that will contain the same quantity?
Ans. 35.44+ rods.

2. I have a circular field 100 rods in circumference; what must be the side of a square field that shall contain the same area?
Ans. 28.2+ rods.

ART. 344. To find the side of a square inscribed in a given circle.

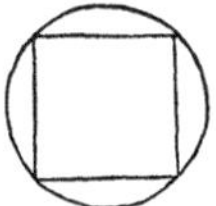

A square is said to be inscribed in a circle when each of its angles touches the circumference or periphery of the circle.

QUESTIONS. — Art. 342. What is the rule for finding the area of a circle, when the diameter is given? When the circumference is given? When the diameter and circumference are both given? — Art. 343. What is the first rule for finding the side of a square equal in area to a given circle? What the second? — Art. 344. When is a square said to be inscribed in a circle? What is the first rule for finding the side of a square inscribed in a circle? The second?

RULE 1. — *Multiply the diameter by* .707106, *and the product is the side of the square inscribed.* Or,

RULE 2. — *Multiply the circumference by* .225079, *and the product is the side of the square inscribed.*

1. What is the thickness of a square stick of timber that may be hewn from a log 30 inches in diameter?

Ans. 21.21+ inches.

2. How large a square field may be inscribed in a circle whose circumference is 100 rods? Ans. 22.5+ rods square.

THE ELLIPSE.

ART. **345.** An ELLIPSE is an oval figure having two diameters, or axes, the longer of which is called the *transverse* and the shorter the *conjugate* diameter.

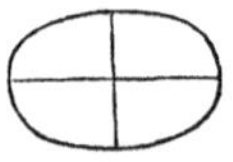

ART. **346.** To find the area of an ellipse.

RULE. — *Multiply the two diameters together, and their product by* .785398; *the last product is the area.*

1. What is the area of an ellipse whose transverse diameter is 14 inches, and its conjugate diameter 10 inches?

Ans. 109.95+ square inches.

2. What is the area of an elliptical table, 8 feet long and 5 feet wide? Ans. 31 square feet, 59+ square inches.

---

§ XLIV. MENSURATION OF SOLIDS.

ART. **347.** A SOLID is a magnitude which has length, breadth, and thickness.

Mensuration of solids includes two operations: first, to find their superficial contents, and, second, their solidities.

THE PRISM.

ART. **348.** A PRISM is a solid whose ends are any plane figures which are equal and similar, and whose sides are parallelograms. It takes particular names, according to the figure

---

QUESTIONS. — Art. 345. What is an ellipse? What is the longer diameter called? The shorter? — Art. 346. What is the rule for finding the area of an ellipse? — Art. 347. What is a solid? What two operations does mensuration of solids include? — Art. 248. What is a prism? What particular names does it take?

of its base or ends, namely, *triangular prism*, *square prism*, *pentagonal prism*, &c.

The *base* of a prism is either end; and of solids in general, the part upon which they are supposed to stand.

All prisms whose bases are parallelograms are comprehended under the general name *parallelopipedons* or *parallelopipeds*.

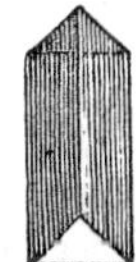

A *triangular prism* is a solid whose base is a triangle.

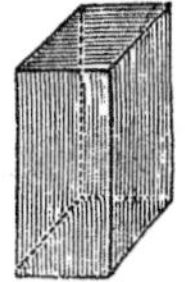

A *square prism* is a solid whose base is a square, and when all the sides are squares it is called a cube.

A *pentagonal prism* is a solid whose base is a pentagon.

ART. **349.** To find the surface of a prism.

RULE. — *Multiply the perimeter of its base by its height, and to this product add the area of the two ends; the sum is the area of the prism.*

1. What are the superficial contents of a triangular prism, the width of whose side is 3 feet, and its length 15 feet?
Ans. 142.79+ square feet.

2. What is the surface of a square prism, whose side is 9 feet wide, and its length 25 feet? Ans. 1062 square feet.

ART. **350.** To find the solidity of a prism.

RULE. — *Multiply the area of the base by the height, and the product is the solidity.*

QUESTIONS. — What is the base of a prism and of solids in general? What is a parallelopiped or parallelopipedon? What is a triangular prism? A square prism? A pentagonal prism? — Art. 349. What is the rule for finding the surface of a prism? — Art. 350. What is the rule for finding the solidity of a prism?

1. What are the contents of a triangular prism, whose length is 20 feet, and the three sides of its triangular end or base 5, 4, and 3 feet? Ans. 120 cubic feet.

2. How many cubic feet are there in a cube, whose sides are 8 feet? Ans. 512 cubic feet.

3. What is the number of cubic feet in a room 30 feet long, 20 feet wide, and 10 feet high? Ans. 6000 cubic feet.

THE CYLINDER.

ART. **351.** A CYLINDER is a round solid, of uniform diameter, with circular ends.

The *axis* of a cylinder is a straight line drawn through it, from the centre of one end to the centre of the other.

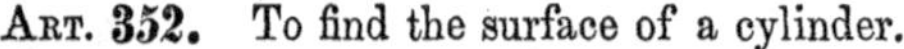

ART. **352.** To find the surface of a cylinder.

RULE. — *Multiply the circumference of the base by the altitude, and to the product add the areas of the two ends; the sum will be the whole surface.*

1. What is the surface of a cylinder, whose length is 4 feet, and the circumference 3 feet? Ans. 13.43+ square feet.

2. John Snow has a roller 12 feet long and 2 feet in diameter; what is its convex surface? Ans. 75.39+ square feet.

ART. **353.** To find the solidity of a cylinder.

RULE. — *Multiply the area of the base by the altitude, and the product will be the solidity.*

1. What is the solidity of a cylinder 8 feet in length and 2 feet in diameter? Ans. 25.13+ cubic feet.

2. What is the solidity of a cylinder, whose diameter is 5 feet, and its altitude 20 feet? Ans. 392.69+ cubic feet.

THE PYRAMID AND CONE.

ART. **354.** A PYRAMID is a solid, standing on a triangular, square, or polygonal base, with its sides tapering uniformly to a point at the top, called the vertex.

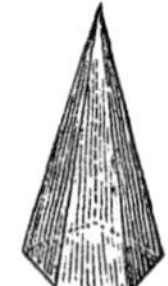

The *slant height* of a pyramid is a line drawn from the vertex to the middle of one of the sides of the base.

QUESTIONS. — Art. 351. What is a cylinder? What is the axis of a cylinder? — Art. 352. What is the rule for finding the surface of a cylinder? — Art. 353. What is the rule for finding the solidity of a cylinder? — Art. 354. What is a pyramid? What is the slant height of a pyramid?

ART. **355.** A CONE is a solid, having a circle for its base, and tapering uniformly to a point, called the vertex.

The *altitude* of a pyramid and of a cone is a line drawn from the vertex perpendicular to the plane of the base; as B C.

The *slant height* of a cone is a line drawn from the vertex to the circumference of the base; as A C.

ART. **356.** To find the surface of a pyramid and of a cone.

RULE. — *Multiply the perimeter or the circumference of the base by half its slant height, and the product is the convex surface.*

1. How many yards of cloth, that is 27 inches wide, will it require to cover the sides of a pyramid whose slant height is 100 feet, and whose perimeter at the base is 54 feet?

Ans. 400 yards.

2. Required the convex surface of a cone, whose slant height is 50 feet, and the circumference at its base 12 feet.

Ans. 300 square feet.

ART. **357.** To find the solidity of a pyramid and of a cone.

RULE. — *Multiply the area of the base by one third of its altitude, and the product will be its solidity.*

1. The largest of the Egyptian pyramids is square at its base, and measures 693 feet on a side. Its height is 500 feet. Now, supposing it to come to a point at its vertex, what are its solid contents, and how many miles in length of wall would it make, 4 feet in height and 2 feet thick?

Ans. 80,041,500 cubic feet; 1894.9 miles in length.

2. What are the solid contents of a cone, whose height is 30 feet, and the diameter of its base 5 feet? Ans. 196.3+ feet.

ART. **358.** A FRUSTUM OF A PYRAMID is the part that remains after cutting off the top, by a plane parallel to the base.

QUESTIONS. — Art. 355. What is a cone? What is the altitude of a pyramid and of a cone? What is the slant height of a cone? — Art. 356. What is the rule for finding the surface of a pyramid and of a cone? — Art. 357. What is the rule for finding the solidity of a pyramid and of a cone? — Art. 358. What is the frustum of a pyramid?

ART. **359.** A FRUSTUM OF A CONE is the part that remains after cutting off the top, by a plane parallel to the base.

ART. **360.** To find the surface of a frustum of a pyramid or of a cone.

RULE. — *Add the perimeters or the circumferences of the two ends together, and multiply this sum by half the slant height. Then add the areas of the two ends to this product, and their sum will be the surface.*

1. There is a square pyramid, whose top is broken off 20 feet slant height from the base. The length of each side at the base is 8 feet, and at the top 4 feet; what is its whole surface?
Ans. 560 square feet.

2. There is a frustum of a cone, whose slant height is 12 feet, the circumference of the base 18 feet, and that of the upper end 9 feet; what is its whole surface?
Ans. 194.22+ square feet.

ART. **361.** To find the solidity of a frustum of a pyramid or of a cone.

RULE. — *Find the area of the two ends of the frustum; multiply these two areas together, and extract the square root of the product. To this root add the two areas, and multiply their sum by one third of the altitude of the frustum; the product will be the solidity.*

1. What is the solidity of the frustum of a square pyramid, whose height is 30 feet, and whose side at the bottom is 20 feet, and at the top 10 feet? Ans. 7000 cubic feet.

2. What are the contents of a stick of timber 20 feet long, and the diameter at the larger end 12 inches, and at the smaller end 6 inches? Ans. 9.162+ feet.

THE SPHERE.

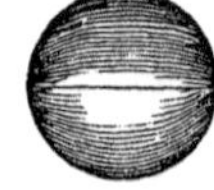

ART. **362.** A SPHERE is a solid, bounded by one continued convex surface, every part of which is equally distant from a point within, called the centre.

The *axis* or *diameter* of a sphere is a line passing through the centre, and terminated by the surface.

QUESTIONS. — Art. 359. What is the frustum of a cone? — Art 360. What is the rule for finding the surface of a frustum of a pyramid or of a cone? — Art. 361. What is the rule for finding the solidity of a frustum of a pyramid or of a cone? — Art. 362. What is a sphere? What is the diameter or axis of a sphere?

ART. **363.** To find the surface of a sphere.

RULE. — *Multiply the diameter by the circumference, and the product will be the surface.*

1. What is the convex surface of a globe, whose diameter is 20 inches? Ans. 1256.6+ square inches.

2. If the diameter of the earth is 8000 miles, what is its convex surface? Ans. 201061888 square miles.

ART. **364.** To find the solidity of a sphere.

RULE. — *Multiply the cube of the diameter by* .523598, *and the product is the solidity.*

1. What is the solidity of a sphere, whose diameter is 20 inches? Ans. 4188.7+ inches.

2. If the diameter of a globe or sphere is 5 feet, how many cubic feet does it contain? Ans. 65.44+ cubic feet.

ART. **365.** To find how large a cube may be cut from any given sphere, or be inscribed in it.

RULE. — *Square the diameter of the sphere, divide the product by* 3, *and extract the square root of the quotient for the answer.*

1. How large a cube may be inscribed in a sphere 10 inches in diameter? Ans. 5.773+ inches.

2. What is the side of a cube that may be cut from a sphere 30 inches in diameter? Ans. 17.32+ feet.

THE SPHEROID.

ART. **366.** A SPHEROID is a solid, generated by the revolution of an ellipse about one of its diameters.

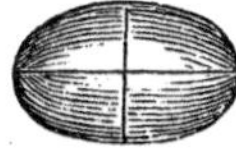

If the ellipse revolves about its *longer* or *transverse* diameter, the spheroid is *prolate*, or *oblong*; if about its *shorter* or *conjugate* diameter, the spheroid is *oblate*, or *flattened*.

ART. **367.** To find the solidity of a spheroid.

RULE 1. — *Multiply the square of the shorter axis by the longer axis, and this product by* .523598, *if the spheroid is prolate, and the product will be its solidity.*

QUESTIONS. — Art. 363. What is the rule for finding the surface of a sphere? — Art. 364. What is the rule for finding the solidity of a sphere? — Art. 365. What is the rule for finding how large a cube can be cut from a given sphere? — Art. 366. What is a spheroid? What is a prolate spheroid? What an oblate spheroid? — Art. 367. What is the rule for finding the solidity of a spheroid?

RULE 2. — *If it is oblate, multiply the square of the longer axis by the shorter axis, and this product by* .523598 ; *the last product will be the solidity.*

1. What is the solidity of a prolate spheroid, whose transverse axis is 30 feet, and the conjugate axis 20 feet?

Ans. 6283.17+ cubic feet.

2. What is the solidity of an oblate spheroid, whose axes are 30 and 10 feet? Ans. 4712.38+ cubic feet.

---

§ XLV. MENSURATION OF LUMBER AND TIMBER.

ART. **368.** ALL rectangular and square lumber and timber, as planks, joists, beams, &c., are usually surveyed by board measure, the board being considered to be 1 inch in thickness. Round timber is sometimes measured by the ton, and sometimes by board measure.

ART. **369.** To find the contents of a board.

RULE. — *Multiply the length of the board, taken in feet, by its breadth, taken in inches, and divide this product by* 12 ; *the quotient is the contents in square feet.*

1. What are the contents of a board 18 inches wide and 16 feet long? Ans. 24 feet.

2. What are the contents of a board 24 feet long and 30 inches wide? Ans. 60 feet.

ART. **370.** To find the contents of joists, beams, &c.

RULE. — *Multiply the depth, taken in inches, by the thickness, and this product by the length, in feet; divide the last product by* 12, *and the quotient is the contents in feet.*

1. What are the contents of a joist 4 inches wide, 3 inches thick, and 12 feet long? Ans. 12 feet.

2. What are the contents of a square stick of timber 25 feet long and 10 inches thick? Ans. $208\frac{1}{3}$ feet.

ART. **371.** To find the contents of round timber.

RULE. — *Multiply the length of the stick, taken in feet, by the square of one fourth the girt, taken in inches; divide this product by* 144, *and the quotient is the contents in cubic feet.*

QUESTIONS. — Art. 368. By what measure are planks, joists, &c., usually surveyed? What is the usual thickness of a board? How is round timber measured? — Art. 369. What is the rule for finding the contents of a board? — Art. 370. What is the rule for finding the contents of joists, &c.?

NOTE 1. — The girt is usually taken about one third the distance from the larger to the smaller end.

NOTE 2. — A ton of timber, estimated by this method, contains $50\frac{92}{100}$ cubic feet.

1. How many cubic feet of timber in a stick, whose length is 50 feet, and whose girt is 60 inches? Ans. $78\frac{1}{8}$ cubic feet.

2. What are the contents of a stick, whose length is 30 feet, and girt 30 inches? Ans. 11.7 + solid feet.

---

§ XLVI. MISCELLANEOUS QUESTIONS.

1. WHAT number is that, to which if $\frac{1}{8}$ be added, the sum will be $7\frac{1}{2}$? Ans. $7\frac{3}{8}$.

2. What number is that, from which if $3\frac{2}{7}$ be taken, the remainder will be $4\frac{1}{4}$? Ans. $7\frac{15}{28}$.

3. What number is that, to which if $3\frac{2}{7}$ be added, and the sum divided by $5\frac{3}{7}$, the quotient will be 5? Ans. $23\frac{6}{7}$.

4. From $\frac{7}{11}$ of a mile take $\frac{7}{9}$ of a furlong. Ans. 4fur. 12rd. 8ft. 8in.

5. John Swift can travel 7 miles in $\frac{8}{9}$ of an hour, but Thomas Slow can travel only 5 miles in $\frac{7}{11}$ of an hour. Both started from Danvers at the same time for Boston, the distance being 12 miles. How much sooner will Swift arrive in Boston than Slow? Ans. $12\frac{36}{77}$ seconds.

6. If $\frac{5}{8}$ of a ton cost $49, what will 1cwt. cost? Ans. $3.92.

7. How many bricks 8 inches long, 4 inches wide, and 2 inches thick, will it take to build a wall 40 feet long, 20 feet high, and 2 feet thick? Ans. 43200 bricks.

8. How many bricks will it take to build the walls of a house, which is 80 feet long, 40 feet wide, and 25 feet high, the wall to be 12 inches thick; the brick being of the same dimensions as in the last question? Ans. 159300 bricks.

9. How many tiles, 8 inches square, will cover a floor 18 feet long, and 12 feet wide? Ans. 486 tiles.

10. If it cost $18.25 to carry 11cwt. 3qr. 19lb. 46 miles, how much must be paid for carrying 83cwt. 2qr. 11lb. 96 miles? Ans. 266.70\frac{2019}{4577}$.

11. A merchant sold a piece of cloth for $24, and thereby lost 25 per cent.; what would he have gained had he sold it for $34? Ans. $6\frac{1}{4}$ per cent.

12. Bought a hogshead of molasses, containing 120 gallons for $30; but 20 gallons having leaked out, for what must I sell the remainder per gallon to gain $10? Ans. $0.40.

13. Bought a quantity of goods for $128.25, and having kept them on hand 6 months, for what must I sell them to gain 6 per cent.? Ans. $140.02.

14. If a sportsman spends $\frac{1}{3}$ of his time in smoking, $\frac{1}{4}$ in "gunning," 2 hours per day in *loafing*, and 6 hours in eating, drinking, and sleeping, how much remains for useful purposes? Ans. 2 hours.

15. If a lady spend $\frac{1}{4}$ of her time in sleep, $\frac{1}{5}$ in making *calls*, $\frac{1}{6}$ at her toilet, $\frac{1}{7}$ in reading novels, and 2 hours each day in receiving visits, how large a portion of her time will remain for improving her mind, and for domestic employments? Ans. $3\frac{27}{35}$ hours per day.

16. If $5\frac{3}{5}$ ells English cost $15.16, what will $71\frac{3}{4}$ yards cost? Ans. $155.39.

17. If a staff 4 feet long cast a shadow $5\frac{3}{5}$ feet, what is the height of a steeple whose shadow is 150 feet? Ans. $107\frac{1}{7}$ feet.

18. Borrowed of James Day $150 for six months; afterwards I lent him $100; how long shall he keep it to compensate him for the sum he lent me? Ans. 9 months.

19. A certain town is taxed $6045.50; the valuation of the town is $293275.00; there are 150 polls in the town, which are taxed $1.20 each. What is the tax on a dollar, and what does A pay, who has 4 polls, and whose property is valued at $3675? Ans. $0.02. A's tax $78.30.

20. D. Sanborn's garden is $23\frac{4}{7}$ rods long, and $13\frac{4}{7}$ rods wide, and is surrounded by a good fence $7\frac{4}{9}$ feet high. Now, if he shall make a walk around his garden within the fence, $7\frac{5}{12}$ feet wide, how much will remain for cultivation? Ans. 1A. 3R. 7p. $85\frac{1381}{1764}$ft.

21. J. Ladd's garden is 100 feet long and 80 feet wide; he wishes to enclose it with a ditch, to be dug outside, 4 feet wide; how deep must it be dug, that the soil taken from it may raise the surface one foot? Ans. $5\frac{15}{47}$ feet.

22. How many yards of paper, that is 30 inches wide, will it require to cover the walls of a room that is $15\frac{1}{2}$ feet long, $11\frac{1}{4}$ feet wide, and $7\frac{3}{4}$ feet high? Ans. $55\frac{17}{60}$ yards.

23. Charles Carleton has agreed to plaster the above room, walls and ceiling, at 10 cents per square yard; what will be his bill? Ans. $$6.54\frac{4}{9}$.

24. What is the interest of $17.86, from Feb. 9, 1850, to Oct. 29, 1852, at $7\frac{1}{4}$ per cent.? Ans. $3.52+.

25. Required the superficial surface of the largest cube that can be inscribed in a sphere 30 inches in diameter.

Ans. 1800 inches.

26. What is due, on the following note, at compound interest, Oct. 29, 1862?

$1000. *Salem, N. H., Oct.* 29, 1856.

For value received, I promise to pay Luther Emerson, Jr., or order, on demand, one thousand dollars with interest.

EMERSON LUTHER

Attest, ADAMS AYER.

On this note are the following endorsements:

Jan. 1, 1857,	was received	$125.00,
June 5, 1857,	do.	$316.00,
Sept. 25, 1857,	do.	$417.00,
April 1, 1858,	do.	$100.00,
July 7, 1858,	do.	$ 50.00.

Ans. $53.79.

27. How many cubic inches are contained in a cube that may be inscribed in a sphere 40 inches in diameter?

Ans. 12316.8+ inches.

28. The dimensions of a bushel measure are $18\frac{1}{2}$ inches wide and 8 inches deep; what should be the dimensions of a similar measure that would contain 4 quarts?

Ans. $9\frac{1}{4}$ inches wide, 4 inches deep.

29. A gentleman willed $\frac{1}{3}$ of his estate to his wife, and $\frac{1}{4}$ of the remainder to his oldest son, and $\frac{1}{6}$ of the residue, which was 151.33\frac{1}{3}$, to his oldest daughter; how much of his estate is left to be divided among his other heirs?

Ans. 756.66\frac{2}{3}$.

30. A man bequeathed $\frac{1}{4}$ of his estate to his son, and $\frac{1}{5}$ of the remainder to his daughter, and the residue to his wife; the difference between his son and daughter's portion was $100; what did he give his wife? Ans. $600.00.

31. Sold a lot of shingles for $50, and by so doing I gained $12\frac{1}{2}$ per cent.; what was their value? Ans. 44.44\frac{4}{9}$.

32. If $\frac{3}{11}$ of a yard cost $5.00, what quantity will $17.50 purchase? Ans. $\frac{21}{22}$ yard.

33. John Savory and Thomas Hardy traded in company; Savory put in for capital $1000; they gained $128.00; Hardy received for his share of the gains $70; what was his capital?

Ans. 1206.89\frac{13}{29}$.

34. E. Fuller lent a certain sum of money to C. Lamson, and at the end of 3 years, 7 months, and 20 days, he received interest and principal $1000; what was the sum lent?

Ans. 820.79\frac{251}{731}$.

35. Lent $88 for 18 months, and received for interest and principal $97.57; what was the per cent.? Ans. 7$\frac{1}{4}$ per cent.

36. When $\frac{3}{5}$ of a gallon cost $87, what cost 7$\frac{1}{4}$ gallons?

Ans. $1051.25.

37. When $71 are paid for 18$\frac{3}{7}$ yards of broadcloth, what cost 5 yards? Ans. 19.26\frac{46}{129}$.

38. How many yards of cloth, at $4.00 per yard, must be given for 18 tons 17cwt. 3qr. of sugar, at $9.50 per cwt.?

Ans. 897$\frac{5}{32}$ yards.

39. How much grain, at $1.25 per bushel, must be given for 98 bushels of salt, at $0.45 per bushel?

Ans. 35$\frac{7}{25}$ bushels.

40. A person, being asked the time of day, replied that $\frac{1}{7}$ of the time passed from noon was equal to $\frac{1}{11}$ of the time to midnight. Required the time. Ans. 40 minutes past 4.

41. On a certain night, in the year 1852, rain fell to the depth of 3 inches in the town of Haverhill; the town contains about 20,000 square acres. Required the number of hogsheads of water fallen, supposing each hogshead to contain 100 gallons, and each gallon 282 cubic inches.

Ans. 13346042hhd. 55gal. 1qt. 0pt. 2$\frac{19}{47}$gi.

42. If the sun pass over one degree in 4 minutes, and the longitude of Boston is 71° 4′ west, what will be the time at Boston, when it is 11h. 16m. A. M. at London?

Ans. 6h. 31m. 44sec. A. M.

43. When it is 2h. 36m. A. M. at the Cape of Good Hope, in longitude 18° 24′ east, what is the time at Cape Horn, in longitude 67° 21′ west? Ans. 8h. 53m. P. M.

44. Yesterday my longitude, at noon, was 16° 18′ west; today I perceive by my watch, which has kept correct time, that the sun is on the meridian at 11h. 36m.; what is my longitude?

Ans. 10° 18′ west.

45. Sound, uninterrupted, will pass 1142 feet in 1 second; how long will it be in passing from Boston to London, the distance being about 3000 miles? Ans. 3h. 51m. 10+sec.

46. The time which elapsed between seeing the flash of a gun and hearing its report was 10 seconds; what was the distance?

Ans. 2 miles 860 feet.

47. J. Pearson has tea, which he barters with M. Swift, at

10 cents per lb. more than it costs him, against sugar, which costs Swift 15 cents per pound, but which he puts at 20 cents per pound; what was the first cost of the tea?

Ans. \$0.30 per lb.

48. Q and Y barter; Q makes of 10 cents $12\frac{1}{2}$ cents; Y makes of 15 cents 19 cents; which makes the most per cent., and how much? Ans. Y makes $1\frac{2}{3}$ per cent. more than Q.

49. A certain individual was born in 1786, September 25, at 23 minutes past 3 o'clock, A. M.; how many minutes old will he be July 4, 1844, at 30 minutes past 5 o'clock, P. M., reckoning 365 days for a year, excepting leap years, which have 366 days each? Ans. 30,386,287 minutes.

50. The longitude of a certain star is 3s. 14° 26′ 14″, and the longitude of the moon at the same time is 8s. 19° 43′ 28″; how far will the moon have to move in her orbit to be in conjunction with the star? Ans. 6s. 24° 42′ 46″.

51. From a small field, containing 3A. 1R. 23p. 200ft., there were sold 1A. 2R. 37p. 30yd. 8ft.; what quantity remained?

Ans. 1A. 2R. 25p. 21yd. 5ft. 36in.

52. What part of $\frac{3}{4}$ of an acre is $\frac{5}{8}$ of an acre?

Ans. $\frac{20}{27}$.

53. A thief was brought before a certain judge, and it was proved that he had stolen property to the value of 1£. 19s $11\frac{3}{4}$d. He was sentenced either to one year's imprisonment in the county jail, or to pay 1£. 19s. $11\frac{3}{4}$d. for the value of every pound he had stolen; required the amount of the fine.

Ans. 3£. 19s. 11d. $0\frac{1}{960}$qr.

54. My chaise having been injured by a very bad boy, I am obliged to sell it for \$68.75, which is 40 per cent. less than its original value; what was the cost? Ans. \$$114.58\frac{1}{3}$.

55. Charles Webster's horse is valued at \$120, but he will not sell him for less than \$134.40; what per cent. does he intend to make? Ans. 12 per cent.

56. Three merchants, L. Emerson, E. Bailey, and S. Curtiss, engage in a cotton speculation. Emerson advanced \$3600, Bailey \$4200, and Curtiss \$2200. They invested their whole capital in cotton, for which they received \$15000 in bills on a bank in New Orleans. These bills were sold to a Boston broker at 15 per cent. below par; what is each man's net gain?

Ans. Emerson \$990.00, Bailey \$1155.00, Curtiss \$605.00.

57. Bought a box made of plank, $3\frac{1}{2}$ inches thick. Its length on the outside is 4ft. 9in., its breadth 3ft. 7in., and its height

2ft. 11in. How many square feet did it require to make the box, and how many cubic feet will it hold?

Ans. $70\frac{5}{24}$ square feet, $29\frac{1}{6}$ cubic feet.

58. How many bricks will it require to construct the walls of a house, 64 feet long, and 32 feet wide, and 28 feet high? The walls are to be 1ft. 4in. thick, and there are also three doors 7ft. 4in. high, and 3ft. 8in. wide; also 14 windows 3 feet wide and 6 feet high, and 16 windows 2ft. 8in. wide and 5ft. 8in. high. Each brick is to be 8 inches long, and 4 inches wide, and 2 inches thick. Ans. 167,480 bricks.

59. John Brown gave to his three sons, Benjamin, Samuel, and William, $1000, to be divided in the proportion of $\frac{1}{3}$, $\frac{1}{4}$, and $\frac{1}{5}$, respectively; but William, having received a fortune by his wife, resigns his share to his brothers. It is required to divide the whole sum between Benjamin and Samuel.

Ans. Benjamin \$571.42$\frac{6}{7}$; Samuel \$428.57$\frac{1}{7}$.

60. Peter Webster rented a house for 1 year to Thomas Bailey, for $100; at the end of four months Bailey rented one half of the house to John Bricket, and at the end of eight months it was agreed by Bricket and Bailey to rent one third of the house to John Dana. What share of the rent must each pay?

Ans. Bailey \$61$\frac{1}{9}$, Bricket \$27$\frac{7}{9}$, and Dana \$11$\frac{1}{9}$.

61. I have a plank $42\frac{1}{4}$ feet in length, 24 inches wide, and 3 inches thick; required the side of a cubical box that can be made from it. Ans. 48 inches.

62. D. Small purchased a horse for 10 per cent. less than his value, and sold him for 16 per cent. more than his value, by which he gained $21.84; what did he pay for the horse?

63. Minot Thayer sold broadcloth at $4.40 per yard, and by so doing he lost 12 per cent.; whereas, he ought to have gained 10 per cent.; for what should the cloth have been sold per yard?

64. A gentleman has five daughters, Emily, Jane, Betsey, Abigail, and Nancy, whose fortunes are as follows. The first two and the last two have $19,000; the first four, $19,200; the last four, $20,000; the first and the last three, $20,500; the first three and the last, $21,300. What was the fortune of each?

Ans. Emily has $5,000; Jane, $4,500; Betsey, $6,000; Abigail, $3,700; and Nancy, $5,800.

65. I have a fenced garden, 12 rods square. How many trees may be set on it, whose distance from each other shall be one rod, and no tree to be within half a rod of the fence?

Ans. 152 trees.

THE END.

www.ingramcontent.com/pod-product-compliance
Lightning Source LLC
LaVergne TN
LVHW020221110826
845151LV00003B/782

* 9 7 8 1 4 2 5 5 3 1 9 9 7 *